VOLUME FOUR HUNDRED AND SIXTY-TWO

METHODS IN ENZYMOLOGY

Non-Natural Amino Acids

METHODS IN ENZYMOLOGY

Editors-in-Chief

JOHN N. ABELSON AND MELVIN I. SIMON

Division of Biology
California Institute of Technology
Pasadena, California, USA

Founding Editors

SIDNEY P. COLOWICK AND NATHAN O. KAPLAN

VOLUME FOUR HUNDRED AND SIXTY-TWO

METHODS IN
ENZYMOLOGY
Non-Natural Amino Acids

EDITED BY

TOM W. MUIR
Laboratory of Synthetic Protein Chemistry
The Rockefeller University
New York, NY, USA

JOHN N. ABELSON
Division of Biology
California Institute of Technology
Pasadena, California, USA

AMSTERDAM • BOSTON • HEIDELBERG • LONDON
NEW YORK • OXFORD • PARIS • SAN DIEGO
SAN FRANCISCO • SINGAPORE • SYDNEY • TOKYO
ELSEVIER
Academic Press is an imprint of Elsevier

AP

Academic Press is an imprint of Elsevier
525 B Street, Suite 1900, San Diego, CA 92101-4495, USA
30 Corporate Drive, Suite 400, Burlington, MA 01803, USA
32 Jamestown Road, London NW1 7BY, UK

First edition 2009

Notice
No responsibility is assumed by the publisher for any injury and/or damage to persons or property as a matter of products liability, negligence or otherwise, or from any use or operation of any methods, products, instructions or ideas contained in the material herein. Because of rapid advances in the medical sciences, in particular, independent verification of diagnoses and drug dosages should be made

For information on all Academic Press publications
visit our website at elsevierdirect.com

ISBN: 978-0-12-374310-7
ISSN: 0076-6879

Printed and bound in United States of America
09 10 11 12   10 9 8 7 6 5 4 3 2 1

Working together to grow
libraries in developing countries

www.elsevier.com | www.bookaid.org | www.sabre.org

ELSEVIER   BOOK AID
             International   Sabre Foundation

CONTENTS

Contributors

Farhana B. Abu Bakar
Department of Biological Sciences, NUS MedChem Program of the Office of Life Sciences, National University of Singapore, Singapore

Christian F. W. Becker
Technische Universität München, Department of Chemistry, Protein Chemistry Group, Garching, Germany

Champak Chatterjee
Laboratory of Synthetic Protein Chemistry, Rockefeller University, New York, USA

Souvik Chattopadhaya
Department of Biological Sciences, NUS MedChem Program of the Office of Life Sciences, National University of Singapore, Singapore

Nam Ky Chu
Technische Universität München, Department of Chemistry, Protein Chemistry Group, Garching, Germany

Kevin M. Clark
Department of Biochemistry, University of Illinois at Urbana-Champaign, Urbana, Illinois, USA

Philip A. Cole
Department of Pharmacology and Molecular Sciences, The Johns Hopkins University School of Medicine, Baltimore, Maryland, USA

David Cowburn
New York Structural Biology Center, New York, USA

Claus Czeslik
Fakultät Chemie – Chemische Biologie, Technische Universität Dortmund, Dortmund, Germany

Jordan J. Devereaux
Program in Chemical Biology, Department of Physiology and Pharmacology, Oregon Health and Sciences University, Portland, Oregon, USA

Wilfred A. van der Donk
Departments of Chemistry and Biochemistry, University of Illinois at Urbana-Champaign, Urbana, Illinois, USA

Daniel Garbe
Fakultät Chemie – Chemische Biologie, Technische Universität Dortmund, Dortmund, Germany

Petra Janning
Max-Planck-Institut für Molekulare Physiologie, Dortmund, Germany

Alexander G. Komarov
Program in Chemical Biology, Department of Physiology and Pharmacology, Oregon Health and Sciences University, Portland, Oregon, USA

Kellie M. Linn
Program in Chemical Biology, Department of Physiology and Pharmacology, Oregon Health and Sciences University, Portland, Oregon, USA

Dongsheng Liu
New York Structural Biology Center, New York, USA

Yi Lu
Departments of Chemistry and Biochemistry, University of Illinois at Urbana-Champaign, Urbana, Illinois, USA

Christina Ludwig
Fakultät Chemie – Chemische Biologie, Technische Universität Dortmund, Dortmund, Germany

Robert K. McGinty
Laboratory of Synthetic Protein Chemistry, Rockefeller University, New York, USA

Henning D. Mootz
Fakultät Chemie – Chemische Biologie, Technische Universität Dortmund, Dortmund, Germany

Tom W. Muir
Laboratory of Synthetic Protein Chemistry, Rockefeller University, New York, USA

Ronald T. Raines
Departments of Chemistry and Biochemistry, University of Wisconsin-Madison, Madison, Wisconsin, USA

Dirk Schwarzer
Current address: Leibniz-Institut für Molekulare Pharmakologie, Berlin, Germany, and Fakultät Chemie – Chemische Biologie, Technische Universität Dortmund, Dortmund, Germany

Mohammad R. Seyedsayamdost
Department of Chemistry, Massachusetts Institute of Technology, Cambridge, Massachusetts, USA

JoAnne Stubbe
Departments of Biology and Chemistry, Massachusetts Institute of Technology, Cambridge, Massachusetts, USA

Lawrence M. Szewczuk
Department of Pharmacology and Molecular Sciences, The Johns Hopkins University School of Medicine, Baltimore, Maryland, USA

Annie Tam
Departments of Chemistry and Biochemistry, University of Wisconsin-Madison, Madison, Wisconsin, USA

Mary Katherine Tarrant
Department of Pharmacology and Molecular Sciences, The Johns Hopkins University School of Medicine, Baltimore, Maryland, USA

Francis I. Valiyaveetil
Program in Chemical Biology, Department of Physiology and Pharmacology, Oregon Health and Sciences University, Portland, Oregon, USA

Rong Xu
New York Structural Biology Center, New York, USA

Shao Q. Yao
Departments of Chemistry and Biological Sciences, NUS MedChem Program of the Office of Life Sciences, National University of Singapore, Singapore

Joachim Zettler
Fakultät Chemie – Chemische Biologie, Technische Universität Dortmund, Dortmund, Germany

Xingang Zhang
Department of Chemistry, University of Illinois at Urbana-Champaign, Urbana, Illinois, USA

Preface

Proteins are the most versatile of nature's macromolecules. There seems to be no limit to what proteins can do, whether it is to interact with other biomolecules, in a sense acting as molecular Velcro, or to catalyze biochemical reactions involved in everything from secondary metabolism to the remodeling of chromatin. Understanding how proteins work at the molecular level is at the heart of biochemistry. Indeed, the pursuit of this fundamental question has helped spawn entire areas of study, including structural biology, protein engineering, and of course enzymology. The more we learn about protein function using the tools provided by these venerable fields, the more we appreciate the extraordinary complexity of their inner workings. Despite the enormous progress that has been made, it is clear that additional approaches are required if we are to further penetrate these seemingly Byzantine structure-activity relationships—just as a watchmaker requires extremely fine tools to put together (or fix) an analog timepiece, so too the protein biochemist needs precise methods by which to tweak the chemical structure of his or her favorite protein to figure out what the various cogs and gears (i.e. amino acids) are doing. Often, site-directed mutagenesis is too blunt an instrument for the question at hand; it is easy to perform, but it represents a compromise over what one might really want to do, namely alter the physiochemical properties of an amino acid in a subtle manner that minimizes collateral damage (i.e. unwanted secondary effects) to the protein. Clearly, the ability to incorporate nonnatural amino acids into proteins would go a long way toward correcting this deficiency. The past several years have seen an explosion of research directed at this problem. This volume offers a snapshot of the major developments in this area and highlights both the chemistry-driven techniques that have been devised, as well as the type of problems in protein biochemistry to which they are now applied.

Broadly speaking, two approaches have emerged in recent years that enable the generation of proteins containing unnatural amino acids; protein semisynthesis and nonsense suppression mutagenesis. The former approach involves building the target protein from premade polypeptide fragments, one of which is generated by recombinant DNA expression methods and others by chemical synthesis. The recombinant building block can be extremely large, while the synthetic piece can contain any number or type of noncoded element(s). Thus, by linking the two together we have the best of both worlds, size, and chemical diversity. Several of the chapters in this volume review the various approaches available to link the protein fragments together and offer detailed practical guides to their use. Also covered in this volume is the nonsense suppression methodology. This approach allows for an unnatural amino acid to be incorporated site specifically into a fully recombinant protein via ribosomal synthesis. In the past few years, methods have been developed that allow for incorporation of a wide range of unnatural amino acids into proteins in cells. This *in vivo* methodology relies on the generation (by directed evolution methods) of mutant tRNA synthetases that aminoacylate suppressor tRNAs with the unnatural amino acid directly in living cells. Importantly, the suppressor tRNA/aminoacyl-tRNA synthesase pair is orthogonal to the host cell aminoacylation apparatus, thereby ensuring the fidelity of the unnatural amino acid mutagenesis.

Do we really need two entirely different approaches to the same problem? The answer is a resounding yes. This is because the two strategies are complementary in terms of what they can and cannot do. The suppressor mutagenesis approach has the big advantage of being easy to perform (at least once you have the orthogonal tRNA/aminoacyl-tRNA synthesase pair). Indeed, it is no more difficult than standard mutagenesis and can, in principle, yield large amounts of mutant protein using bacterial overexpression technologies. Also, one can mutate any residue in the protein simply by replacing the corresponding codon in the gene with an amber codon. In contrast, semisynthesis involves *in vitro* manipulations, which can be technically cumbersome by comparison. Moreover, these same technical issues mean that it is much easier to incorporate unnatural amino acids near (within ≈ 50 residues) the N- or C-terminus of a large protein than in the middle. Semisynthesis allows for a much wider range of unnatural amino acids to be introduced into the protein than suppressor mutagenesis; it is possible to incorporate almost any building block into a synthetic peptide, whereas there are significant restrictions on what amino acids the ribosome will accept. Furthermore, only a single unnatural unit can be introduced efficiently using the suppressor mutagenesis approach, whereas any number of unnatural units can be incorporated at once by semisynthesis; indeed, the synthetic building block can be nonpeptidic (e.g. DNA or a glass slide). Another often–overlooked strength of semisynthesis is that it allows NMR–active isotopes to be introduced into a single amino acid, or stretch of amino acids, in a protein, thereby allowing targeted spectroscopic studies to be performed. Thus, semisynthesis and suppressor mutagenesis have different strengths and weaknesses. It is also worth noting that there is no technical reason why they cannot be used in combination; that is to generate a protein containing nonnatural elements introduced by both strategies. Many otherwise intractable problems could yield to this integrated approach.

Many of the chapters in the book describe applications of these protein engineering technologies to specific biochemical problems that exploit their respective strengths, some of which I have alluded to here. Indeed, one of the messages that I hope readers take from reading this volume is that the approaches are robust enough to tackle even the most challenging of problems, whether those be studying conductance of an ion channel or teasing apart proton–coupled electron transfer reactions in a multi–subunit enzyme. Thus, I think the future of this field will be increasingly application driven rather than focused on technology development per se. Hopefully, the reader will gain some inspiration from one or more of these articles.

TOM W. MUIR

METHODS IN ENZYMOLOGY

CHAPTER ONE

PROTEIN PHOSPHORYLATION BY SEMISYNTHESIS: FROM PAPER TO PRACTICE

Lawrence M. Szewczuk,* Mary Katherine Tarrant,* *and* Philip A. Cole*

Contents

* Department of Pharmacology and Molecular Sciences, The Johns Hopkins University School of Medicine, Baltimore, Maryland, USA

Methods in Enzymology, Volume 462
ISSN 0076-6879, DOI: 10.1016/S0076-6879(09)62001-2

Abstract

Deconvolution of specific phosphorylation events can be complicated by the reversibility of modification. Protein semisynthesis with phosphonate analogues offers an attractive approach to functional analysis of signaling pathways. In this technique, N- and C-terminal synthetic peptides containing nonhydrolyzable phosphonates at target residues can be ligated to recombinant proteins of interest. The resultant semisynthetic proteins contain site specific, stoichiometric phosphonate modifications and are completely resistant to phosphatases. Control of stoichiometry, specificity, and reversibility allows for complex signaling systems to be broken down into individual events and discretely examined. This chapter outlines the general methods and considerations for designing and carrying out phosphoprotein semisynthetic projects.

1. OVERVIEW OF PROTEIN PHOSPHORYLATION

Since the discovery of cell surface receptors and second messengers, there has been intensive study of the mechanisms of cell signal transduction. Information flow from the plasma membrane to the nucleus and then back again involves an array of enzymes, adaptor proteins, lipids, and other small molecules whose structural interactions govern cell growth, differentiation, and movement. Among the many cell-signaling pathway regulatory mechanisms, reversible protein phosphorylation stands out as of critical importance. Protein kinases and phosphatases are members of large superfamilies in the human genome, and a large number of these enzymes have been implicated as drug targets for diseases including cancer, diabetes, immune disorders, inflammatory conditions, and cardiovascular conditions (Blume-Jensen and Hunter, 2001; Cohen, 2002; Hunter, 2000). Protein phosphorylation catalyzed by the eukaryotic protein kinase superfamily members are typically Ser/Thr or Tyr selective, and there are approximately 400 protein Ser/Thr kinase (PSK) and 100 protein Tyr kinase (PTKs) human genes (Manning *et al.*, 2002). The protein tyrosine phosphatase (PTP) superfamily has about 100 members, whereas the protein Ser/Thr phosphatase (PSP) family numbers about 15 (Alonso *et al.*, 2004). It has been estimated that 25% of cellular proteins undergo phosphorylation and are thus substrates of one or more kinase and phosphatase (Cohen, 2000). Upon phosphorylation, protein structural interactions can be affected in several ways. In many cases, the addition of a phosphate to a protein side chain can inhibit or promote intra- or intermolecular protein-protein interactions. A number of well-characterized domains have been implicated in selectively binding to pSer/pThr and pTyr, respectively (see Table 1.1). The first of these discovered was the Src homology 2 (SH2) domain, which binds pTyr motifs in proteins. There are about 100 SH2 domains found in

Table 1.1   Phosphoprotein binding domains

Ligand	Domains
Phospho-tyrosine	Src homology 2 domain (SH2)
	Phosphotyrosine binding domain (PTB)
Phospho-serine/threonine	14-3-3 protein
	WW domain
	Forkhead-associated domain (FHA)
	BRCA1 C-terminal (BRCT) domain
	Polo-box domain (PBD)

human genes (Yaffe, 2002). In addition, phosphotyrosine-binding (PTB) domains also bind pTyr-containing sequences. A number of proteins or protein domains that have been identified as maintaining pSer/pThr binding include 14-3-3 adaptor proteins, Polo-Box domains, WW domains, BRCA1 C-terminal (BRCT) domains, and forkhead-associated (FHA) domains (Yaffe and Elia, 2001). Unraveling the complexity of protein phosphorylation functionality is daunting. In general, the identification of specific kinase-substrate interactions and cellular phosphorylation actions for individual cases remains a major challenge. A variety of novel approaches have been developed in recent years to attempt to analyze protein kinases, phosphatases, and the function of individual phosphate modifications (Johnson and Hunter, 2005; Shogren-Knaack *et al.*, 2001; Qiao *et al.*, 2006; Williams and Cole, 2001; Zhang, 2005). In this review, we discuss the steps required to successfully use protein semisynthesis as a tool to investigate protein phosphorylation, and we outline the practical uses of these semisynthetic proteins.

2. INVESTIGATING PROTEIN PHOSPHORYLATION WITH PHOSPHOMIMETICS

Researchers have been searching for ways to investigate protein phosphorylation since the discovery that this modification can reversibly alter the activity of an enzyme through the combined action of kinases and phosphatases (Fischer and Krebs, 1955; Krebs and Fischer, 1955). Unfortunately, it is this reversibility that presents a challenging problem in *in vivo* studies of protein phosphorylation and attempts to elucidate complex signaling cascades. Although various techniques have been used to address this problem, all have limitations. These techniques are discussed subsequently.

2.1. Thiophosphate substitution as a phosphomimetic

One way to deal with the reversibility of phosphorylation is to enzymatically phosphorylate the protein of interest using ATPγS and catalytic amounts of kinase (Cole *et al.*, 1994; Eckstein, 1983). This resulting thiophosphate linkage is more stable to phosphatases and therefore of greater potential utility in *in vivo* assays. However, with this approach, it is difficult to control the site specificity and stoichiometry of the modification (Tailor *et al.*, 1997), and thiophosphates are not fully phosphatase resistant (Cho *et al.*, 1993).

2.2. Using amino acid substitution as a phosphomimetic

A genetic approach to deal with reversibility of phosphorylation has been to mutate the phosphorylated residue to either a Glu or an Asp in hopes that the addition of a negative charge will mimic the phosphorylated residue (Fig. 1.1B). This approach was taken by Thorsness and Koshland (1987) in the early years of site-directed mutagenesis; they showed that Asp was a good mimic for pSer in isocitrate dehydrogenase. In that case, the mutation led to the predicted complete inactivation of the enzyme. Since this early report, there have been numerous examples of this approach being both successful (Hao *et al.*, 1996; Potter and Hunter, 1998) and unsuccessful (Huang and Erikson, 1994; Zheng *et al.*, 2003). Phosphates and carboxylates obviously differ in number of oxygen atoms available for hydrogen bonding, geometry, numbers of negative charges at neutral pH, and size. Moreover, there is no natural isostere for pTyr (Zisch *et al.*, 2000).

2.3. Incorporation of alternative genetically encoded phosphomimetics

More recently, the Schultz group described a general method to expand the number of genetically encoded amino acids in *Escherichia coli* by using unique tRNA/aminoacyl-tRNA synthetase pairs (Wang *et al.*, 2001). To date, more than 30 unnatural amino acids have been incorporated into proteins in *E. coli*, yeast, and mammalian cells (Xie and Schultz, 2005). One such amino acid, *p*-carboxymethyl-phenylalanine (Cmp), acts as a pTyr mimetic in the same way that Glu and Asp act as pThr/pSer mimetics (Fig. 1.1B). Although this methodology expands the number of genetically available phosphomimetics to include a pTyr mimetic, it has not yet been used for phosphate-containing amino acids, presumably because of their limited ability to traverse the cell membrane or wall.

A

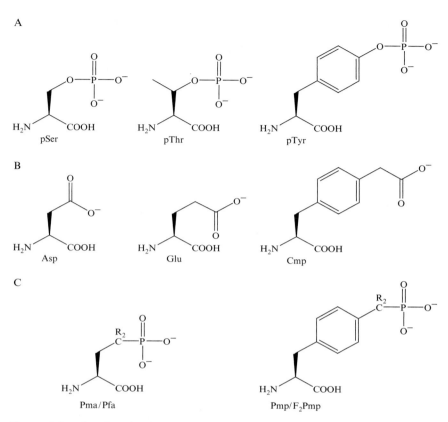

B

C

Figure 1.1 Phosphorylated amino acids and their mimetics. (A) Naturally occurring phosphorylated amino acids, (B) genetically encoded phosphomimetics, and (C) nonhydrolyzable phosphonates, where R = H or F.

3. PHOSPHONATE ANALOGUES AND PROTEIN SEMISYNTHESIS

Nonhydrolyzable phosphonate analogues offer a useful tool for investigating phosphorylated proteins *in vivo*, as proteins bearing these mimetics are completely resistant to phosphatase activity (Fig. 1.1C) (Berkowitz *et al.*, 1996; Chen *et al.*, 1995; Desmarais *et al.*, 1999; Zheng *et al.*, 2003, 2005). Therefore, by incorporating phosphonate analogues into proteins at positions that are normally phosphorylated, it is possible to identify novel interacting proteins by pull–down assays (e.g., phosphatases and phosphopeptide binding domains) and to evaluate protein stability and cellular localization by microinjection assays (Lu *et al.*, 2001; Schwarzer *et al.*, 2006; Zheng *et al.*, 2003, 2005). However, to incorporate the phosphonate analogues into proteins, the proteins must be amenable to semisynthetic technology. In these procedures, N- and C-terminal peptides (up to 50 amino acids in length) containing the phosphonate substitution(s) can be synthetically prepared using a standard Fmoc solid-phase peptide synthesis (SPPS) strategy and ligated to a recombinant portion of the protein of interest (Dawson *et al.*, 1994; Erlanson *et al.*, 1996; Evans *et al.*, 1998; Muir *et al.*, 1998). In this manner, homogenous, stoichiometrically labeled phosphoproteins can be generated. The marriage of phosphonate analogue technology and protein semisynthetic technology makes it possible to determine which protein domains and enzymes can interact with phosphorylated proteins in live cells and aids in the elucidation of complex signal transduction cascades.

The development of native chemical ligation (NCL) revolutionized the possibility of linking large unprotected peptide fragments in a chemoselective and mild fashion (Dawson *et al.*, 1994). In this methodology, ^{1}Cys-containing peptides can be efficiently ligated to $^\alpha$thioester peptides by initial transthioesterification, followed by rearrangement to a stable peptide bond. Extension of NCL for use in protein semisynthesis requires the generation of ^{1}Cys groups or $^\alpha$thioester functionality in recombinant proteins. N-terminal Cys production is most readily achieved using selective proteases (Erlanson *et al.*, 1996; He *et al.*, 2003; Tolbert *et al.*, 2005; Zheng *et al.*, 2003), whereas $^\alpha$thioesters can be generated using modified inteins (Chong *et al.*, 1997; Evans *et al.*, 1998). Expressed protein ligation (EPL) has been used to describe the technique of protein semisynthesis in which a ^{1}Cys-containing peptide is ligated by NCL to a recombinant protein fragment containing a C-terminal thioester (Evans *et al.*, 1998; Muir *et al.*, 1998). In EPL, the recombinant protein thioester fragment is typically generated via a C-terminal intein-chitin binding domain (CBD) fusion protein, whereby the intein catalyzes an *N*- to *S*-acyl shift at the junction of the fusion protein. In the presence of a thiol [e.g., sodium 2-mercaptoethanesulfonate (MESNA), thiophenol, or mercaptophenylacetic acid], this intermediate is trapped as an $^\alpha$thioester and intercepted with a ^{1}Cys synthetic peptide. The newly formed thioester-linked species spontaneously rearranges to yield a native peptide bond at neutral pH. As a result, a semisynthetic protein is generated that contains a chemoselectively ligated synthetic peptide at the C-terminus (Fig. 1.2) (Evans *et al.*, 1998; Muir *et al.*, 1998).

4. METHODS

This section aims to guide investigators in designing and carrying out a phosphoprotein semisynthetic project. In doing so, we refer to specific published examples that have been successful in our laboratory.

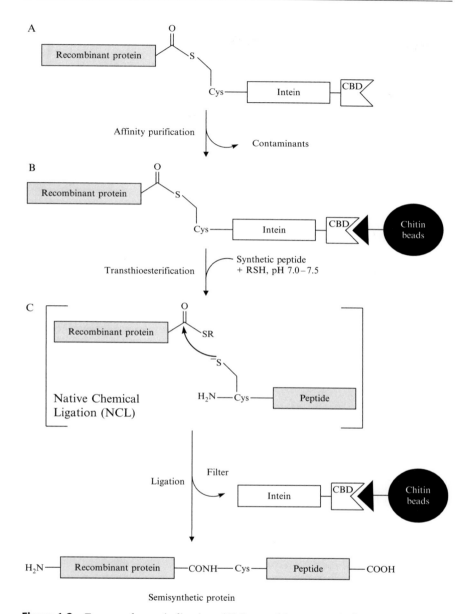

Figure 1.2   Expressed protein ligation. (A) Recombinant protein for EPL is expressed as a C-terminal intein–CBD fusion protein and subjected to affinity purification on chitin beads. The modified inteins used for semisynthesis catalyze an *N*- to *S*-acyl shift resulting in the formation of a thioester-linked fusion protein. (B) The protein thioester is trapped by small-molecule thiol (RSH = sodium 2-mercaptoethansulfonate, thiophenol, or mercaptophenylacetic acid) as a more reactive $^{\alpha}$thioester for semisynthesis.

4.1. Choosing a protein for semisynthesis

Many but not all proteins are attractive candidates for protein semisynthesis. At least three major factors should be considered. First, the target residue (or residues) for modification should be within approximately 50 residues of either the N-terminus or the C-terminus. This reflects a limitation of protein semisynthesis in that there is no easy method to incorporate unnatural amino acids into the central regions of proteins, and SPPS becomes increasingly difficult for very long peptides. Second, in choosing the candidate protein, it is more favorable to focus on C-terminal modifications because recombinant protein $^{\alpha}$thioesters for semisynthesis are easily made by inteins, whereas investigating N-terminal modifications requires the synthesis of a peptide $^{\alpha}$thioester, which can be difficult when using Fmoc SPPS. Third, protein expression and solubility should be considered. It is easiest to work with a protein that is soluble and expresses well in *E. coli*; however, many proteins have poor solubility when expressed recombinantly in prokaryotic systems, and semisynthesis can be achieved only with insect cell expression techniques. It can be difficult to predict how a protein fragment of interest will behave when it is truncated for semisynthesis and expressed as either a C-terminal intein-CBD fusion protein or N-terminal fusion protein for semisynthesis. Breaking into the middle of an independently folded domain is often problematic and may result in poor solubility. However, this does not mean that the semisynthetic project is not feasible. As long as the protein can be refolded to its native state, the project is likely to succeed (e.g., K^{+} channel, LMW–PTP) (Schwarzer *et al.*, 2006; Valiyaveetil *et al.*, 2002). Dealing with soluble versus insoluble proteins for semisynthesis is discussed further in later sections.

4.2. Choosing the peptide ligation site

When ligating a synthetic peptide to a recombinant protein by NCL, it is necessary to have a Cys residue at the ligation site. Depending on whether C-terminal or N-terminal semisynthesis is used, the Cys residue will need to be included in either the synthetic peptide or the recombinant protein, respectively. Ideally, a native Cys residue will be present in close proximity, interior to the target amino acid for modification. Often, this is not the case, and investigators must select where to insert the ligation site. This is somewhat more complex in N-terminal semisynthesis, as many of the

---

(C) In the native chemical ligation reaction, the $^{\alpha}$thioester intermediate is intercepted by a ^{1}Cys synthetic peptide and the newly formed thioester-linked species spontaneously rearranges to yield a native peptide bond at pH 7.0–7.5. The semisynthetic protein is eluted from the column and now contains a chemoselectively ligated synthetic peptide at the C-terminus.

protocols for synthesizing peptide $^\alpha$thioesters by Fmoc SPPS can result in epimerization at the alpha carbon of the thioester. In such cases, we prefer to use a Gly residue for the $^\alpha$thioester so that epimerization is not a problem. There are several factors that influence selection of the ligation junction. First is the secondary structure of the protein. If a crystal structure of the protein is available, potential ligation sites can be examined closely to inspect whether truncation of the protein is likely to disrupt the secondary or tertiary structure. It is preferable to place the ligation site in a nondisruptive (i.e., flexible) region. Second, if a native Cys residue is not available for ligation, then conservative mutation of a residue to Cys is desirable. We recommend replacement by Ser first, Ala second, and other residues if necessary. Third, the identity of the residue N-terminal to the Cys at the ligation site can influence efficiency of the NCL reaction. Hackeng *et al.* (1999) have published a detailed analysis describing the kinetics of the NCL reaction on a model peptide in which every amino acid was substituted immediately N-terminal to the Cys residue at the ligation site. The authors found that the NCL reaction rates depended on the identity of the residue. The time it took for the reactions to be completed were as follows: HGC (4 h) < FMYAW (9 h) < NSDQEKR (24 h) < LTVIP (48 h). We use these data as guidelines for selecting ligation junctions. Additional information on this topic can be found in the Impact-CN manual available online from New England Biolabs. Fourth, in choosing a site for N-terminal ligation, one has to consider that it is convenient to have a Gly residue adjacent to the $^\alpha$thioester on the synthetic peptide. Ideally, the protein would contain a native Gly-Cys, Gly-Ser, or Gly-Ala sequence interior to the amino acid targeted for modification. If no such sequence exists, one can consider insertion of a Gly-Cys sequence or make nonconservative mutations to yield the Gly-Cys sequence. Amino acid insertion should generally be avoided, but it can be necessary, as described for arylalkylamine-*N*-acetyltransferase (AANAT) (Zheng *et al.*, 2003). Finally, after the ligation site has been chosen, a full-length protein harboring the mutation(s) should be purified and compared to wild-type protein to make sure that no deleterious effects arise from the mutation(s).

4.3. Design of recombinant constructs for semisynthesis

Once the candidate protein and ligation site have been selected and the mutant protein has been validated as behaving identical to the wild-type protein, the semisynthetic project can proceed. The truncated candidate protein must be subcloned into an appropriate vector for ligation. For C-terminal EPL, our lab generally subclones the C-terminally truncated candidate protein into both commercially available Impact-CN vectors pTYB2 or pTXB1 (New England Biolabs) which contain the *Saccharomyces cerevisiae* VMA1 intein and *Mycobacterium xenopi* GyrA intein, respectively.

The VMA1 intein is larger than the GyrA intein and cannot be refolded, which can be limiting (Valiyaveetil et al., 2002). However, because it is impossible to predict which fusion protein will yield good expression, soluble protein, and ligation efficiency, it is advisable to start with both vectors. After the subcloning is complete, expression levels should be checked on a small scale to determine whether the fusion protein is soluble or insoluble. The vector(s) that yield soluble protein are preferred; however, if both vectors yield insoluble fusion protein, then only the GyrA fusion can be carried forward.

For N-terminal ligation, a cleavable fusion partner for the N-terminus of the protein must be chosen. It is important to realize that cleavage of the N-terminal tag must yield a ^{1}Cys recombinant protein fragment for the NCL reaction. Proteases typically used for this task are factor Xa (Erlanson et al., 1996; Zheng et al., 2003), TEV protease (Tolbert et al., 2005), and SUMO protease. Alternatively, one may rely on bacterial methionine aminopeptidase to expose an N-terminal Cys after translation; however, it is impossible to predict to what extent this will occur in vivo (He et al., 2003). The GST–factor Xa vectors (pGEX) and factor Xa protease are commercially available from GE Healthcare Lifesciences (formerly Amersham Biosciences), a His$_6$ tag and the TEV protease recognition sequence (ENLYFQ/G) can be introduced easily by PCR and His$_6$ tagged TEV protease can be produced recombinantly, and the Champion pET-SUMO expression system is commercially available from Invitrogen. Although TEV and SUMO proteases are more specific than factor Xa, one should recognize that cleavage with any protease can lead to the formation of unwanted degradation products. After the desired N-terminally truncated candidate protein is subcloned into the appropriate vector, expression levels should be checked on a small scale to determine whether the fusion protein is soluble or insoluble. Larger-scale protein purification is then necessary to work out conditions for specific proteolytic cleavage to yield the ^{1}Cys recombinant protein fragment and its subsequent purification for use in N-terminal NCL.

4.4. Peptide synthesis

Once it can be established that the recombinant protein fragment is accessible for semisynthesis, peptide synthesis can be pursued. The use of either native phosphorylated amino acids or nonhydrolyzable mimics depends on the project's goals. For example, there is little benefit to using a phospho-mimetic, which is a greater experimental investment, if the semisynthetic protein will not be exposed to phosphatases in subsequent experiments (Huse et al., 2001; Muir et al., 1998; Ottessen et al., 2004). With the exception of the pTyr mimetic phosphonomethylene phenylalanine (Pmp, Advance ChemTech), we are unaware of a commercial source for nonhydrolyzable phosphonates for use in Fmoc SPPS; therefore, these unnatural amino acids must be synthesized as needed. In general, the preferred phosphonates are phosphono-difluoromethylene alanine (Pfa), a pSer/pThr mimetic, and phosphono-difluoromethylene phenylalanine (F$_2$Pmp), a pTyr mimetic (Fig. 1.1C). Although the corresponding CH$_2$ phosphonates are a reasonable option and have been used successfully by our group, the presence of the halogens act to lower the second pK_a of the phosphonate to be similar to that observed for the natural phosphorylated amino acid. A discussion of the synthetic schemes for these analogues is beyond the scope of this manuscript; however, we have had success with the procedures of Berkowitz et al. (1996) (Pfa) and Guo et al. (2002) (F$_2$Pmp).

When embarking on a new Fmoc SPPS project, our lab follows standard protocols. For incorporation of phosphonates, which are typically limiting, we use a single coupling reaction with 1.2 eq of Fmoc amino acid for 4 to 10 h, followed by N-terminal capping with acetic anhydride and normal peptide extension. N-terminal Cys peptide acids for EPL are typically assembled on the Wang resin, but when synthesizing $^\alpha$thioesters for N-terminal NCL, the weak acid labile 2-Chlorotrityl (2-Cl-Trt) resin can be used. Although there are now several methods for synthesizing $^\alpha$thioesters by Fmoc SPPS (Botti et al., 2004; Gross et al., 2005), our lab has had success using the method described by Futaki et al. (1997). The thioester peptides/protein fragments need to be stored at low pH (<5) and temperature to prevent hydrolysis.

4.5. C-terminal semisynthesis (EPL) on a soluble protein

C-terminal EPL on a soluble protein is probably the simplest of the semi-synthetic methods. As a starting point for C-terminal EPL, three HEPES-based buffers are typically used in our lab:

1. Cell lysis buffer: 25 mM HEPES (pH 7 to 7.5), 150 mM NaCl, 1 mM MgSO$_4$, 5% glycerol, 5% ethylene glycol (protease inhibitors as needed)
2. Chitin column buffer: 50 mM HEPES (pH 7 to 7.5), 250 mM NaCl, 1 mM EDTA, 0.1% triton X-100 (protease inhibitors as needed)
3. Ligation buffer: 50 mM HEPES (pH 7 to 7.5), 250 mM NaCl, 1 mM EDTA (protease inhibitors as needed)

To perform NCL, one of several thiols can be selected. The major options include MESNA, thiophenol, and mercaptophenylacetic acid (MPAA), and we usually compare the efficiency of each for a given case (Johnson and Kent, 2006). These thiols will intercept the recombinant protein $^\alpha$thioester to yield a more reactive $^\alpha$thioester for C-terminal EPL. Ligations in the presence of MESNA can be somewhat slower than in the thiophenol or MPAA reactions (Ayers et al., 1999; Johnson and Kent, 2006). Thiophenol can be problematic because of its low solubility in aqueous buffers (generally 0.5 to 2% v/v is used) and consequently its presence may have deleterious effects on protein stability and solubility. However, MPAA offers a compromise between MESNA and thiophenol, as it is quite soluble and forms a very reactive $^\alpha$thioester (Johnson and Kent, 2006). Because MESNA can induce unexpected enzymatic properties (Zhang *et al.*, 2003), it is advisable to rule out unwanted effects of the selected thiol on a protein of interest prior to modification.

For a protein that expresses well in *E. coli* (i.e., $\geq$5 mg/L of culture), C-terminal EPL is generally conducted by starting with the cell paste from 1 L of culture. Cells are resuspended in ice-cold lysis buffer, lysed via double pass on a French press (16,000–18,000 psi), and clarified by centrifugation at 25,000 $\times$ *g* for 15 min (4°C). The clarified soluble protein is then double loaded onto a pre-equilibrated 5-mL chitin column (chitin available from New England Biolabs) at approximately 1 mL/min (4°C). The column is washed with 20 vol of chitin column buffer to remove impurities, followed by 10 vol of ligation buffer to remove detergent (4°C). At this point, the column is allowed to warm up to room temperature for 30 min. For EPL initiation, two solutions are needed:

1. 5 mL ligation buffer + 200 mM thiol
2. 5 mL ligation buffer + 200 mM thiol + 1 to 2 mM ^{1}Cys-peptide for ligation.

Because NCL is a bimolecular reaction, the rate of ligation should generally be proportional to the amount of peptide used. The solutions for initiation are prepared at room temperature and the pH must be readjusted to 7.0 to 7.5 on addition of thiol and synthetic peptide. For longer ligation times, the solutions should be purged with N_2 to prevent the oxidation of thiol. To initiate EPL, 5 mL of solution A (thiol solution) is quickly passed through the column and the flow-through discarded. This is followed immediately by 5 mL of solution B (peptide solution), but the flow-through is collected and loaded back onto the top of the column, both ends of the column are sealed, and the column is purged with N_2 and incubated at room temperature for the duration of the ligation reaction. Adding agitation (e.g., rotation in three dimensions) and increasing the temperature may improve or reduce ligation efficiency and depends on the protein of interest.

Typically, C-terminal EPL takes between 24 h and 96 h to reach greater than 80% completion depending on the thiol used. Reaction progress can be monitored via SDS-PAGE by pipetting some solution off the top of the column and running a gel. One can generally distinguish the bands corresponding to the ligated and unligated semisynthetic protein (this is easy when the mass difference is $\geq$1000 Da). Although MALDI-TOF MS is not strictly quantitative, it can be used to monitor reaction progress; however, the protein must be desalted before analysis. Desalting is usually achieved by C_4 zip-tip purification, but sometimes the high concentrations of thiol or salt cannot be removed, thus rendering MALDI-TOF MS analysis impossible at this stage.

On completion of C-terminal EPL, the protein is eluted from the column with ligation buffer (no thiol), and protein-containing fractions are pooled and dialyzed into storage buffer overnight to remove thiol. The semisynthetic protein is then concentrated and subjected to size exclusion chromatography to remove excess ^{1}Cys peptide. A 1-L preparation of a protein that expresses well can often yield 3 to 5 mg of semisynthetic protein. If expression is poor, the initial amount of culture can be scaled up for the desired production.

4.6. C-terminal EPL on an insoluble protein

Conducting protein semisynthesis on an insoluble protein requires a similar approach but should be undertaken only if the target protein can easily be refolded. As mentioned earlier, investigators are limited to using the smaller GyrA intein because it can be denatured and refolded to a functional state (Schwarzer *et al.*, 2006; Valiyaveetil *et al.*, 2002). In our experience, the GyrA intein is functional at up to 3 M urea. It should be noted that the CBD has lower affinity for chitin under these denaturing conditions. Therefore, it is beneficial to determine the optimum urea concentration to maintain fusion protein solubility, intein functionality, and chitin affinity.

Typically, the insoluble protein from washed inclusion bodies is resuspended in 8 M urea buffer and then the GyrA intein is refolded by dialysis down to $\leq$3 M urea for intein mediated protein $^\alpha$thioester formation. The refolded soluble fusion protein is then loaded onto pre-equilibrated chitin beads (when affinity permits) and extensively washed or used as is. The $^\alpha$thioester protein is generated by treating the intein fusion protein with 200 mM thiol (MESNA, thiophenol, or MPAA) in the absence of ^{1}Cys peptide. If the $^\alpha$thioester protein precipitates after cleavage from the intein fusion protein, it can be solubilized with a mixture of 50% trifluoroethanol (TFE) and 0.1% trifluoroacetic acid (TFA) in H_2O and dried on a lyophilizer. This lyophilized protein mixture will contain the desired protein $^\alpha$thioester and may contain the uncleaved fusion protein as well as the intein–CBD that results from cleavage. This mixture can be used directly for ligation to ^{1}Cys peptide by simply mixing 1 to 10 eq of synthetic ^{1}Cys peptide with recombinant protein $^\alpha$thioester in the presence of 200 mM thiol (MESNA, thiophenol, or MPAA) in pH 7.0 to 7.5 buffer. Additives such as sodium dodecylsulfate (SDS) can also be used to completely solubilize the mixture of proteins. It should be noted that if SDS is used, it must be removed after ligation. For this purpose, we have used Extracti-Gel D Detergent Removing Gel (Pierce). Optimizing the level of agitation and temperature of the ligation reaction may result in better yields of semisynthetic protein. Upon completion of peptide ligation, the semisynthetic protein can either be refolded and purified by a subsequent chromatography step (i.e., affinity, ion exchange, or size-exclusion chromatography) or kept in the denatured state and purified by RP–HPLC on a C4 column followed by refolding (Schwarzer *et al.*, 2006).

4.7. N-terminal semisynthesis

As has been discussed, in this complementary semisynthetic approach, the $^{\alpha}$thioester for ligation is generated synthetically and the ^{1}Cys protein for ligation is generated by proteolytic cleavage of a recombinant protein. Ligation is afforded by mixing 1 to 10 eq of synthetic peptide $^{\alpha}$thioester with recombinant ^{1}Cys protein in the presence of 200 mM thiol (MESNA, thiophenol, or MPAA) in pH 7.0 to 7.5 buffer (Zheng *et al.*, 2003). Typically this is done in a small volume so that, on completion of the ligation reaction, the mixture can be directly subjected to size-exclusion chromatography to remove excess synthetic peptide $^{\alpha}$thioester and small-molecule thiol. Again, this procedure can be carried out on an insoluble protein provided that the ^{1}Cys protein can be generated under denaturing conditions and the target protein can easily be refolded.

4.8. Purification of the semisynthetic protein

If the ligation reaction does not go to $\geq 95\%$ completion, the semisynthetic protein must be further purified to remove the unligated protein and prevent its interference with biological evaluation. This separation may be problematic if there is not a great enough difference in charge, size, or affinity for a ligand on generation of the semisynthetic protein. In some cases, a biotin tag can be included on the synthetic peptide during Fmoc SPPS as a handle for purification, but this approach should be used with caution because it introduces another variable into the evaluation of the semisynthetic protein. With any semisynthetic project, the mass of the final purified semisynthetic protein should be confirmed by MALDI–TOF MS.

5. PRACTICAL USES OF SEMISYNTHETIC PHOSPHOPROTEINS

5.1. Kinetic analysis of phosphonylated enzymes

The combination of protein semisynthesis, phosphonate analogue technology, and kinetic analysis yields a powerful tool that can be used to dissect complex signaling cascades. The combined approaches enable the investigator to determine the effect of stoichiometric phosphorylation on enzyme activity, which often provides unique insights. The methods proved especially valuable in analyzing the protein tyrosine phosphatases (PTPs) SHP-1, SHP-2, and low molecular weight (LMW) PTP.

Both SHP-1 and SHP-2 are cytosolic PTPs that share a common domain structure that includes two SH2 domains followed by the PTP domain and a C-terminal tail. SHP-1 plays a role as a negative regulator of cell signaling in cells of hematopoietic lineage (Zhang et al., 2000), whereas SHP-2 is critical in the early development of many species, including mammals (Feng, 1999). The C-terminal tail of each PTP contains two sites of tyrosine phosphorylation that have been proposed to be a possible regulatory mechanism (Bennett et al., 1994; Feng et al., 1993; Vogel et al., 1993). This hypothesis could not be tested in the absence of phosphonate analogues because of the inherent tendency of the PTPases to autodephosphorylate. EPL was used to incorporate nonhydrolyzable phosphonates at these positions, thus allowing for the effect of phosphorylation of each position to be determined. With respect to SHP-1, phosphonylation with F_2Pmp at Tyr^{536} and Tyr^{564} led to 8-fold and 1.6-fold stimulation of PTPase activity, respectively. These findings formed the basis for a structural model in which $pTyr^{536}$ and $pTyr^{564}$ engage the SH2 domains intramolecularly to relieve basal inhibition (Zhang et al., 2003). The findings with SHP-2 (Tyr^{542} and Tyr^{580}) were similar to those with SHP-1, thus allowing a similar model for regulation to be proposed (Lu et al., 2001, 2003). Semisynthesis with SHP-2 was extended to include bis-phosphonylated enzyme, where it was shown that the Pmp groups have close to additive effects on PTPase activity, thereby suggesting dual occupancy of the SH2 domains. Furthermore, catalytically inactive forms of phosphorylated SHP-2 were generated for use in a systematic analysis of intermolecular autodephosphorylation. In these studies, autodephosphorylation was shown to be both dependent on the surrounding sequence and competitive with the intramolecular binding to the SH2 domains (Lu et al., 2003).

LMW-PTP is a nonclassical PTPase involved in regulating growth factor responses and in reorganizing the cytoskeleton (Chiarugi et al., 1995; Nimnual et al., 2003; Raugei et al., 2002). A potential mechanism for regulating LMW-PTPase is tyrosine phosphorylation at positions 131 and 132, which are close to the active site. Investigating the roles of these residues has been difficult because of the intrinsic tendency of the enzyme to autodephosphorylate. Previous groups addressed this problem by using ATPγS-mediated phosphorylation of LMW-PTP; however, this approach can be problematic, as described earlier. Nevertheless, these teams reported stimulation of PTPase activity on tail phosphorylation (Bucciantini et al., 1999; Tailor et al., 1997). Our lab revisited this problem using a combination of protein semisynthesis and phosphonate analogue technology to produce site-specific, stoichiometrically phosphonylated LMW-PTP for detailed kinetic analysis (Schwarzer et al., 2006). With physiological phosphopeptide substrates derived from the PDGF receptor and p190RhoGap, it was shown that tail phosphorylation led to a decrease in PTPase activity and that the effect of the *bis*-phosphonylation was almost additive. This represents, to our knowledge, the first example of a PTPase being inhibited by tyrosine phosphorylation. In this manner, phosphorylation of LMW-PTP would act to remove a negative influence on PDGF receptor stimulation.

5.2. Microinjection of phosphonylated enzymes

Microinjection provides the investigator with a way to introduce semisynthetic proteins into a cell. This is achieved by applying direct pressure to the cell membrane with a glass Femtotip. The advantages of this technique are its versatility (almost anything can be injected into the cell with no limit on size or charge), temporal control, and the ability to target delivery to either the cytoplasm or the nucleus. With microinjection, one can rapidly increase the level of a modified kinase or phosphatase and follow its downstream effects via reporter construct. Protein levels and stability can be directly quantitated by immunofluorescence, and cellular localization can be mapped with confocal microscopy. The main disadvantages of microinjection are that it is time consuming, requires concentrated proteins (>1 mg/mL) and fixed cells, and is not amenable to high-throughput analysis. Other disadvantages of the technique include the small sample size (only about 100 cells can be microinjected per time point) and the expertise and apparatus needed for the experiments.

To perform microinjection experiments, one needs an inverted microscope, either a phase contrast scope (for flat specimens like those in tissue cell culture) or differential-interference contrast (DIC) scope (for rounded cells and nuclear injection). In addition, both a micromanipulator (to control where the needle is) and a microinjector (or air regulator to control the pressure of the fluid into the needle) are required. The needles used for microinjection (Femtotips) and the microloaders used to fill them are available from Eppendorf.

Before beginning microinjection experiments, an appropriate cell line must be selected. This can depend on the protein or signal pathway being studied and on the robustness of the cell line. Effects observed in some cell lines may be exaggerated or not observed at all in other cell lines. Cells should be plated (seeded) at least 12 h prior to microinjection, though 24 h is optimal. Most cell types are very sensitive to plasma membrane damage during the early hours after seeding, and microinjection works best at 70 to 80% confluency. Some cell lines are very sensitive to pH change that occurs while a dish is outside the CO_2 incubator. For this reason, cells cannot be microinjected for longer than 15 min. To circumvent this problem, bicarbonate-free media (Leibovitz L-15 medium) or a microscope fitted with a CO_2 incubator can be used.

Proper sample preparation can be a critical aspect of the microinjection experiments. Typical protein concentrations used are 1 to 5 mg/mL. Although this can be varied, samples that are too concentrated are likely to clog the needle tip, and samples that are too dilute may be difficult to detect by immunofluorescence. The sample needs to be diluted in a physiological buffer. An example of an appropriate sample buffer is 20 mM Tris pH 7.5, 20 mM NaCl, 1 mM MgCl$_2$, 150 mM KCl (if the protein is sensitive to oxidation, 1 to 5 mM 2-mercaptoethanol can be included). Each sample needs to be clarified via high-speed centrifugation prior to loading into the Femtotip to remove any particulate matter that can clog the tip (Ridley, 1995).

Adherent cells should be microinjected with the tip at 30 to 40° angle, so the tip penetrates the cell by moving in a vertical z–axis movement. The volume injected into cells depends on the pressure and is usually femtoliters to nanoliters. Typically pressures of 5 to 12 kPa are used and injection times are between 0.2 s and 1 s. Ideally injection volume is 10% of the cell volume (Minaschek et al., 1989). Injections can be performed manually, in which the operator controls the movement of the tip and length of injection. Alternatively, one can use automatic injection. In this case, the Z–line (or vertical movement of the needle into the cell) and length of injection are fixed. This provides more consistency in the volumes being injected into the cells, but it can be troublesome if the cells are not flat or the dish is not level. A tip diameter of ≈0.3 μm is optimal for microinjection of mammalian cells. Increases in tip diameter result in increases in delivery rate, which can cause more cell damage. Decreasing the tip diameter can result in frequent clogging of the tip and loss of sample (once a tip is clogged, it can no longer be used and sample cannot be recovered). For nuclear injection, a smaller shoulder angle on the tip is necessary to limit the delivery of protein into the cytoplasm above the nucleus.

During the course of a microinjection, one needs to work quickly and keep track of the number of cells that were injected (typically 50 to 100 per time point). To practice the microinjection technique, FITC-labeled dextran (available from Sigma at 5 mg/mL) can be used. This provides a quick way to determine whether one's microinjection technique is successful. Because dextran is an inert polymer of high molecular weight, it will remain in the cell compartment into which it was injected. If the microscope is fitted with a fluorescent lamp and appropriate filter, the cells can be visualized immediately following the injection. One should not attempt to microinject a cell more than once. If it is desired that two or more proteins are injected into the same cell, they should be mixed together at the desired ratio before sample preparation and then injected together (Lamb et al., 1996).

Immediately after injections are complete, the dishes with the cells should be returned to the incubator. Following suitable incubation times, the cells can be fixed in 4% formaldehyde/phosphate buffered saline (PBS) solution. After washing with 10 mM glycine in PBS, the cells are permeabilized with

0.2% Triton-X100. At this point, the cells can be immunostained with the primary antibody followed by a fluorophore-conjugated secondary antibody. Blocking with 10 mg/mL BSA or other blocking agents is recommended to prevent nonspecific binding of the antibodies. Dilutions for the antibodies will most likely need to be optimized. If the signal is poor, it may be necessary to increase the protein concentration in the injected sample. The stained cells then can be analyzed and quantitated by fluorescence microscopy. Our lab has used microinjection to investigate the roles of phosphorylation on the cellular stability of AANAT and LMW-PTP, the cellular localization of SHP-2 and LMW-PTP, and the downstream effects of SHP-2 (Lu *et al.*, 2001; Schwarzer *et al.*, 2006; Zheng *et al.*, 2003, 2005).

The penultimate enzyme in the melatonin biosynthetic pathway is AANAT (Klein *et al.*, 1997). The diurnal rise and fall of melatonin is governed by the rhythmic phosphorylation of this enzyme at two sites (Thr^{31} and Ser^{205}) (Ganguly *et al.*, 2005). Microinjection experiments were key in elucidating the role of phosphorylation in regulation of AANAT activity. The results of studies in which AANAT modified with Pma at Thr^{31} was injected into cells provided the first direct evidence that Thr^{31} phosphorylation controls AANAT cellular stability (Zheng *et al.*, 2003). Semisynthetic AANAT-Pma^{31} injected into CHO cells and measured by immunofluorescent staining had 3-fold greater signal than that of unmodified AANAT when assessed at 1 h after cellular microinjection. After 4 h, AANAT-Pma^{31} continued to show substantial fluorescent signal, whereas the signal from the unmodified enzyme was no longer present. This finding confirmed that phosphorylation at Thr^{31} protects against proteolytic degradation. Note that the corresponding Glu replacement was unable to mimic this effect, consistent with our understanding of 14-3-3/phosphopeptide interaction (Zheng *et al.*, 2003). Phosphorylation at Ser^{205} was also investigated and yielded the same result; phosphorylation of the residue was important for cellular stability of AANAT (Zheng *et al.*, 2005). The findings, coupled with earlier studies that identified 14-3-3 proteins as ligands for phosphorylated AANAT, helped confirm the 14-3-3/phosphorylation-dependent model for AANAT regulation (Ganguly *et al.*, 2001). In contrast, similar studies conducted with LMW-PTP indicated that phosphorylation did not play a role in regulating the cellular stability of the enzyme (Schwarzer *et al.*, 2006).

Phosphorylation may play a role in changing the cellular localization of a protein. In these cases, microinjection followed by confocal microscopy can be used to follow protein movement. With respect to the two proteins that our lab has investigated via this technique (SHP-2 and LMW-PTP), there has been no change in cellular localization on protein modification (Lu *et al.*, 2001; Schwarzer *et al.*, 2006). By contrast, a change in cellular localization on phosphorylation was observed with the semisynthetically prepared Smad2-MH protein (Hahn and Muir, 2004). Here, protein semisynthesis was used to introduce two photocaged (discussed later) phosphoserines at the C-terminus of the protein. The caged Smad2 was bound to SARA (Smad anchor for receptor activation) in the cytosol, but on uncaging by brief irradiation, it localized to the nucleus and promoted gene transcription (Hahn and Muir, 2004).

Microinjection can be combined with reporter plasmid technology to provide a functional assay for the injected protein by monitoring downstream effects. This methodology was useful in elucidating the role of SHP-2 phosphorylation *in vivo*. Co-injection of phosphonylated SHP-2 and the reporter construct 5xSRE-CAT, which allows the investigator to demonstrate the ability of the microinjected protein to transcriptionally activate the serum response element (SRE), indicated that single phosphorylation at Tyr^{542} was sufficient to activate the MAP kinase pathway (Lu *et al.*, 2001). This result was previously unattainable because of the ability of SHP-2 to autodephosphorylate.

5.3. Pull-down assays using phosphonylated enzymes as bait

Phosphorylation on proteins can directly lead to the formation of multimolecular signaling complexes through specific interactions between phosphoprotein binding domains and phosphorylated proteins. Enzymes bearing the nonhydrolyzable phosphonate modification can be used as bait to pulldown novel interacting proteins from cell lysates because of their stability to phosphatases. In addition, pull-down assays can be used to demonstrate specific binding between two purified proteins.

As was previously discussed, the phosphorylation of Thr^{31} and Ser^{205} of AANAT leads to its cellular stabilization. Through the use of pull-down studies, it could be demonstrated that the phosphorylation of these residues leads to the recruitment of 14-3-3, which in turn protects AANAT from degradation *in vivo* (Zheng *et al.*, 2005). Similar pull-down assays were performed with semisynthetic SHP-2 and demonstrated that phosphorylation at Tyr^{542} was sufficient to enable binding to Grb2, an SH2 domain containing adaptor protein (Lu *et al.*, 2001).

 # 6. FUTURE OF PROTEIN SEMISYNTHESIS IN SIGNALING

Where does protein phosphorylation by semisynthesis go from here? We have seen in the previous examples that incorporation of phosphonates into native proteins and their subsequent analysis in activity assays, microinjection assays, and pull-down assays gives the investigator access to data that is otherwise not available because of the reversible nature of the modification. In this manner, it is possible to deconvolute the specific roles of phosphorylation in complex systems (e.g., regulation of activity, stability, cellular localization, and protein–protein interactions) and piece together signaling cascades one phosphorylation event at a time.

Currently, our lab is expanding on the notion of phosphorylation by protein semisynthesis to include ATP linkage via a thiophosphate at sites of protein phosphorylation (Parang *et al.*, 2001). In this manner, a high-affinity kinase ligand can be generated. This approach has recently been used to convert Src into a high affinity Csk ligand. Src-ATP was a fairly potent inhibitor of the Csk kinase activity, with a K_i of 100 nM, and could be used as bait to selectively pull-down recombinant Csk from a 1% spiked mammalian cell lysate (Shen and Cole, 2003). The studies outline a general strategy for identifying unknown kinases that might be responsible for the phosphorylation of a specific target protein and add another tool to be used in the dissection of signaling cascades. The utility of ATP linkage is not limited to identifying kinases. The high-affinity protein–ATP conjugates could be used to facilitate X-ray crystal structure determination of the kinase with its physiological substrate by stabilizing the complex.

Nonhydrolyzable phosphomimetics such as Pfa and Pmp are not the only phosphorylated amino acid substitutions that can be incorporated into proteins through semisynthesis. Photolabile caged analogues of phosphorylated residues allowing temporal and spatial control provide further tools for uncovering the regulation of signaling pathways (Rothman *et al.*, 2005). Caged amino acids can be introduced into the cell in an inert form by microinjection and then switched on with a short burst of light in a spatially well-defined fashion (Lawrence, 2005). A caged thiophospho-Thr[197] variant of PKA was prepared by semisynthesis and on uncaging of this protein 85 to 90% of activity was restored (Zou *et al.*, 2002). As previously discussed, Smad2 could be prepared with two photocaged phosphoserines at the C-terminus and on irradiation became activated (Hahn and Muir, 2004).

The advantages of protein semisynthesis are its versatility with respect to the protein of interest, the type of modification to be investigated, and the position of the modified amino acid within the protein. Therefore, the combination of this technology with unnatural amino acids and traditional experimental approaches yields a level of experimental control that was previously elusive. Such control of stoichiometry, specificity, and reversibility allows complex signaling systems to be broken down to individual events and discretely examined. Relatively unexplored is the use of semisynthesis to investigate other reversible posttranslational modifications including glycosylation, acetylation, methylation, and ubiquitination (He *et al.*, 2003; Shogren-Knaak *et al.*, 2006; Thompson *et al.*, 2004). In the future, protein semisynthesis in combination with new technologies is likely to play an increasingly important role in the elucidation of mechanisms in complex biological systems.

ACKNOWLEDGMENTS

We are grateful for the insightful comments of many of our colleagues, most of whose names are noted in the references. We thank the National Institutes of Health for financial support.

REFERENCES

Ayers, B., Blaschke, U. K., Camarero, J. A., Cotton, G. J., Holford, M., and Muir, T. W. (1999). Introduction of unnatural amino acids into proteins using expressed protein ligation. *Biopolymers* **51,** 343–354.

Alonso, A., Sasin, J., Bottini, N., Friedberg, I., Freidberg, I., Ostermann, A., Godzik, A., Hunter, T., Dixon, J., and Mustelin, T. (2004). Protein tyrosine phosphatases in the human genome. *Cell* **117,** 699–711.

Bennett, A. M., Tang, T. L., Sugimoto, S., Walsh, C. T., and Neel, B. G. (1994). Protein-tyrosine-phosphatase SHPTP2 couples platelet-derived growth factor receptor beta to Ras. *Proc. Natl. Acad. Sci. USA* **91,** 7335–7339.

Berkowitz, D. B., Eggen, M., Shen, Q., and Shoemaker, R. K. (1996). Ready access to fluorinated phosphonate mimics of secondary phosphatases: Synthesis of the (alpha,alpha-Difluoroalkyl)phosphonate analogues of L-phosphoserine, L-phosphoallothreoine, and L-phosphothreoine. *J. Org. Chem.* **61,** 4666–4675.

Blume-Jensen, P., and Hunter, T. (2001). Oncogenic kinase signalling. *Nature* **411,** 355–365.

Botti, P., Villain, M., Manganiello, S., and Gaertner, H. (2004). Native chemical ligation through in situ O to S acyl shift. *Org. Lett.* **6,** 4861–4864.

Bucciantini, M., Chiarugi, P., Cirri, P., Taddei, L., Stefani, M., Raugei, G., Nordlund, P., and Ranponi, G. (1999). The low Mr phosphotyrosine protein phosphatase behaves differently when phosphorylated at Tyr 131 or Tyr 132 by Src kinase. *FEBS Lett.* **465,** 73–78.

Chen, L., Wu, L., Otaka, A., Smyth, M. S., Roller, P. P., Burke, T. R., den Hertog, J., and Zhang, Z. Y. (1995). Why is phosphonodifluoromethyl phenylalanine a more potent inhibitory moiety than phosphonomethyl phenylalanine toward protein-tyrosine phosphatases? *Biochem. Biophys. Res. Commun.* **216,** 976–984.

Chiarugi, P., Cirri, P., Raugei, G., Camici, G., Dolfi, F., Berti, A., and Ramponi, G. (1995). PDGF receptor as a specific *in vivo* target for low M(r) phosphotyrosine protein phosphatase. *FEBS Lett.* **373,** 49–53.

Cho, H., Krishnaraj, R., Itoh, M., Kitas, E., Bannwarth, W., Saito, H., and Walsh, C. T. (1993). Substrate specificities of catalytic fragments of protein tyrosine phosphatases (HPTPb, LAR, and CD45) toward phosphotyrosylpeptide substrates and thiophosphorylated peptides as inhibitors. *Protein Sci.* **2,** 977–984.

Chong, S., Mersha, F. B., Comb, D. G., Scott, M. E., Landry, D., Vence, L. M., Perler, F. B., Benner, J., Kucera, R. B., Hirvonen, C. A., Pelletier, J. J., Paulus, H., et al. (1997). Single-column purification of free recombinant proteins using a self-cleavable affinity tag derived from a protein splicing element. *Gene* **192,** 271–281.

Cohen, P. (2000). The regulation of protein function by multisite phosphorylation. *Trends Biochem. Sci.* **25,** 596–601.

Cohen, P. (2002). Protein kinases: The major drug targets of the twenty-first century? *Nat. Rev. Drug Discov.* **1,** 309–315.

Cole, P. A., Burn, P., Takacs, B., and Walsh, C. T. (1994). Evaluation of the catalytic mechanism of recombinant human Csk (C-terminal Src kinase) using nucleotide analogs and viscosity effects. *J. Biol. Chem.* **269,** 30880–30887.

Dawson, P. E., Muir, T. W., Clark-Lewis, I., and Kent, S. B. (1994). Synthesis of proteins by native chemical ligation. *Science* **266,** 776–779.

Desmarais, S., Friesen, R. W., Zamboni, R., and Ramachandran, C. (1999). [Difluoro (phosphono)methyl]phenylalanine-containing peptide inhibitors of protein kinase phosphatases. *Biochem. J.* **337,** 219–223.

Eckstein, F. (1983). Phosphorothioate analogues of nucleotides-tools for the investigation of biochemical processes. *Angew. Chem. Int. Ed. Engl.* **22,** 423–439.

Erlanson, D. A., Chytil, M., and Verdine, G. L. (1996). The leucine zipper domain controls the orientation of AP-1 in the NFAT.AP-1.DNA complex. *Chem. Biol.* **3,** 981–991.

Evans, T. C., Benner, J., and Xu, M. Q. (1998). Semisynthesis of cytotoxic proteins using a modified protein splicing element. *Protein Sci.* **7,** 2256–2264.

Feng, G. S. (1999). Shp-2 tyrosine phosphatase: Signaling one cell or many. *Exp. Cell Res.* **253,** 47–54.

Feng, G. S., Hui, C. C., and Pawson, T. (1993). SH2-containing phosphotyrosine phosphatase as a target of protein-tyrosine kinases. *Science* **259,** 1607–1611.

Fischer, E. H., and Krebs, E. G. (1955). Conversion of phosphorylase b to phophorylase a in muscle extracts. *J. Biol. Chem.* **216,** 121–132.

Futaki, S., Sogawa, K., Maruyama, J., Asahara, T., Niwa, M., and Hojo, H. (1997). Preparation of peptide thioesters using Fmoc-solid-phase peptide synthesis and its application to the construction of a template-assembled synthetic protein (TASP). *Tetrahedron Lett.* **38,** 6237–6240.

Ganguly, S., Gastel, J. A., Weller, J. L., Schwartz, C., Jaffe, H., Namboodiri, M. A., Coon, S. L., Hickman, A. B., Rollag, M., Obsil, T., Beasverger, P., Ferry, G., *et al.* (2001). Role of a pineal cAMP-operated arylalkylamine N-acetyltransferase/14-3-3-binding switch in melatonin synthesis. *Proc. Natl. Acad. Sci. USA* **98,** 8083–8088.

Ganguly, S., Weller, J. L., Ho, A., Chemineau, P., Malpaux, B., and Klein, D. C. (2005). Melatonin synthesis: 14-3-3-dependent activation and inhibition of arylalkylamine N-acetyltransferase mediated by phosphoserine-205. *Proc. Natl. Acad. Sci. USA* **102,** 1222–1227.

Gross, C. M., Lelievre, D., Woodward, C. K., and Barany, G. (2005). Preparation of protected peptidyl thioester intermediates for native chemical ligation by Nalpha-9-fluorenymethoxycarbonyl (Fmoc) chemistry: Considerations of side-chain and backbone anchoring strategies, and compatible protection for N-terminal cysteine. *J. Pept. Res.* **65,** 395–410.

Guo, X. L., Shen, K., Wong, F., Lawrence, D. S., and Zhang, Z. Y. (2002). Probing the molecular basis for potent and selective protein-tyrosine phosphatase 1B inhibition. *J. Biol. Chem.* **277,** 41014–41022.

Hackeng, T. M., Griffin, J. H., and Dawson, P. E. (1999). Protein synthesis by native chemical ligation: Expanded scope by using straightforward methodology. *Proc. Natl. Acad. Sci. USA* **96,** 10068–10073.

Hahn, M. E., and Muir, T. W. (2004). Photocontrol of Smad2, a multiphosphorylated cell-signaling protein, through caging of activating phosphoserines. *Angew. Chem. Int. Ed. Engl.* **43,** 5800–5803.

Hao, M., Lowy, A. M., Kapoor, M., Deffie, A., Liu, G., and Lozano, G. (1996). Mutation of phosphoserine 389 effects p53 function *in vivo. J. Biol. Chem.* **271,** 29380–29385.

He, S., Bauman, D., Davis, J. S., Loyola, A., Nishioka, K., Gronlund, J. L., Reinberg, D., Meng, F., Kelleher, N., and McCafferty, D. G. (2003). Facile synthesis of site-specifically acetylated and methylated histone proteins: Reagents for evaluation of the histone code hypothesis. *Proc. Natl. Acad. Sci. USA* **100,** 12033–12038.

Huang, W., and Erikson, R. L. (1994). Constitutive activation of Mek1 by mutation of serine phosphorylation sites. *Proc. Natl. Acad. Sci. USA* **91,** 8960–8963.

Hunter, T. (2000). Signaling – 2000 and Beyond. *Cell* **100,** 113117.

Huse, M., Muir, T. W., Xu, L., Chen, Y. G., Kuriyan, J., and Massague, J. (2001). The TGF beta receptor activation process: An inhibitor- to substrate-binding switch. *Mol. Cell* **8,** 671–682.

Johnson, E. C., and Kent, S. B. (2006). Insights into the mechanism and catalysis of the native chemical ligation reaction. *J. Am. Chem. Soc.* **128,** 6640–6646.

Johnson, S. A., and Hunter, T. (2005). Kinomics: Methods for deciphering the kinome. *Nat. Methods* **2,** 17–25.

Klein, D. C., Coon, S. L., Roseboom, P. H., Weller, J. L., Bernard, M., Gastel, J. A., Zatz, M., Iuvone, P. M., Rodriguez, I. R., Begay, V., Falcon, J., Cahill, G. M., *et al.* (1997). The melatonin rhythm-generating enzyme: Molecular regulation of serotonin N-acetyltransferase in the pineal gland. *Recent Prog. Horm. Res.* **52,** 307–357.

Krebs, E. G., and Fischer, E. H. (1955). Phosphorylase activity of skeletal muscle extracts. *J. Biol. Chem.* **216,** 113–120.

Lamb, N. J., Gauthier-Rouviere, C., and Fernandez, A. (1996). Microinjection strategies for the study of mitogenic signaling in mammalian cells. *Front. Biosci.* **1,** 19–29.

Lawrence, D. S. (2005). The preparation and *in vivo* applications of caged peptides and proteins. *Curr. Opin. Chem. Biol.* **9,** 570–575.

Lu, W., Gong, D., Bar-Sagi, D., and Cole, P. A. (2001). Site-specific incorporation of a phosphotyrosine mimetic reveals a role for tyrosine phosphorylation of SHP-2 in cell signalling. *Mol. Cell* **8,** 759–769.

Lu, W., Shen, K., and Cole, P. A. (2003). Chemical dissection of the effects of tyrosine phosphorylation of SHP-2. *Biochemistry* **42,** 5461–5468.

Manning, G., Whyte, D. B., Martinez, R., Hunter, T., and Sudarsanam, S. (2002). The protein kinase complement of the human genome. *Science* **298,** 1912–1934.

Minaschek, G., Bereiter-Hahn, J., and Bertholdt, G. (1989). Quantitation of the volume of liquid injected into cells by means of pressure. *Exp. Cell Res.* **183,** 434–442.

Muir, T. W., Sondhi, D., and Cole, P. A. (1998). Expressed protein ligation: A general method for protein engineering. *Proc. Natl. Acad. Sci. USA* **95,** 6705–6710.

Nimnual, A. S., Taylor, L. J., and Bar-Sagi, D. (2003). Redox-dependent downregulation of Rho by Rac. *Nat. Cell Biol.* **5,** 236–241.

Ottesen, J. J., Huse, M., Sekedat, M. D., and Muir, T. W. (2004). Semisynthesis of phosphovariants of Smad2 reveals a substrate preference of the activated TGF beta RI kinase. *Biochemistry* **43,** 5698–5706.

Parang, K., Till, J. H., Ablooglu, A. J., Kohanski, R. A., Hubbard, S. R., and Cole, P. A. (2001). Mechanism-based design of a protein kinase inhibitor. *Nat. Struct. Biol.* **8,** 37–41.

Potter, L. R., and Hunter, T. (1998). Phosphorylation of the kinase homology domain is essential for activation of the A-type natriuretic peptide receptor. *Mol. Cell Biol.* **18,** 2164–2172.

Qiao, Y., Molina, H., Pandey, A., Zhang, J., and Cole, P. A. (2006). Chemical rescue of a mutant enzyme in living cells. *Science* **311,** 1293–1297.

Raugei, G., Ramponi, G., and Chiarugi, P. (2002). Low molecular weight protein tyrosine phosphatases: small, but smart. *Cell Mol. Life Sci.* **59,** 941–949.

Ridley, A. J. (1995). Microinjection of Rho and Rac into Quiescent Swiss 3T3 Cells. *Methods Enzymol.* **256,** 313–320.

Rothman, D. M., Shults, M. D., and Imperiali, B. (2005). Chemical approaches for investigating phosphorylation in signal transduction networks. *Trends Cell Biol.* **15,** 502–510.

Schwarzer, D., Zhang, Z., Zheng, W., and Cole, P. A. (2006). Negative regulation of a protein tyrosine phosphatase by tryrosine phosphorylation. *J. Am. Chem. Soc.* **128,** 4192–4193.

Shen, K., and Cole, P. A. (2003). Conversion of a tyrosine kinase protein substrate to a high affinity ligand by ATP linkage. *J. Am. Chem. Soc.* **125,** 16172–16173.

Shogren-Knaak, M. A., Alaimo, P. J., and Shokat, K. M. (2001). Recent advances in chemical approaches to the study of biological systems. *Annu. Rev. Cell Dev. Biol.* **17,** 405–433.

Shogren-Knaak, M., Ishii, H., Sun, J. M., Pazin, M. J., Davie, J. R., and Peterson, C. L. (2006). Histone H4-K16 acetylation controls chromatin structure and protein interactions. *Science* **311**, 844–847.

Tailor, P., Gilman, J., Williams, S., Coutur, C., and Mustelin, T. (1997). Regulation of the low molecular weight phosphotyrosine phosphatase by phosphorylation at tyrosines 131 and 132. *J. Biol. Chem.* **272**, 5371–5374.

Thompson, P. R., Wang, D., Wang, L., Fulco, M., Pediconi, N., Zhang, D., An, W., Ge, Q., Roeder, R. G., Wong, J., Levrero, M., Sartorelli, V., *et al.* (2004). Regulation of p300 HAT domain via a novel activation loop. *Nat. Struct. Mol. Biol.* **11**, 308–315.

Thorsness, P.E, and Koshland, D. E. (1987). Inactivation of isocitrate dehydrogenase by phosphorylation is mediated by the negative charge of the phosphate. *J. Biol. Chem.* **262**, 10422–10425.

Tolbert, T. J., Franke, D., and Wong, C. H. (2005). A new strategy for glycoprotein synthesis: Ligation of synthetic glycopeptides with truncated proteins expressed in E. coli as TEV protease cleavable fusion proteins. *Bioorg. Med. Chem.* **13**, 909–915.

Valiyaveetil, F. I., MacKinnon, R., and Muir, T. W. (2002). Semisynthesis and folding of the potassium channel KcsA. *J. Am. Chem. Soc.* **124**, 9113–9120.

Vogel, W., Lammers, R., Huang, J., and Ullrich, A. (1993). Activation of a phosphotyrosine phosphatase by tyrosine phosphorylation. *Science* **259**, 1611–1614.

Wang, L., Brock, A., Herberich, B., and Schultz, P. G. (2001). Expanding the genetic code of *Escherischia coli*. *Science* **292**, 498–500.

Williams, D. M., and Cole, P. A. (2001). Kinase Chips Hit the Proteomics Era. *Trends Biochem. Soc.* **26**, 271–273.

Xie, J., and Schultz, P. G. (2005). Adding amino acids to the genetic repertoire. *Curr. Opin. Chem. Biol.* **9**, 548–554.

Yaffe, M. B. (2002). Phosphotyrosine-binding domains in signal transduction. *Nat. Rev. Mol. Cell Biol.* **3**, 177–186.

Yaffe, M. B., and Elia, A. E. (2001). Phosphoserine/threonine-binding domains. *Curr. Opin. Cell Biol.* **13**, 131–138.

Zhang, Z. Y. (2005). Functional studies of protein tyrosine phosphatases with chemical approaches. *Biochim. Biophys. Acta* **1754**, 100–107.

Zhang, J., Somani, A. K., and Siminovitch, K. A. (2000). Roles of the SHP-1 tyrosine phosphatase in the negative regulation of cell signalling. *Semin. Immunol.* **12**, 361–378.

Zhang, Z., Shen, K., Lu, W., and Cole, P. A. (2003). The role of C-terminal tyrosine phosphorylation in the regulation of SHP-1 explored via expressed protein ligation. *J. Biol. Chem.* **278**, 4668–4674.

Zheng, W., Schwarzer, D., LeBeau, A., Weller, J. L., Klein, D. C., and Cole, P. A. (2005). Cellular stability of serotonin N-acetyltransferase conferred by phosphonodifluoro-methylene alanine (Pfa) substitution for Ser-205. *J. Biol. Chem.* **280**, 10462–10467.

Zheng, W., Zhongsen, Z., Ganguly, S., Weller, J. L., Klein, D. C., and Cole, P. A. (2003). Cellular stabilization of the melatonin rhythm enzyme induced by nonhydrolyzable phosphonate incorporation. *Nat. Struct. Biol.* **10**, 1054–1057.

Zisch, A. H., Pazzagli, C., Freeman, A. L., Schneller, M., Hadman, M., Smith, J. W., Ruoslahti, E., and Pasquale, E. B. (2000). Replacing two conserved tyrosines of the EphB2 receptor with glutamic acid prevents binding of SH2 domains without abrogating kinase activity and biological responses. *Oncogene* **19**, 177–187.

Zou, K., Cheley, S., Givens, R. S., and Bayley, H. (2002). Catalytic subunit of protein kinase A caged at the activating phosphothreonine. *J. Am. Chem. Soc.* **124**, 8220–8229.

CHAPTER TWO

PROTEIN ENGINEERING WITH THE TRACELESS STAUDINGER LIGATION

Annie Tam *and* Ronald T. Raines

Contents

Abstract

The engineering of proteins can illuminate their biological function and improve their performance in a variety of applications. Within the past decade, methods have been developed that facilitate the ability of chemists to manipulate proteins in a controlled manner. Here, we present the traceless Staudinger ligation as a strategy for the convergent chemical synthesis of proteins. This reaction unites a phosphinothioester and an azide to form an amide bond with no residual atoms. An important feature of this reaction is its ability to ligate peptides at noncysteine residues, thereby overcoming a limitation of alternative strategies. Attributes of the traceless Staudinger ligation are discussed, and an overall comparison of known reagents for effecting the reaction is presented.

Departments of Chemistry and Biochemistry, University of Wisconsin-Madison, Madison, Wisconsin, USA

Methods in Enzymology, Volume 462
ISSN 0076-6879, DOI: 10.1016/S0076-6879(09)62002-4

General methods are elaborated for the synthesis of the most efficacious phosphinothiol for mediating the traceless Staudinger ligation, as well as for the preparation of phosphinothioester and azide fragments and the ligation of peptides immobilized on a solid support. Together, this information facilitates the use of this emerging method to engineer proteins.

1. INTRODUCTION

The advent of recombinant DNA technology and site-directed muta-genesis has made facile the substitution of one amino acid for another at any site within a protein (Smith, 1994). For protein chemists, however, there remains a major barrier—the genetic code, which only tolerates the intro-duction of 20 amino acids. Methods that overcome this limitation but still rely on the ribosome are limited to the substitution of a subset of α-amino acids and α-hydroxyacids.

Driven by the desire to achieve complete flexibility in the manipulation of primary structure, protein chemists are developing methods that enable nonnatural amino acids and artificial modules to be incorporated into proteins. The most popular such method is "native chemical ligation," which was developed by Kent and coworkers as a means to join large peptide fragments (Kent, 2003). In native chemical ligation, the thiolate of an N-terminal cysteine residue of one peptide reacts with a C-terminal thioester of a second peptide, forming an amide bond after rapid $S \rightarrow N$ acyl group transfer. The ligation also works with selenocysteine—the rare "21st" amino acid—in the place of cysteine (Hondal and Raines, 2002; Hondal et al., 2001). An extension of native chemical ligation, "expressed protein ligation," employs an engineered intein to access a polypeptide containing the C-terminal thioester (Muir, 2003). Although these methods have produced landmark results, both require a cysteine residue at the ligation juncture. Cysteine is uncommon, comprising <2% of all protein residues. The introduction of a new cysteine residue can be detrimental, as its high nucleophilicity and propensity to oxidize leads to undesirable side reactions. Accordingly, many natural proteins can be neither synthesized nor modified by a ligation method that relies on cysteine residues.

2. TRACELESS STAUDINGER LIGATION

Emerging strategies for the unconstrained engineering of proteins avoid the requisite cysteine residues (Nilsson et al., 2005). Here, we describe one such strategy—the Staudinger ligation—which is based on the Stau-dinger reaction (Staudinger and Meyer, 1919). In the Staudinger reaction, a phosphine is used to reduce an azide to an amine: $PR_3 + N_3R' + H_2O \rightarrow O = PR_3 + H_2NR' + N_2(g)$. This reaction occurs via a stable intermediate, an iminophosphorane ($R_3P^{+}-^{-}NR'$, also known less precisely as an "aza-ylide"), which has a nucleophilic nitrogen. Vilarrasa and others showed that this nitrogen can be acylated, both in intermolecular (i.e., three-component) and intramolecular (i.e., two-component) ligations (Bosch *et al.*, 1995; Velasco *et al.*, 2000). Hydrolysis of the resulting amidophos-phonium salt gives an amide and a phosphine oxide. Bertozzi and coworkers showed that the phosphine itself can serve as the acyl group donor in a two-component ligation (Saxon and Bertozzi, 2000).

To apply the Staudinger reaction to peptide synthesis, we developed the use of a phosphinothiol to unite a thioester and azide, as shown in Fig. 2.1 (Nilsson *et al.*, 2000, 2001). This phosphinothiol is bifunctional, having a thiol group that can be tethered to the C-terminus of a peptide fragment, and a phosphino group that can react with a peptide fragment that has an azido group at its N-terminus to form an iminophosphorane intermediate. Attack of the iminophosphorane nitrogen on the conjoined thioester carbon leads first to a tetrahedral intermediate, and then to an amidophos-phonium salt (Soellner *et al.*, 2006a). Hydrolysis of the amidophosphonium salt releases a phosphine oxide and produces a native amide bond between the two peptides. Significantly, no extraneous atoms remain in the amide product—the reaction is "traceless" (Nilsson *et al.*, 2000). This attribute is a strict requirement for the use of the Staudinger ligation in the chemical synthesis of proteins or other molecules. It is noteworthy that the traceless Staudinger ligation mediated by a phosphinothiol couples the energetics of native chemical ligation with that of the Staudinger reaction (which is highly exergonic), resulting in an enormous thermodynamic driving force for the overall transformation (Nilsson *et al.*, 2005).

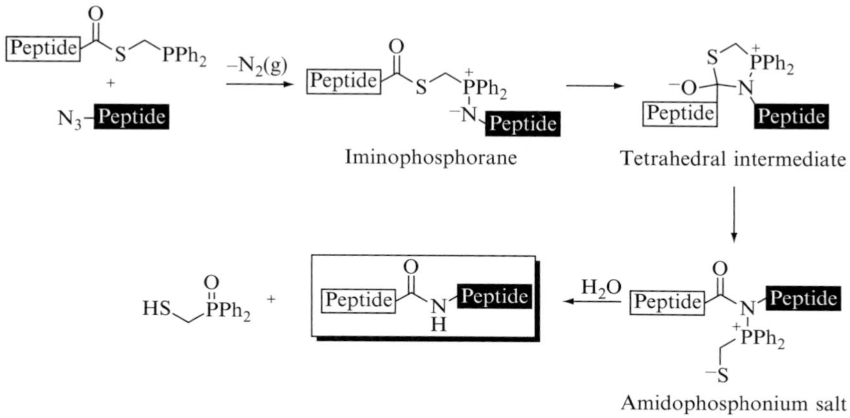

Figure 2.1   Putative mechanism for the traceless Staudinger ligation of two peptides.

The kinetics of the traceless Staudinger ligation have been characterized by using a sensitive and continuous assay based on ^{13}C NMR spectroscopy (Soellner et al., 2006a). In this assay, a phosphinothioester is allowed to react with a ^{13}C$^{\alpha}$-labeled azide in a deuterated solvent, and the course of the reaction is monitored over time. Significantly, intermediates do not accumulate, indicating that the rate-limiting step is the association of the phosphinothioester and the azide. For the reaction of AcGlySCH$_2$PPh$_2$ and N$_3$^{13}CH$_2$C(O)NHBn at room temperature, $t_{1/2} = 7$ min. The traceless Staudinger ligation proceeds without detectable (<0.5%) epimerization of the α-carbon of the azido acid (Soellner et al., 2002). This attribute is crucial for its application in protein chemistry, as all 20 proteinogenic amino acids except glycine have a stereogenic center at their α-carbon. The reaction of phosphinothioesters (but not phosphinoesters) with azides is also chemoselective in the presence of the functional groups in native proteins, and unprotected peptide fragments can be ligated with no undesirable side reactions (Soellner et al., 2006a). These attributes endow the Staudinger ligation with broad utility.

The traceless Staudinger ligation has been applied to the assembly of a protein from constituent peptides (Nilsson et al., 2003a), as well as the site-specific immobilization of peptides and proteins to a surface (Gauchet et al., 2006; Soellner et al., 2003). Variations of the Staudinger ligation have also been used in the synthesis of glycopeptides (Bianchi and Bernardi, 2006; Bianchi et al., 2005; He et al., 2004; Liu et al., 2006) and biomolecular labeling experiments in vitro (Grandjean et al., 2005; Tsao et al., 2005) and in vivo (Dube et al., 2006), and for drug delivery (Azoulay et al., 2006). As with auxiliary-mediated ligations (Nilsson et al., 2005), steric hindrance at the ligation junction (as in nonglycyl couplings) diminishes the ligation yield. Phosphinothiols that mediate the efficient coupling of nonglycyl amino acids are, however, now known (Soellner et al., 2006b; Tam et al., 2008).

3. CHOICE OF COUPLING REAGENT

Several coupling reagents have been used in the traceless Staudinger ligation, with varied success. These compounds include phosphinomethanethiol **I** (Nilsson et al., 2001), phosphinothiophenol **II** (Nilsson et al., 2000), phosphinomethanol **III** (Saxon et al., 2000), phosphinoethanethiol **IV** (Han and Viola, 2004), and phosphinophenol **V** (Saxon et al., 2000). The efficacy of these coupling reagents in a model reaction between its AcGly(thio)ester and ^{13}C$^{\alpha}$-labeled N$_3$GlyNHBn in a wet organic solvent has been compared directly (Soellner et al., 2006b), and the key results are listed in Table 2.1. Traceless Staudinger ligations mediated by reagents

Table 2.1 Effect of coupling reagent on the rate and product distribution of the Staudinger ligation

Phosphino(thio)ester (AcGlyR) + N_3—^{13}C...NHBn $\xrightarrow[\text{(6:1)}]{k_2 \quad DMF/D_2O}$ ...product with ^{13}C...NHBn

Coupling reagent (HR)	k_2 ($\times 10^{-3}$ M^{-1} s^{-1})	Yield (%)
HS—PPh$_2$ **I**	7.7 ± 0.3	95
HS—(phenyl)—PPh$_2$ **II**	1.04 ± 0.05	38
HO—PPh$_2$ **III**	0.12 ± 0.01	11
HS—PPh$_2$ **IV**	0.65 ± 0.01	39
HO—(phenyl)—PPh$_2$ **V**	7.43 ± 0.03	99

II, **III**, and **IV** are sluggish compared to those by reagents **I** and **V**. Furthermore, coupling reagents **II**, **III**, and **IV** also display low-ligation yields. The low rate and yield with **II** and **IV** could be due to the increased size of the ring that is formed during the nucleophilic attack of the imino-phosphorane nitrogen on the thioester (e.g., to produce the tetrahedral intermediate in Fig. 2.1). Reagent **III** enabled a direct comparison of an ester and thioester reagent, and highlights the advantage of a good leaving group (thiolate vs alkoxide) in mediating the traceless Staudinger ligation. Finally, phosphinophenol **V** gave amide yields and reaction rates nearly indistinguishable from phosphinomethanethiol **I**. Although Staudinger ligation with **V** requires the formation of a six- rather than a five-membered ring during $S \rightarrow N$ acyl group transfer (Fig. 2.1), the conjugate base of **V** is a somewhat better leaving group than that of **I**. Upon further investigation, ligations mediated by **V** were found to suffer a decrease in amide yield in the presence of the functional groups found in proteinogenic amino acids. This result is presumably due to the aryl ester of **V** being more electrophilic than the thioester of **I**, increasing its susceptibility to nonspecific acyl transfer reactions (e.g., with the ε-amino group of a lysine residue). On the contrary, Staudinger ligations performed with **I** can be performed on unprotected peptide fragments (Gauchet *et al.*, 2006; Liu *et al.*, 2006; Soellner *et al.*, 2003).

Because of its high reaction rate, high ligation yields, and chemoselectivity, (diphenylphosphino)methanethiol (**I**) is the most efficacious of known reagents for mediating the traceless Staudinger ligation (Soellner *et al.*, 2006b). Thiol-based reagents (e.g., **I**) have another intrinsic advantage over hydroxyl-based reagents (e.g., **V**). The thiol-based reagents react readily with thioester fragments generated by expressed protein ligation or other methods to form phosphinothioesters poised for a traceless Staudinger ligation.

Phosphinothiol **I** can be prepared from diphenylphosphine–borane complex and other commercial materials by two routes, designated as **a** (Soellner *et al.*, 2002) and **b** (He *et al.*, 2004) in Fig. 2.2, both with overall yields of 55%. A precursor that is common to both routes, phosphine–borane complex **X**, is stable to air and moisture and can be stored on the

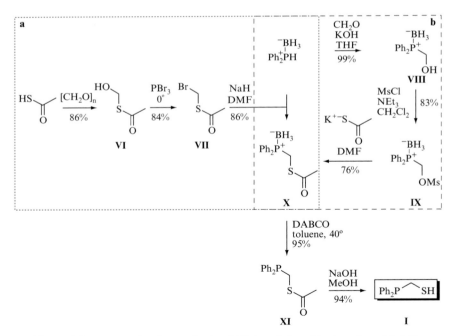

Figure 2.2 Routes for the synthesis of (diphenylphosphino)methanethiol (**I**).

shelf at room temperature for months without any sign of oxidation or decomposition. Phosphine–borane complex **X** is also available from a commercial vendor (Sigma–Aldrich product #670359). Although fully deprotected phosphinothiol **I** is stable under Ar(g) for several days, it is best when prepared freshly from phosphine–borane complex **X**.

3.1. Experimental procedure: Synthesis of phosphinothiol I

Route a. In route **a** of Fig. 2.2 (Soellner *et al.*, 2002), a P–C bond is made by alkylation of a diphenylphosphine–borane complex with agent **VII**, which was known previously (Farrington *et al.*, 1989). Thioacetic acid (50 g, 0.65 mol) and paraformaldehyde (20 g) are mixed and heated at 100 °C for 2 h under Ar(g). The reaction mixture becomes clear and light yellow, which indicates that the reaction is complete. Distillation under a high vacuum (bp 36 °C at 0.1 mm Hg) gives the $AcSCH_2OH$ (**VI**) as a colorless oil (typical yield: 59 g, 0.65 mmol, 86%). $AcSCH_2OH$ (**VI**, 59 g, 0.56 mol) is cooled under Ar(g) in an ice bath, and PBr_3 (50.5 g, 0.19 mol) is added dropwise slowly such that the reaction temperature does not exceed 8 °C. After the complete addition of PBr_3, the reaction mixture is stirred for an additional 30 min in an ice bath, and then allowed to warm to room temperature. The reaction mixture is poured over an ice/water mixture (100 ml), and extracted with ether (3 × 100 ml). The organic extracts are dried over anhydrous $MgSO_4$(s), filtered, and concentrated under reduced pressure. The residue is distilled under a high vacuum (bp 53 °C at 0.1 mm Hg) to give alkylating agent **VII** as a colorless oil (typical yield: 0.80 g, 0.47 mmol, 84%). Spectral data should be as reported previously (Farrington *et al.*, 1989).

Diphenylphosphine–borane complex (10.33 g, 51.6 mmol) is dissolved in dry DMF under Ar(g) and cooled to 0 °C. NaH (1.24 g, 51.6 mmol) is added slowly, and the mixture is stirred at 0 °C until bubbling ceases. Alkylating agent **VII** (8.73 g, 51.6 mmol) is then added, and the mixture is allowed to warm to room temperature and stirred for 12 h. The product is concentrated under reduced pressure, and the residue is purified by flash chromatography (silica gel, 10% (v/v) EtOAc in hexanes). Phosphine–borane complex **X** is isolated as a colorless oil (typical yield: 12.8 g, 44.4 mmol, 86%), and can be stored under air in a flask or bottle for extended periods in this form. ^1H NMR (300 MHz, $CDCl_3$) δ 7.74–7.67 (m, 4 H), 7.54–7.41 (m, 6 H), 3.72 (d, $J = 6$ Hz, 2 H), 2.23 (s, 3 H), 1.51–0.53 (broad m, 3 H) ppm; ^{13}C NMR (75 MHz, $CDCl_3$) δ 192.94, 132.26 (d, $J = 9.2$ Hz), 131.61 (d, $J = 2.3$ Hz), 128.71 (d, $J = 10.2$ Hz), 127.43 (d, $J = 55.4$ Hz), 29.87, 23.59 (d, $J = 35.5$ Hz) ppm; ^{31}P NMR (121 MHz, $CDCl_3$) δ 19.40 (d, $J = 59.3$ Hz) ppm; typical MS (ESI) m/z 311.0806 ($MNa^+ = 311.0807$).

Route b. In route **b** of Fig. 2.2 (He *et al.*, 2004), a P–C bond is made by addition of a diphenylphosphine–borane complex to formaldehyde. Diphenylphosphine–borane complex (2.45 g, 12.2 mmol) is dissolved in THF (7 ml). Formaldehyde (37% (v/v) in H_2O, 7.16 ml) is added to the solution, followed by potassium hydroxide (825 mg, 14.7 mmol). The resulting bilayered solution is stirred overnight, and then concentrated under reduced pressure. The residue is dissolved in ethyl acetate (10 ml), and the layers are separated. The organic extracts are washed with brine, dried over anhydrous $MgSO_4$(s), filtered, and concentrated under reduced pressure. The crude oil is purified by flash chromatography (silica gel, 50% (v/v) CH_2Cl_2 in hexanes) to give phosphine–borane complex **VIII** as a colorless oil (typical yield: 2.81 g, 12.2 mmol, 99%). ^1H NMR ($CDCl_3$, 300 MHz) δ 7.73–7.68 (m, 4 H), 7.52–7.41 (m, 6 H), 4.40 (broad s, 2 H), 2.38 (broad s, 1 H), 1.50–0.50 (broad m, 3 H) ppm; ^{13}C NMR ($CDCl_3$, 75 MHz) δ 132.89 (d, $J = 8.9$ Hz), 131.79, 129.10 (d, $J = 10.8$ Hz), 126.88 (d, $J = 54.5$ Hz), 60.47 (d, $J = 41.4$ Hz) ppm; ^{31}P NMR ($CDCl_3$, 121 MHz) δ 17.61 (d, $J = 58.9$ Hz) ppm; typical MS (ESI) m/z 253.0927 ($MNa^+ = 253.0930$).

Triethylamine (2.56 ml, 18.35 mmol) is added to a solution of phosphine–borane complex **VIII** (2.81 g, 12.2 mmol) in CH_2Cl_2 (36 ml), and the reaction mixture is cooled to 0 °C with an ice bath. Methanesulfonyl chloride (1.33 ml, 17.1 mmol) is added dropwise, and the resulting solution is allowed to warm to room temperature slowly (e.g., overnight). The solution is washed with 0.1 N HCl and brine, and the combined organic extracts are dried over anhydrous $MgSO_4$(s), filtered, and concentrated under reduced pressure. The residue is purified by flash chromatography (silica gel, 30% (v/v) ethyl acetate in hexanes) to give phosphine–borane complex **IX** as a pale yellow oil (typical yield: 3.14 g, 10.16 mmol, 83% yield). ^1H NMR ($CDCl_3$, 300 MHz) δ 7.76–7.71 (m, 4 H), 7.59–7.48 (m, 6 H), 4.90 (d, $J = 1.90$ Hz, 2 H), 2.87 (s, 3 H), 1.50–0.50 (broad m, 3 H) ppm; ^{13}C NMR ($CDCl_3$, 75 MHz) δ 133.09 (d, $J = 9.50$ Hz), 132.48, 129.343 (d, $J = 10.0$ Hz), 125.32 (d, $J = 58.3$ Hz), 64.68 (d, $J = 37.8$ Hz), 37.65 ppm; ^{31}P NMR ($CDCl_3$, 121 MHz) δ 18.87 (d, $J = 57.8$ Hz) ppm; typical MS (ESI) m/z 331.0719 ($MNa^+ = 331.0705$).

Potassium thioacetate (1.4 g, 12.2 mmol) is added to a solution of phosphine–borane complex **IX** (3.14 g, 10.16 mmol) in anhydrous DMF (50 ml) under Ar(g). The resulting solution is stirred overnight, and then concentrated under reduced pressure. The residue is dissolved in ethyl acetate (25 ml), and the resulting solution is washed with water and brine. The combined organic extracts are dried over anhydrous $MgSO_4$(s), filtered, and concentrated under reduced pressure. The residue is purified by flash chromatography (silica gel, 30% (v/v) CH_2Cl_2 in hexanes) to give phosphine–borane complex **X** as a colorless oil (typical yield: 2.22 g, 7.7 mmol, 76%). Spectral data should be as reported for route **a**.

Phosphine–borane complex **X** (4.00 g, 13.9 mmol) is dissolved in toluene (140 ml) under Ar(g). 1,4-Diazabicylo[2.2.2]octane (DABCO) (1.56 g, 13.9) is added, and the mixture is heated at 40 °C for 4 h. The product is concentrated under reduced pressure, dissolved in CH_2Cl_2 (50 ml), and washed with both 1 N HCl (20 ml) and saturated brine (20 ml). The organic layer is dried over $MgSO_4(s)$, and concentrated under reduced pressure. Phosphine **XI** is isolated as a colorless oil (typical yield: 3.62 g, 13.2 mmol, 95%) and is used without further purification. ^1H NMR ($CDCl_3$, 500 MHz) δ 7.43–7.40 (m, 4 H), 7.33–7.30 (m, 6 H), 3.50 (d, $J = 4$ Hz, 2 H), 2.23 (s, 3 H) ppm; ^{13}C NMR ($CDCl_3$, 125 MHz) δ 194.01, 136.42 (d, $J = 13.6$ Hz), 132.28 (d, $J = 19.4$ Hz), 128.69, 128.11 (d, $J = 6.8$ Hz), 29.83, 25.41 (d, $J = 23.4$ Hz) ppm; ^{31}P NMR ($CDCl_3$, 202 MHz) δ −15.11 ppm; typical MS (ESI) m/z 274.06 ($MH^+ = 275.0$, fragments at 233.0, 199.2, 121.2).

Phosphine **XI** (17.27 g, 63.0 mmol) is dissolved in anhydrous methanol (0.40 l), and Ar(g) is bubbled through the solution for 1 h. Sodium hydroxide (5.04 g, 126 mmol) is then added, and the mixture is stirred under Ar(g) for 2 h. The product is concentrated under reduced pressure, and then dissolved in methylene chloride (0.30 l). The resulting solution is washed with 2 N HCl (2 × 0.10 l) and brine (0.10 l). The organic layer is dried over $MgSO_4(s)$, filtered, and concentrated under reduced pressure. The residue is purified by flash chromatography (alumina, 25% ethyl acetate in hexanes). (Diphenylphosphino)methanethiol (**I**) is isolated as a colorless oil (typical yield: 10.8 g, 46.6 mmol, 74%). ^1H NMR ($CDCl_3$, 300 MHz) δ 7.41–7.38 (m, 4 H), 7.33–7.26 (m, 6 H), 3.02 (d, $J = 7.8$ Hz, 2 H), 1.38 (t, $J = 7.5$ Hz, 1 H) ppm; ^{13}C NMR ($CDCl_3$, 75 MHz) δ 132.54 (d, $J = 17.1$ Hz), 128.86, 128.36, 128.14, 20.60 (d, $J = 21.7$ Hz) ppm; ^{31}P NMR ($CDCl_3$, 121 MHz) δ −7.94 ppm; typical MS (ESI) m/z 232.05 ($MH^+ = 233.0$, fragments at 183.0, 155.0, 139.0, 91.2).

4. PREPARATION OF THE AZIDO FRAGMENT

Three methods have been described for preparing a peptide with an N-terminal azido group. These methods are listed in Table 2.2. Two of these methods involve a protected peptide on a solid support; the third involves an unprotected peptide in solution.

N1. A synthetic azido acid (or peptide) is coupled to the N-terminus of a synthetic peptide on a solid support (Nilsson *et al.*, 2003a; Soellner *et al.*, 2003). Side-chain functional groups are protected from side reactions (as indicated by the triangles).

N2. The N-terminal amino group of a peptide on solid support is converted into an azide by diazo transfer from, for example, triflyl azide in

Table 2.2 Strategies for preparation of the azide fragment

the presence of divalent copper ions (Rijkers *et al.*, 2002). Side-chain functional groups are protected from reaction.

N3. A protease-catalyzed peptide condensation reaction is used to intro-duce azido dipeptides to the N-terminus of an unprotected peptide fragment (Liu *et al.*, 2006). A large excess (10 equiv.) of synthetic azido dipeptides is needed, along with the protease subtilisin.

4.1. Experimental procedure: Strategy N1

The azido derivatives of amino acids can be prepared by a method described previously (Lundquist and Pelletier, 2001). In our example, azido glycine is prepared by partially dissolving sodium azide (20.56 g, 317 mmol) by stir-ring in DMSO (880 ml) for 1.5 h. Bromoacetic acid (20.96 g, 151 mmol) is added to this slurry, and the remaining NaN_3 dissolves within minutes. The reaction mixture is stirred overnight at room temperature, before diluting with H_2O (1.0 l) and adjusting the pH to 2.5 with concentrated HCl. The desired azido glycine is extracted with EtOAc (2 × 1 l). The organic extracts are dried over anhydrous $MgSO_4$(s), and then concentrated under reduced pressure to yield azido glycine as a pale oil (typical yield: 11.1 g, 110 mmol, 73%). Spectral data should be as reported previously (Lundquist and Pelletier, 2001).

A desired $(n − 1)$ peptide fragment (in our example, RNase A fragment 113–124: NPYVPVHFDASV (Nilsson *et al.*, 2003a)) is synthesized by solid–phase peptide synthesis on a Novasyn® TGA resin loaded with FmocValOH (110 mg, 22 μmol) by using standard methods on an automated synthesizer. The resin containing the $(n - 1)$ peptide fragment is swollen in DMF (1 ml) for 1 h. Azido glycine (10.1 mg, 100 μmol), PyBOP (52 mg, 100 μmol), HOBT (14 mg, 100 μmol), and diisopropylethylamine (DIEA, 35 μl, 200 μmol) are dissolved in DMF (4 ml), and this mixture is added to the resin. The resin is agitated for 2 h by bubbling Ar(g) through the slurry. The resin is filtered, and this coupling protocol is repeated to ensure maximal coupling. After the second coupling, the resin containing the azido peptide is rinsed and dried under Ar(g).

5. Preparation of the Phosphinothioester Fragment

Many methods are known for installing a phosphinothioester at the C–terminus of a synthetic peptide (or module). Five of these methods are listed in Table 2.3. All peptide fragments synthesized via solid–phase peptide synthesis have the potential of incorporating nonnatural amino acids and synthetic modules anywhere within the peptide fragment.

C1. A peptide fragment is synthesized on a sulfonamide-linker ("safety-catch") resin (Backes and Ellman, 1999). After activation of the fully loaded resin with iodoacetonitrile, treatment with an excess of phosphinothiol **I** liberates the thioester fragment.

C2. A peptide fragment is synthesized on an acid-sensitive resin (e.g., NovaSyn TGA resin or 2-chlorotrityl resin) and liberated with 1% (v/v) TFA, which leaves intact the amino acid protecting groups. The C–terminus is then activated (e.g., with DCC, PyBOP, or NHS) and coupled with phosphinothiol **I**.

C3. A peptide fragment is assembled by standard Fmoc chemistry on a 4-hydroxymethyl-phenylacetamidomethyl (PAM) or 4-hydroxy-methylbenzoic acid (HMBA) resin (Sewing and Hilvert, 2001). The ester linkage is activated for cleavage by $AlMe_3$ in the presence of an excess of phosphinothiol **I**. Epimerization at the C-terminal residue can occur, limiting this strategy to peptide fragments with a C-terminal glycine.

C4. A peptide fragment is assembled by standard Fmoc chemistry on an ester-linked (acid-stable) resin, which is loaded with N-4,5-dimethoxy-2-mercaptobenzyl (Dmmb)–Ala. The Dmmb group undergoes an $N \rightarrow S$ acyl group shift under acidic conditions, and the resulting thioester can undergo transthioesterification (Kawakami *et al.*, 2005).

36

Table 2.3 Strategies for preparation of the phosphinothioester fragment

Strategy	Route

C4

C5

C5. A polypeptide with a C-terminal intein and resin-binding domain is produced by recombinant DNA technology. Transthioesterification with water-soluble phosphinothiol **XII** liberates the peptide from the resin, simultaneously forming the C-terminal phosphinothioester (Tam and Raines, 2009; Tam *et al.*, 2007).

5.1. Experimental procedure: Strategy C1

First, a peptide is synthesized on resin. In our example (Nilsson *et al.*, 2003a), FmocGlu(O*t*Bu)OH is loaded onto 4-sulfamylbutyryl resin as described previously (Backes and Ellman, 1999). 4-Sulfamylbutyryl resin (1 g, 1.12 mmol) is swollen in $CHCl_3$ (25 ml) for 1 h. DIEA (1.56 ml, 8.96 mmol) and FmocGlu(O*t*Bu)OH (1.91 g, 4.48 mmol) are added to the resin. The reaction mixture is cooled to $-20\ ^\circ C$ under a flow of Ar(g). After 20 min, PyBOP (2.33 g, 4.48 mmol) is added to the solution and the resulting mixture is stirred, allowing the temperature to warm slowly to room temperature over a period of 8 h. The resin is filtered immediately and rinsed with $CHCl_3$. It is important to terminate the reaction after 8 h so as to minimize epimerization (Backes and Ellman, 1999). The coupling protocol is repeated to ensure maximal loading. After the second coupling is complete, the resin is filtered, rinsed with $CHCl_3$, and dried under Ar(g).

Fmoc-deprotection is achieved by swelling the resin in DMF. A solution of piperidine in DMF (30% (v/v), 10 ml) is then added to the resin, and agitated for 2 h. The resin is filtered, and rinsed with DMF (10 × 5 ml) and CH_2Cl_2 (10 × 5 ml).

To couple the subsequent amino acid, FmocCys(Trt)OH (2.62 g, 4.48 mmol), PyBOP (2.33 g, 4.48 mmol), and HOBT (0.605 g, 4.48 mmol) are dissolved in DMF (10 ml). DIEA (1.56 ml, 8.96 mmol) is added to the mixture, and the resulting solution is added to the resin described above. After agitating for 3 h, the resin is filtered, and rinsed with DMF (5 × 10 ml) and CH_2Cl_2 (5 × 10 ml).

The linker between a resin and its pendant synthetic peptide (in our example, RNase A fragment 110–111) is then activated with iodoaceto-nitrile as follows (Nilsson *et al.*, 2003a). The resin is swollen in CH_2Cl_2. A solution of iodoacetonitrile (3.4 ml, 46.8 mmol), DIEA (3.2 ml, 18.7 mmol), and NMP (75 ml) is filtered through a plug of basic alumina, and added to the resin. The resin is agitated for 18 h, filtered, and washed with NMP (5 × 10 ml) and CH_2Cl_2 (5 × 10 ml).

The phosphinothioester is liberated by incubating the above resin (1.0 g, 1.12 mmol peptide loading) with a solution of phosphinothiol **I** (2.1 g, 9.0 mmol) in DMF (15 ml) for 12 h under Ar(g). The resin is filtered, and rinsed with DMF (5 × 10 ml) and CH_2Cl_2 (5 × 10 ml), and the filtrate is concentrated under reduced pressure. The residue is purified by flash chromatography (silica gel, 30% (v/v) EtOAc in hexanes) to yield FmocCys (Trt)Glu(OtBu)SCH$_2$PPh$_2$ (typical yield: 0.71 g, 0.72 mmol, 64% based on a 1.12-mmol resin loading). ^{1}H NMR (CDCl$_3$, 300 MHz) δ 7.75–7.70 (m, 2 H), 7.57–7.55 (m, 2 H), 7.42–7.14 (m, 29 H), 6.68 (d, $J = 6.6$ Hz, 1 H), 5.13 (d, $J = 8.1$ Hz, 1 H), 4.56–4.50 (m, 1 H), 4.36–4.34 (m, 2 H), 4.19–4.17 (m, 1 H), 3.81–3.80 (m, 1 H), 3.44–3.38 (m, 2 H), 2.78–2.68 (m, 1 H), 2.61–2.57 (m, 1 H), 2.27–2.23 (m, 2 H), 2.11–1.95 (m, 1 H), 1.83–1.70 (m, 1 H), 1.37 (s, 9 H) ppm; ^{13}C NMR (CDCl$_3$, 75 MHz) δ 198.14, 171.89, 170.19, 155.81, 144.17, 143.59, 143.46, 141.10, 136.49 (d, $J = 14$ Hz), 132.69 (d, $J = 4.2$ Hz), 132.44 (d, $J = 4.1$ Hz), 129.41, 128.98, 128.38 (d, $J = 6.6$ Hz), 127.93, 127.59, 126.94, 126.74, 124.93, 119.80, 80.74, 67.21, 67.00, 58.50, 53.83, 46.89, 31.00, 27.85, 27.23, 25.45 (d, $J = 24.8$ Hz) ppm; ^{31}P NMR (CDCl$_3$, 121 MHz) δ −14.51 ppm; typical MS (ESI) m/z 1007.3340 (MNa$^+$ = 1007.3371).

5.2. Experimental procedure: Strategy C5

A water-soluble phosphinothiol can effect the traceless Staudinger ligation in purely aqueous medium in moderate yields, thereby integrating the traceless Staudinger ligation with expressed protein ligation (Tam et al., 2007). Incubation of the phosphinothiol and the chitin-bound peptide expressed via rDNA technology, and direct elution from the chitin resin yields the C-terminal phosphinothioester, which can then be used in Staudinger ligation with an azido peptide fragment.

Proteins and peptide fragments can be produced with rDNA methods in which the fragment is fused with the Mxe intein and a chitin-binding domain (CBD) (Arnold et al., 2002). In our example (Tam et al., 2007), this method is performed on Met(−1)RNase A–Gly–intein–CBD fusion protein to generate its C-terminal phosphinothioester. The desired plasmid is transformed into E. coli BL21(DE3) cells. Luria–Bertani (LB) medium (5 ml) containing ampicillin (0.10 mg/ml) is inoculated with a single colony and grown for 16 h at 37 °C. The cells are collected by centrifugation (2000×g for 2 min), and resuspended in LB medium (4 ml). Four 4–l flasks each containing 1 l of LB medium with ampicillin (0.10 mg/ml) are then inoculated with the resuspended cells (1 ml to each flask) from the 16 h culture. Cultures are grown with shaking at 37 °C until OD = 0.5 at 600 nm. Gene expression is then induced by the addition of isopropyl β-D-thiogalactopyranoside (IPTG; to 0.5 mM), and the cultures are grown for an additional 3–4 h at 25 °C. The lower temperature prevents the formation of inclusion bodies. Cells are harvested by centrifugation, and the cell pellet is stored at −20 °C.

Frozen cells are thawed and suspended in lysis and column buffer (LCB), which is 20 mM 3-(N-morpholino)propanesulfonic acid (MOPS)–NaOH buffer (pH 6.8) containing NaCl (0.5 M), ethylenediaminetetraacetic acid (EDTA; 0.1 mM), Triton X-100 (0.1%, w/w). Cells are lysed by sonication, and the lysate is subjected to centrifugation at 15,000×*g* for 30 min. The supernatant is applied slowly to an LCB–equilibrated column of chitin resin (New England Biolabs, Ipswich, MA). Approximately 6 ml of chitin resin is needed for 1 g of cells. The loaded resin is washed thoroughly with LCB (8 column–volumes), and LCB containing 0.5 *M* NaCl (2 column–volumes).

Intein-mediated cleavage is induced by incubating the resin with degassed cleavage buffer, which is 50 m*M* MOPS–NaOH buffer (pH 6.8) containing NaCl (0.5 *M*), EDTA (0.1 m*M*), and a water-soluble thiol such as 2-mercaptoethanesulfonic acid (MESNA) for 14 h under Ar(g). The thiol effects the transthioesterification of the fusion protein to form a C-terminal thioester of the protein, which is eluted from the resin with 0.5 *M* NaCl (2 ml). The peptide thioester is precipitated by the addition to 1% (v/v) of an aqueous solution of sodium deoxycholate (NaDOC) (1%, v/v) and by the addition to 2% (v/v) of an aqueous solution of trichloroacetic acid (TCA, 50%, w/v). After mixing, the precipitate is collected by centrifugation (5000×*g* for 5 min), decanted, and resuspended in acetone to remove small-molecule additives. MALDI mass spectrometry can be used to confirm the identity of the peptide thioester. After dissolving the peptide thioester in the appropriate solvent/buffer, transthioesterification can be performed with a phosphinothiol to generate the C-terminal phosphinothioester. The resulting peptide phosphinothioester can be isolated by the above precipitation procedure using NaDOC and TCA.

6. PROTEIN ASSEMBLY BY ORTHOGONAL CHEMICAL LIGATIONS

Perhaps the most well-characterized protein, bovine pancreatic ribonuclease (RNase A; Raines, 1998), has been used to evaluate the efficacy of some of the strategies above. The 124 amino acids of RNase A were assembled by using a variety of sequential and convergent amide-bond forming reactions, including the traceless Staudinger ligation, as depicted in Fig. 2.3 (Nilsson *et al.*, 2003a). The enzyme thus created is remarkable in that its peptide bonds were synthesized by four distinct processes, two of which are sequential (mRNA translation by a ribosome and solid-phase peptide synthesis) and two of which are convergent (native chemical ligation and traceless Staudinger ligation).

6.1. Experimental procedure: Traceless Staudinger ligation on a solid phase

The resin-bound azido peptide (RNase A fragment 112–124: GNPYVPVHFDASV, 180 mg, 25 μmol) as synthesized with Strategy N1 is swollen in DMF for 1 h. The C-terminal phosphinothioester of RNase A

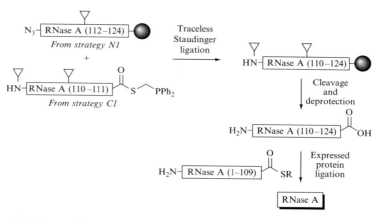

Figure 2.3 Route for the assembly of RNase A with solid-phase peptide synthesis, Staudinger ligation, and expressed protein ligation.

fragment 110–111 is synthesized with Strategy C1 as FmocCys(Trt)Glu (OtBu)SCH$_2$PPh$_2$ (99 mg, 100 μmol), dissolved in 10:1 DMF/H$_2$O (1.5 ml), and added to the swollen resin. The slurry is agitated gently for 12 h, after which the solvent is removed by filtration, and the resin is rinsed with DMF (5 × 10 ml) and CH$_2$Cl$_2$ (5 × 10 ml). The resin is dried under high vacuum and then treated with a cleavage cocktail (38:1:1 TFA/H$_2$O/ ethanedithiol, 2 ml) for 2 h. The resin is filtered, and added to ice-cold diethyl ether (20 ml) to precipitate the deprotected peptide, RNase A fragment 110–124. The peptide is purified by reverse-phase HPLC and can be analyzed by MALDI mass spectrometry. The ligated peptide can be elaborated further with orthogonal-ligation methods. In our example, expressed protein ligation with the C-terminal thioester of RNase A fragment 1–109 gives full-length RNase A, as shown.

7. PROSPECTUS

The traceless Staudinger ligation has joined the repertoire of ligation methods for the convergent synthesis of proteins. This method has been used along with others to assemble an entire protein. A putative strategy for the assembly of proteins is depicted in Fig. 2.4 (Nilsson et al., 2003b). Here, a target protein is divided into shorter fragments, and the ultimate C-terminal fragment is attached to a solid support. This immobilized fragment is capped with an α-azido acid and then reacted with a protected C-terminal phosphinothioester peptide fragment. The cycle is repeated until all fragments have been added. Deprotection and folding of the nascent

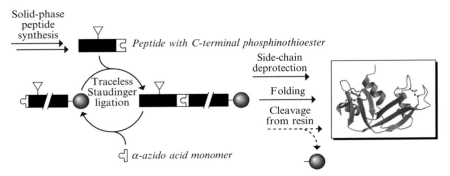

Figure 2.4 Strategy for the chemical synthesis of proteins by iterative cycles of solid-phase peptide synthesis and solid-phase Staudinger ligation.

polypeptide while still attached to the solid support (to avoid aggregation) yields a functional protein. The protein can be left attached to the resin for high-throughput assays or liberated for structure–function analyses in solution. The entire process is amenable to automation. Most notably, nonnatural amino acids or synthetic modules can be substituted for native ones, affording otherwise inaccessible proteins for otherwise unattainable goals.

7.1. Experimental procedure: General

All chemicals and reagents are available from Aldrich Chemical (Milwaukee, WI), with the exception of Fmoc-protected amino acids and alkanesulfonamide safety-catch resins, which are available from Novabiochem (San Diego, CA). Solution-phase reactions are monitored by thin-layer chromatography and visualized by UV light or staining with I_2. Flash chromatography is performed with columns of silica gel 60, 230–400 mesh (Silicycle, Québec City, Québec, Canada). HPLC purification is performed on a C18 reverse-phase column.

The term "concentrated under reduced pressure" refers to the removal of solvents and other volatile materials using a rotary evaporator at water-aspirator pressure (<20 mm Hg) while maintaining the water-bath temperature below 40 °C. The term "high vacuum" refers to a vacuum (≤0.1 mm Hg) achieved by a mechanical belt-drive oil pump.

Peptide synthesis is performed by standard Fmoc-protection strategies using an automated synthesizer with HATU activation. Phosphorus-31 NMR spectra are proton-decoupled and referenced against an external standard of deuterated phosphoric acid. Mass spectra are obtained with electrospray ionization (ESI) or matrix-assisted laser desorption ionization (MALDI) techniques.

ACKNOWLEDGMENTS

We are grateful to L. L. Kiessling for contributive discussions. Research on protein chemistry in the authors' laboratory is supported by Grant GM044783 (NIH).

REFERENCES

Arnold, U., Hinderaker, M. P., Nilsson, B. L., Huck, B. R., Gellman, S. H., and Raines, R. T. (2002). Protein prosthesis: A semisynthetic enzyme with a β-peptide reverse turn. *J. Am. Chem. Soc.* **124**, 368–369.

Azoulay, M., Tuffin, G., Sallem, W., and Florent, J. C. (2006). A new drug-release method using the Staudinger ligation. *Bioorg. Med. Chem. Lett.* **16**, 3147–3149.

Backes, B. J., and Ellman, J. A. (1999). An alkanesulfonamide "safety-catch" linker for solid-phase synthesis. *J. Org. Chem.* **64**, 2322–2330.

Bianchi, A., and Bernardi, A. (2006). Traceless Staudinger ligation of glycosyl azides with triaryl phosphines: Stereoselective synthesis of glycosyl amides. *J. Org. Chem.* **71**, 4565–4577.

Bianchi, A., Russo, A., and Bernardi, A. (2005). Neo-glycoconjugates: Stereoselective synthesis of α-glycosyl amides via Staudinger ligation reactions. *Tetrahedron Asymmetry* **16**, 381–386.

Bosch, I., Urpí, F., and Vilarrasa, J. (1995). Epimerization-free peptide formation from carboxylic-acid anhydrides and azido derivatives. *Chem. Commun.* 91–92.

Dube, D. H., Prescher, J. A., Quang, C. N., and Bertozzi, C. R. (2006). Probing mucin-type O-linked glycosylation in living animals. *Proc. Natl. Acad. Sci. USA* **103**, 4819–4824.

Farrington, G. K., Kumar, A., and Wedler, F. C. (1989). A convenient synthesis of diethyl (mercaptomethyl)phosphonate. *Org. Prept. Proced. Int.* **21**, 390–392.

Gauchet, C., Labadie, G. R., and Poulter, C. D. (2006). Regio- and chemoselective covalent immobilization of proteins through unnatural amino acids. *J. Am. Chem. Soc.* **128**, 9274–9275.

Grandjean, C., Boutonnier, A., Guerreiro, C., Fournier, J. M., and Mulard, L. A. (2005). On the preparation of carbohydrate-protein conjugates using the traceless Staudinger ligation. *J. Org. Chem.* **70**, 7123–7132.

Han, S., and Viola, R. E. (2004). Splicing of unnatural amino acids into proteins: A peptide model study. *Protein Pept. Lett.* **11**, 107–114.

He, Y., Hinklin, R. J., Chang, J. Y., and Kiessling, L. L. (2004). Stereoselective N-glycosylation by Staudinger ligation. *Org. Lett.* **6**, 4479–4482.

Hondal, R. J., and Raines, R. T. (2002). Semisynthesis of proteins containing selenocysteine. *Methods Enzymol.* **347**, 70–83.

Hondal, R. J., Nilsson, B. L., and Raines, R. T. (2001). Selenocysteine in native chemical ligation and expressed protein ligation. *J. Am. Chem. Soc.* **123**, 5140–5141.

Kawakami, T., Sumida, M., Nakamura, K., Vorherr, T., and Aimoto, S. (2005). Peptide thioester preparation based on an N–S acyl shift reaction mediated by a thiol ligation auxiliary. *Tetrahedron Lett.* **46**, 8805–8807.

Kent, S. (2003). Total chemical synthesis of enzymes. *J. Pept. Sci.* **9**, 574–593.

Liu, L., Hong, Z. Y., and Wong, C. H. (2006). Convergent glycopeptide synthesis by traceless Staudinger ligation and enzymatic coupling. *ChemBioChem* **7**, 429–432.

Lundquist, J. T., and Pelletier, J. C. (2001). Improved solid-phase peptide synthesis method utilizing α-azide-protected amino acids. *Org. Lett.* **3**, 781–783.

Muir, T. W. (2003). Semisynthesis of proteins by expressed protein ligation. *Annu. Rev. Biochem.* **72**, 249–289.

Nilsson, B. L., Kiessling, L. L., and Raines, R. T. (2000). Staudinger ligation: A peptide from a thioester and azide. *Org. Lett.* **2**, 1939–1941.

Nilsson, B. L., Kiessling, L. L., and Raines, R. T. (2001). High-yielding Staudinger ligation of a phosphinothioester and azide to form a peptide. *Org. Lett.* **3**, 9–12.

Nilsson, B. L., Hondal, R. J., Soellner, M. B., and Raines, R. T. (2003a). Protein assembly by orthogonal chemical ligation methods. *J. Am. Chem. Soc.* **125**, 5268–5269.

Nilsson, B. L., Soellner, M. B., and Raines, R. T. (2003b). Protein assembly to mine the human genome. *In* "Chemical Probes in Biology (NATO ASI Series)" (M. P. Schneider, ed.), pp. 359–369. Kluwer Academic, Boston, MA.

Nilsson, B. L., Soellner, M. B., and Raines, R. T. (2005). Chemical synthesis of proteins. *Annu. Rev. Biophys. Biomol. Struct.* **34**, 91–118.

Raines, R. T. (1998). Ribonuclease A. *Chem. Rev.* **98**, 1045–1066.

Rijkers, D. T. S., van Vugt, H. H. R., Jacobs, H. J. F., and Liskamp, R. M. J. (2002). A convenient synthesis of azido peptides by post-assembly diazo transfer on the solid phase applicable to large peptides. *Tetrahedron Lett.* **43**, 3657–3660.

Saxon, E., and Bertozzi, C. R. (2000). Cell surface engineering by a modified Staudinger reaction. *Science* **287**, 2007–2010.

Saxon, E., Armstrong, J. I., and Bertozzi, C. R. (2000). A "traceless" Staudinger ligation for the chemoselective synthesis of amide bonds. *Org. Lett.* **2**, 2141–2143.

Sewing, A., and Hilvert, D. (2001). Fmoc-compatible solid-phase peptide synthesis of long C-terminal peptide thioesters. *Angew. Chem. Int. Ed.* **40**, 3395–3396.

Smith, M. (1994). Nobel lecture. Synthetic DNA and biology. *Biosci. Rep.* **14**, 51–66.

Soellner, M. B., Nilsson, B. L., and Raines, R. T. (2002). Staudinger ligation of α-azido acids retains stereochemistry. *J. Org. Chem.* **67**, 4993–4996.

Soellner, M. B., Dickson, K. A., Nilsson, B. L., and Raines, R. T. (2003). Site-specific protein immobilization by Staudinger ligation. *J. Am. Chem. Soc.* **125**, 11790–11791.

Soellner, M. B., Nilsson, B. L., and Raines, R. T. (2006a). Reaction mechanism and kinetics of the traceless Staudinger ligation. *J. Am. Chem. Soc.* **128**, 8820–8828.

Soellner, M. B., Tam, A., and Raines, R. T. (2006b). Staudinger ligation of peptides at non-glycyl residues. *J. Org. Chem.* **71**, 9824–9830.

Staudinger, H., and Meyer, J. (1919). New organic compounds of phosphorus. III. Phosphinemethylene derivatives and phosphinimines. *Helv. Chim. Acta* **2**, 635–646.

Tam, A., and Raines, R. T. (2009). Coulombic effects on the traceless Staudinger ligation in water. *Bioorg. Med. Chem.* **17**, 1055–1063.

Tam, A., Soellner, M. B., and Raines, R. T. (2007). Water-soluble phosphinothiols for traceless Staudinger ligation and integration with expressed protein ligation. *J. Am. Chem. Soc.* **129**, 11421–11430.

Tam, A., Soellner, M. B., and Raines, R. T. (2008). Electronic and steric effects on the rate of the traceless Staudinger ligation. *Org. Biomol. Chem.* **6**, 1173–1175.

Tsao, M. L., Tian, F., and Schultz, P. G. (2005). Selective Staudinger modification of proteins containing p-azidophenylalanine. *ChemBioChem* **6**, 2147–2149.

Velasco, M. D., Molina, P., Fresneda, P. M., and Sanz, M. A. (2000). Isolation, reactivity and intramolecular trapping of phosphazide intermediates in the Staudinger reaction of tertiary phosphines with azides. *Tetrahedron* **56**, 4079–4084.

CHAPTER THREE

Replacement of Y_{730} and Y_{731} in the $\alpha2$ Subunit of *Escherichia coli* Ribonucleotide Reductase with 3-Aminotyrosine Using an Evolved Suppressor tRNA/tRNA-Synthetase Pair

Mohammad R. Seyedsayamdost* *and* JoAnne Stubbe*,†

Contents

* Department of Chemistry, Massachusetts Institute of Technology, Cambridge, Massachusetts, USA
† Department of Biology, Massachusetts Institute of Technology, Cambridge, Massachusetts, USA

Methods in Enzymology, Volume 462
ISSN 0076-6879, DOI: 10.1016/S0076-6879(09)62003-6

Abstract

Since the discovery of the essential tyrosyl radical ($Y \bullet$) in *E. coli* ribonucleotide reductase (RNR), a number of enzymes involved in primary metabolism have been found that use transient or stable tyrosyl (Y) or tryptophanyl (W) radicals in catalysis. These enzymes engage in a myriad of charge transfer reactions that occur with exquisite control and specificity. The unavailability of natural amino acids that can perturb the reduction potential and/or protonation states of redox-active Y or W residues has limited the usefulness of site-directed muta-genesis methods to probe the attendant mechanism of charge transport at these residues. However, recent technologies designed to site-specifically incorporate unnatural amino acids into proteins have now made viable the study of these mechanisms. The class Ia RNR from *E. coli* serves as a paradigm for enzymes that use amino acid radicals in catalysis. It catalyzes the conversion of nucleotides to deoxynucleotides and utilizes both stable and transient protein radicals. This reaction requires radical transfer from a stable tyrosyl radical ($Y_{122} \bullet$) in the β subunit to an active-site cysteine (C_{439}) in the α subunit, where nucleotide reduction occurs. The distance between the sites is proposed to be >35 Å. A pathway between these sites has been proposed in which transient aromatic amino acid radicals mediate radical transport. To examine the pathway for radical propagation as well as requirements for coupled elec-tron and proton transfers, a suppressor tRNA/aminoacyl-tRNA synthetase (RS) pair has been evolved that allows for site-specific incorporation of 3-aminotyrosine (NH_2Y). NH_2Y was chosen because it is structurally similar to Y with a similar phenolic pK_a. However, at pH 7, it is more easily oxidized than Y by 190 mV (≈ 4.4 kcal/mol), thus allowing it to act as a radical trap. Here we present the detailed procedures involved in evolving an NH$_2$Y-specific RS, assessing its efficiency in NH$_2$Y insertion, generating RNR mutants with NH$_2$Y at selected sites, and determining the spectroscopic properties of NH$_2$Y • and the kinetics of its formation.

1. INTRODUCTION

Long-range electron transfer (ET) reactions are prevalent in biology, notably in photosynthesis and respiration, and in almost all cases occur between metal cofactors that are spaced 10 to 15 Å apart (Gray and Winkler, 1996; Marcus and Sutin, 1985; Moser *et al.*, 1992; Page *et al.*, 1999, 2003). ET theory, articulated by Marcus, has been immensely useful in dissecting the kinetics and thermodynamics of these reactions. More recently, however, the role of amino acid radicals as mediators of charge transport has been recognized (Cukier and Nocera, 1998; Reece *et al.*, 2006). Oxidation of amino acids at neutral pH requires loss of an electron and a proton, implicating proton–coupled electron transfer (PCET). Ribo-nucleotide reductases (RNRs), the topic of this chapter, serve as a paradigm for long-range PCET reactions, in which transient amino acid radicals are involved (Jordan and Reichard, 1998; Stubbe and Riggs-Gelasco, 1998; Stubbe and van der Donk, 1998; Stubbe *et al.*, 2003).

The *E. coli* RNR consists of two homodimeric protein subunits: α2 and β2. The β2 subunit contains a stable diferric tyrosyl radical (Y$_{122}$ •) cofactor, which is essential for catalysis and is buried deeply within the protein (Ehrenberg and Reichard, 1972; Nordlund *et al.*, 1990; Reichard and Ehrenberg, 1983; Sjoberg *et al.*, 1978). The α2 subunit harbors the active site, where a cysteinyl radical (C$_{439}$ •), generated by Y$_{122}$ • on radical transfer from β2, initiates nucleotide reduction (Stubbe, 1990, 1998; Uhlin and Eklund, 1994). This intersubunit radical transfer reaction has been proposed to span a distance of >35 Å on the basis of the Eklund docking model created from the structures of α2 and of β2 (Eklund *et al.*, 2001; Uhlin and Eklund, 1994). A pathway of amino acids that gap the distance between the Y$_{122}$ • in β2 and C$_{439}$ • in α2 has been proposed (Fig. 3.1) (Nordlund *et al.*, 1990; Stubbe *et al.*, 2003; Uhlin and Eklund, 1994). Several properties of the proposed radical transfer pathway in the *E. coli* class Ia RNR have made it difficult to examine with conventional methods. First, replacement of any residue involved in radical propagation by another natural amino acid via site-directed mutagenesis renders the enzyme inactive, thus precluding mechanistic studies (Climent *et al.*, 1992; Ekberg *et al.*, 1996; Stubbe *et al.*, 2003). Second, pre-steady-state kinetic experiments have revealed that radical transfer is preceded by a slow conformational change (Ge *et al.*, 2003). Consequently, to study this process, the rate-limiting step must be changed from the physical step to the radical transfer step. Finally, the

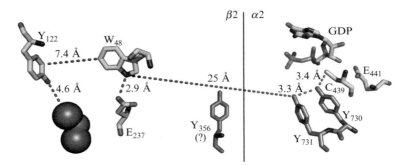

Figure 3.1 The radical transfer pathway in *E. coli* RNR. Residues Y_{122}, W_{48}, E_{237}, and Y_{356} reside in $\beta2$, while Y_{731}, Y_{730}, C_{439}, and E_{441} reside in $\alpha2$. Residues in gray have been shown to be redox-active using DOPA-$\beta2$ and NH$_2$Y-$\alpha2$s (see text). Note that the position of Y_{356} is unknown from structural studies. (See Color Insert.)

system has evolved to avoid buildup of any intermediates during this process, which could result in loss of the essential radical (Licht and Stubbe, 1999). In *E. coli* RNR, the $Y_{122}\bullet$ has a lifetime of several days (Atkin *et al.*, 1973).

Site-specific replacement of the tyrosines within the radical transfer pathway (Fig. 3.1) with appropriately designed unnatural amino acids could potentially allow for investigation of the length of the pathway and the existence of transient radical intermediates. Therefore, we have sought modified tyrosines that can block radical transfer because of their inability to be oxidized to the corresponding radical form, report on radical transfer because of their ease of oxidation relative to tyrosine, or change the rate-limiting step from the conformational change to radical transfer by modulating the radical reduction potential over several hundred mV relative to tyrosine in the physiological pH range.

To incorporate a thermodynamic block into the radical transfer pathway, we previously chose 3–nitrotyrosine (NO$_2$Y) as an analogue because it is substantially more difficult to oxidize than Y (by 200 mV at pH 7) (Yee *et al.*, 2003b). This block in the pathway could potentially allow accumulation of radical intermediates preceding the site of incorporation. NO$_2$Y also functioned as a reporter on the ability of the protein environment to perturb the phenolic pK_a, essential in understanding PCET mechanisms. To insert a radical trap into the pathway, we chose 3,4–dihydroxyphenylalanine (DOPA) because it is substantially easier to oxidize than Y (by 270 mV at pH 7) (Seyedsayamdost and Stubbe, 2006). This thermodynamic trap then served as a readout for the involvement of residue 356 in radical transfer. Finally, we chose a series of fluorotyrosine analogues (F$_n$Ys, n = 1–4) which perturb the phenolic pK_a and modulate the radical reduction potential relative to Y by -40 to $+270$ mV, depending on the reaction pH and the number of fluorines (Seyedsayamdost *et al.*, 2006a). These analogues allowed us to change the rate-limiting step in RNR and to uncouple the electron and proton flows at a given residue (Seyedsayamdost *et al.*, 2006b; Yee *et al.*, 2003a). We recently reported detailed procedures for replacement of Y_{356} in $\beta2$ by expressed protein ligation (EPL) methods (Muralidharan and Muir, 2006; Pellois and Muir, 2006) to site-specifically insert NO_2Y, DOPA, and F_nYs (Seyedsayamdost *et al.*, 2007c).

To study residues involved in radical transfer in $\alpha2$ in a similar fashion, we have considered two methods for the insertion of unnatural amino acids. The first method, semisynthesis of the α subunit of RNR using EPL, is not feasible because of the large size of $\alpha2$ (each monomer is 761 amino acids), its poor solubility, and the position of the residues to be replaced (Y_{730}/Y_{731}, Fig. 3.1). We have therefore employed an alternative method for site-specific insertion of unnatural amino acids. Herein, we detail our approach for inserting the unnatural tyrosine analogue 3-aminotyrosine (NH_2Y) into the $\alpha2$ subunit of RNR using suppressor tRNA/aminoacyl-tRNA synthetase (RS) methodologies (Seyedsayamdost *et al.*, 2007b; Wang and Schultz, 2004; Xie and Schultz, 2005).

2. Site-Specific Insertion of Unnatural Amino Acids using the Suppressor tRNA/RS Method

Schultz and colleagues have developed robust methods for site-specific incorporation of a wide range of unnatural tyrosine and phenylalanine analogues, *in vivo* (Wang and Schultz, 2004; Xie and Schultz, 2005, 2006). In collaboration with the Schultz lab, we have been able to apply this methodology to incorporate NH_2Y into α (Seyedsayamdost *et al.*, 2007a,b). In this method, six residues at or near the active site of the *Methanococcus jannaschii* Tyr-RS were randomized to generate a library of RSs. The resulting library was then subjected to a set of positive and negative selections (see section 4) to find one or several RSs that can selectively charge the cognate *M. jannaschii* tRNA$_{Tyr}$ with NH_2Y and not with Y. This suppressor tRNA, designated mutRNA$_{CUA}$, contains an amber stop codon and was optimized for interacting with the library of RSs (Wang and Schultz, 2001). A number of criteria must be satisfied by the selected *M. jannaschii* mutRNA$_{CUA}$/RS pair to make it suitable in this procedure (Xie and Schultz, 2005). First, the mutRNA$_{CUA}$/RS pair must be orthogonal to the expression host's tRNA/RS pairs; that is, the amber suppressor mutRNA$_{CUA}$ must not be charged by any RSs from the host organism *in vivo*. Second, the selected RS must charge only its cognate mutRNA$_{CUA}$ and not any of the host's tRNAs. Third, *in vivo* RSs must not be able to charge their cognate tRNAs with the unnatural amino acid. Fourth, the unnatural amino acid must be taken up by the host cell and must not be toxic. Finally, the gene of interest, with the amber stop codon at the residue where the unnatural amino acid is to be inserted, must be expressed inside the host. If all of these criteria are met, then the gene of interest and the genes for the evolved mutRNA$_{CUA}$/RS pair specific for the unnatural amino acid are transformed into the host, and the expression of the protein of interest results in the incorporation of the unnatural amino acid with high fidelity and specificity.

Both the expressed protein ligation method (Dawson and Kent, 2000; Giriat and Muir, 2003; Mootz et al., 2003; Muir, 2003; Perler, 2005) and the suppressor tRNA/RS method (Wang and Schultz, 2004; Wang et al., 2006; Zhang et al., 2003) have been used in vitro and in vivo, each with its advantages and disadvantages. While we have used EPL extensively with the β subunit and gained important insights into radical transfer, the tRNA/RS method promises to be more robust for incorporation of unnatural amino acids into both RNR subunits for several reasons. First, it allows incorporation of the unnatural amino acid anywhere in the protein, not just at the C- or N-terminus. Muir and colleagues have developed methods for insertion of probes in the middle of proteins (Cotton et al., 1999); however, refolding of α or β is not an option, and therefore incorporation of unnatural amino acids by these techniques into the center of the protein would not be viable. Second, the β subunit of RNR is a dimer. Therefore, the semisynthesis of $\beta2$ by EPL required separation of the full-length dimeric $\beta2$, the heterdimeric $\beta\beta'$ (composed of a full-length monomer, β, and a truncated monomer, β') and the doubly truncated dimer $\beta'2$. While this separation was possible, it came at the expense of protein yield. Third, expressed protein ligation of $\beta2$ required two additional mutations to provide sufficient quantities of semisynthetic $\beta2$ for physical biochemical methods necessary to study radical transfer in RNR. A serine residue (S$_{354}$) needed to be changed to a cysteine for the ligation reaction, and the residue at the site of the activated thioester-intein construct (V$_{353}$) needed to be replaced with a less bulky amino acid (Gly) to increase the yield of the ligated product (Yee et al., 2003b). With the tRNA/RS method, insertions can be made with a single unnatural point mutation. In addition, the requirement to assemble the essential diferric Y$_{122}$• cofactor of $\beta2$ and the sensitivity of the unnatural amino acids to oxidation made the intein method challenging at every step.

As with every method, there are a number of problems that may be encountered with the suppressor tRNA/RS method. The unnatural amino acid of interest might not be readily taken up into the host cell or may be toxic. In addition, each unnatural amino acid requires evolution of a new RS. As documented subsequently, the suppressor tRNA/RS pair has been highly successful in site-specific insertion of NH$_2$Y into RNR and has allowed us to perform mechanistic studies, which would not have been accessible by other methods.

3. NH$_2$Y, A Y ANALOGUE FOR INVESTIGATING ENZYMATIC PCET REACTIONS

3.1. Overview: Choice of NH$_2$Y

NH$_2$Y was chosen as a probe because its radical reduction potential of 0.64 V (pH 7.0) is 0.19 V lower than that of Y (DeFelippis et al., 1991), which suggests that it might act as a radical trap and directly report on the participation of residues Y$_{730}$ and Y$_{731}$ in radical transfer (Fig. 3.1). In addition, the ease of synthesis of NH$_2$Y (Seagle and Cowgill, 1976) and its kinetic stability to oxidation relative to other low reduction potential radical traps, such as DOPA (Jovanovic, 1994), make it a more practical target. The characterization of the UV–vis and EPR spectroscopic properties of the oxidized NH$_2$Y, the amino tyrosyl radical (NH$_2$Y$\bullet$), will allow NH$_2$Y to serve as a probe for enzymes that are thought to employ transient Y$\bullet$s, or modified Y$\bullet$s in catalysis or in electron transfer (Seyedsayamdost et al., 2007b). The ability to incorporate NH$_2$Y may also be of more general use for site-specific appendage of probes, as recent studies have shown that NH$_2$Y itself can be derivatized with fluorescent dyes (Hooker et al., 2004; Kovacs et al., 2007). Further, we have shown that NH$_2$Y can be used as a redox-active distance probe to determine molecular distances of 15 to 80 Å between two paramagnetic sites within an enzyme or enzyme complex (Bennati et al., 2003; Seyedsayamdost et al., 2007a).

Before performing directed-evolution on the library of RSs to select for an NH$_2$Y-RS, three criteria must be fulfilled by NH$_2$Y: it must be taken up by E. coli cells, it must not be toxic, and it must not be incorporated into proteins by any endogenous RSs. In this section, we outline the procedures used to determine the uptake and toxicity of NH$_2$Y.

3.2. Protocol for assessing uptake and toxicity of NH$_2$Y in E. coli

To assess the uptake of NH$_2$Y by E. coli (Xie and Schultz, 2005), growth of DH10B cells (Invitrogen) was initiated in the absence of antibiotics in two 50-mL Erlenmeyer flasks containing 15 mL of glycerol minimal media + leucine (GMML), which contains final concentrations of 1% (v/v) glycerol, 1× M9 salts (Sigma-Aldrich), 0.05% (w/v) NaCl, 1 mM MgSO$_4$, 0.1 mM CaCl$_2$, and 0.3 mM L-leucine (Sigma-Aldrich). The inclusion of L-leucine is important, as it greatly enhances growth rates. One flask was supplemented with 1 mM NH$_2$Y (Sigma-Aldrich) and 0.1 mM DTT (Mallinckrodt), while the other served as a control. The cultures were grown to saturation, harvested by centrifugation (6000g, 8 min) and each supernatant was transferred to a Falcon tube. The cell pellets were washed twice by resuspension in 1 mL of GMML, the cells isolated by centrifugation (6000g, 8 min), and the supernatant discarded. The washed cell pellets were then resuspended in 0.2 mL of BugBuster Protein Extraction Reagent (Novagen) and shaken at 37 °C for 1 h. Cell debris was pelleted by centrifugation (20,000g, 15 min) and the crude extract filtered using Microcon YM-10 membranes (Millipore). The filtered crude extract was subjected to LC-ESI-MS analysis using a Zorbax SB-C18 column (5 μm, 4.6 × 150 mm, Agilent Technologies) with a linear gradient from 5 to 25% MeCN in 0.1% TFA solution over 8 min at a flow rate of 0.5 mL/min. Under these conditions, NH_2Y elutes at ≈13% MeCN. ESI-MS sampling of the eluate was performed under positive ionization mode. Authentic NH_2Y standard solutions were prepared in water and chromatographed under identical conditions. NH_2Y from the crude extract was identified on the basis of comparison with the retention time of authentic NH_2Y, as well as by ESI-MS results.

3.3. Results

The DH10B *E. coli* cells grew normally in the presence of NH_2Y, indicating that it is not toxic, though the presence of DTT, included in the medium to maintain a reducing environment, reduced the growth rate by 25 to 35% (data not shown). Previous studies with DOPA had indicated that inclusion of DTT in the growth medium significantly increased the lifetime of reduced DOPA (Alfonta *et al.*, 2003). Therefore, DTT was maintained in all our experiments despite the reduction in the growth rate.

The HPLC of crude extracts clearly identified a peak corresponding to NH_2Y, as determined by the retention time and the ESI-MS results, which were identical to those of the authentic standard NH_2Y solution ($[M+H]^+_{exp}$ 197, $[M+H]^+_{obs}$ 196.9). The uptake was not determined as a function of time but rather at culture saturation. We were also able to detect NH_2Y in the supernatant after the cells were harvested by centrifugation, which showed that NH_2Y remains reduced in the culture throughout the time course of the growth. Together, the results showed that NH_2Y is not toxic and is taken up into *E. coli*.

The structure of *E. coli* Tyr-RS was examined to evaluate whether it would be able to bind NH_2Y and consequently charge $tRNA_{Tyr}$ with this analogue. Previous reports had demonstrated that *E. coli* Tyr-RS is unable to adenylate 3-iodo-Tyr (Kiga *et al.*, 2002), suggesting that NH_2Y, which is similar in size, will unlikely be a substrate for Tyr-RS. In support of this hypothesis, the structures of the adenylation domain of *E. coli* Tyr-RS in the presence of Tyr and adenylated Tyr (Kobayashi *et al.*, 2005) show that residues Y_{37}, Q_{179} and Q_{195}, and D_{182} are within 3.4 Å of the phenol C_3

and phenol C$_5$ carbon atoms, respectively, which are ortho to the hydroxyl group at C$_4$. This analysis suggests that NH$_2$Y would be unable to bind *E. coli* Tyr-RS because of unfavorable steric interactions with its 3–amino moiety. Although this conclusion needs to be further verified experimentally by isolating and assaying the *E. coli* Tyr-RS, the foregoing arguments suggest that NH$_2$Y meets the final criterion for commencing directed evolution of a specific NH$_2$Y-RS.

4. DIRECTED EVOLUTION OF NH$_2$Y-RS IN *E. COLI*

In this section, we summarize the general selection process used to isolate NH$_2$Y-RS. The protocols used to select for an NH$_2$Y-RS are similar to those previously described in detail and are summarized in Fig. 3.2 (Wang and Schultz, 2004; Xie and Schultz, 2005). The positive selection is performed in DH10B *E. coli* cells containing the positive selection plasmid, pREP/YC-J17 (Santoro *et al.*, 2002), which carries a tetracycline marker (Tet^R), and plasmid pBK-JYRS (Wang *et al.*, 2001), which contains the library of RSs and a kanamycin marker (Kn^R). The positive selection is

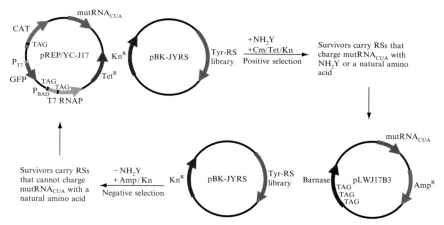

Figure 3.2 General scheme for selection of NH$_2$Y-RS. In the positive selection, suppression of a permissive TAG codon in chloramphenicol acetyl transferase (CAT, the Cm resistance gene) allows synthetases that charge mutRNA$_{CUA}$ with any amino acid to survive. The positive selection may also be performed by monitoring green fluorescence stemming from GFPuv, which contains a T7 promoter and is expressed when the TAG codons in the T7 RNA polymerase gene are suppressed. The synthetases are then carried through a negative selection cycle in which those that suppress the TAG codons in the barnase gene in the absence of NH$_2$Y are eliminated. Therefore, only synthetases that are functional with the host's translation machinery and do not charge mutRNA$_{CUA}$ with a natural amino acid will survive. See text for a description of the features on each plasmid. (See Color Insert.)

based on suppression of an amber stop codon (TAG) at a permissive site in the chloramphenicol acetyl transferase (CAT) gene when DH10B *E. coli* cells are grown in the presence of chloramphenicol (Cm), Kn, Tet, NH_2Y/DTT, and the cognate tRNA on GMML plates. Surviving clones carry RSs that are functional with the host cell's translation machinery and incorporate NH_2Y or natural amino acids into CAT in response to the amber stop codon. This leads to production of CAT and survival of the clones in the presence of 60 μg/mL Cm. In addition, the positive selection plasmid contains a GFPuv gene under the control of the T7 promoter and a T7 RNA polymerase, which contains amber codons at two permissive sites. Suppression of the amber stop codons generates T7 RNA polymerase, which drives the expression of GFPuv resulting in green fluorescence in the desired clones. Therefore, Cm resistance and green fluorescence may be monitored to isolate clones that carry the desired RSs (Xie and Schultz, 2005).

After the positive selection, the cells are scraped from the plates and isolated by centrifugation. The vectors pREP/YC-J17 ($\approx$10 kb) and pBK-JYRS ($\approx$3 kb) are isolated using the Qiagen Miniprep Kit and separated on a 1% agarose gel. Then, pBK-JYRS is extracted using the Qiagen Gel Extraction Kit and transformed into DH10B *E. coli* cells carrying the negative selection plasmid, pLWJ17B3 (Chin *et al.*, 2002; Wang *et al.*, 2003). The negative selection is based on lack of suppression of three amber codons in the barnase gene, which codes for a ribonuclease, when the cells are grown in the absence of NH_2Y in LB. The RSs that can charge the suppressor $mutRNA_{CUA}$ with a natural amino acid generate barnase, which kills the cells. Thus, this selection removes RSs that are not specific for NH_2Y. The surviving clones carry RSs that do not incorporate any natural amino acids in response to the amber stop codon. The RSs are then examined to determine the fidelity and specificity of NH_2Y insertion in a model system, as described in the subsequent section.

Four rounds of positive and three rounds of negative selections were required to obtain RSs that were specific for NH_2Y and did not cross-react with natural amino acids (Seyedsayamdost *et al.*, 2007b). The RS that conferred the highest degree of Cm resistance ($\approx$110 μg/mL) was selected and amplified. The plasmid carrying this RS was designated pBK-NH_2Y-RS (Seyedsayamdost *et al.*, 2007b). While structural studies have not yet been performed on this RS, DNA sequencing revealed a Q_{32} instead of the wild-type Y_{32} (note that this Y corresponds to Y_{37} in the *E. coli* Tyr-RS discussed earlier). In the wild-type *M. jannaschii* RS, the phenol oxygen of Y_{32} is positioned within 3.4 Å of the phenolic C-atom ortho to the hydroxyl group of the Tyr ligand (Kobayashi *et al.*, 2003; Turner *et al.*, 2006; Zhang *et al.*, 2005). In the selected NH_2Y-RS, this residue is Q, which indicates that the binding pocket has been expanded to allow accommodation for, and perhaps favorable H-bonding interactions with, the *o*-NH_2 moiety of NH_2Y.

5. Examination of the Fidelity and Specificity of NH$_2$Y Incorporation in a Protein Expressed in *E. coli*

5.1. Overview: The Z-domain as a model

A variety of expression systems for small proteins have been developed by the Schultz lab to assess the fidelity and specificity of unnatural amino acid incorporation. In our case, the efficiency of NH$_2$Y incorporation using pBK–NH$_2$Y-RS was tested using the C–terminally His-tagged Z-domain of protein A (the Z-domain) (Nilsson *et al.*, 1987; Wang *et al.*, 2003; Zhang *et al.*, 2002). There are several advantages and disadvantages to using small model proteins such as the Z-domain. The advantages are that the Z-domain is small and stable, allowing for robust expression and rapid purification at room temperature. More important, the incorporation of unnatural Y analogues can be assessed using simple analytical methods, such as MALDI-TOF-MS. Even small modifications, such as addition of an NH$_2$ group (Δ_{MW} = 15 Da) can be easily detected using this method. Further, trypsin digests result in a small set of peptides that can be sequenced using MS methods. In contrast, the Z-domain is a poor surrogate for target enzymes, which are typically larger, less robust, and/or multimeric in nature. Therefore, incorporation of an unnatural amino acid into the Z-domain does not guarantee that it will also occur in the target protein, and thus expression of the latter must be optimized. In addition, the Z-domain is subject to multiple forms of posttranslational modification that complicates MS analysis (Wang *et al.*, 2003). Nevertheless, it provides a rapid test for the efficiency of the evolved RS.

5.2. Protocol for incorporation of NH$_2$Y into the Z-domain

The vectors used to incorporate NH$_2$Y into the Z-domain were pBK–NH$_2$Y-RS and pLEIZ (Wang *et al.*, 2003; Zhang *et al.*, 2002). The latter contains the Z-domain gene with an amber stop codon at residue 7 and is under the control of a T7 promoter. It also contains the cognate mutRNA$_{CUA}$ and a Cm^R marker. The plasmids were cotransformed into BL21(DE3)-competent cells (Invitrogen) using the manufacturer's instructions and plated on LB/agar plates containing 50 μg/mL Kn and 35 μg/mL Cm. All subsequent cultures were grown in the presence of Kn (50 μg/mL) and Cm (35 μg/mL) at 37 °C. A single colony from the plate was inoculated into 5 mL of 2YT medium (Becton Dickinson) and grown to saturation ($\approx$13 h). One mL of this saturated culture was diluted into 25 mL 2YT medium and grown to saturation overnight ($\approx$11 h). Ten mL of that culture was then diluted into each of 2 $\times$ 1 L Erlenmeyer flasks containing 250 mL of GMML. When the $OD_{600 \text{ nm}}$ reached 0.65 (9 h), one of the cultures was supplemented with final concentrations of 1 mM NH_2Y and 0.1 mM DTT. The other culture was supplemented only with DTT (0.1 mM). Fifteen minutes after addition of NH_2Y/DTT (or DTT), IPTG (Sigma-Aldrich) was added to each culture to a final concentration of 1 mM. After 5 h, cells were harvested by centrifugation. The Z-domain produced from these cultures grown in the presence and absence of NH_2Y was then purified by Ni^{2+} affinity chromatography (Novagen) under denaturing conditions, per manufacturer's instructions. After purification, the Z-domain was analyzed by SDS-PAGE and MALDI-TOF MS. For MALDI-TOF MS, the Z-domain was exchanged into water by dialysis at 4 °C using 1-kDa molecular weight cutoff membranes (SpectroPor), and mass spectra were subsequently obtained under positive ionization mode at the Scripps Center for Mass Spectrometry.

5.3. Results

The SDS-PAGE analysis of the expression and purification of Z-domain in the presence and absence of NH_2Y is shown in Fig. 3.3A. The results show that expression of the Z-domain in the presence of NH_2Y/DTT in GMML resulted in suppression of the amber stop codon. Typically, 5 mg of Z-domain

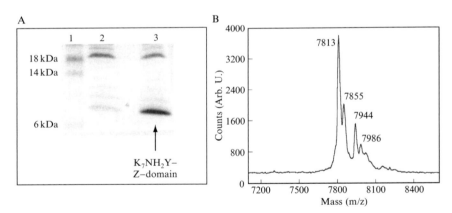

Figure 3.3 SDS-PAGE and MALDI-TOF MS analysis of K_7NH_2Y-Z-domain. (A) SDS gel of purified Z-domain after expression in the absence (lane 2) or presence (lane 3) of NH_2Y. The arrow designates the band corresponding to K_7NH_2Y-Z-domain. Protein ladder and MWs are shown in lane 1. (B) MALDI-TOF MS of purified K_7NH_2Y-Z-domain obtained under positive ionization mode. For the four main peaks in the spectrum, m/z $[M+H]^+$ are indicated. These correspond to N-terminally cleaved Met form of K_7NH_2Y-Z-domain, $[M+H]^+_{exp} = 7814$; its acetylated form, $[M+H]^+_{exp} = 7856$; full-length K_7NH_2Y-His-Z-domain, $[M+H]^+_{exp} = 7945$; and its acetylated form, $[M+H]^+_{exp} = 7987$. (See Color Insert.)

per liter of culture were obtained after purification. MALDI-TOF MS analysis of the sample yielded four major peaks (Fig. 3.3B) with MW = 7812, 7854, 7943, and 7985 Da, corresponding to K$_7$NH$_2$Y-Z-domain minus the first Met (MW$_{exp}$= 7813 Da), its acetylated form (MW$_{exp}$= 7855 Da), full-length K$_7$NH$_2$Y-Z-domain (MW$_{exp}$ = 7944 Da), and its acetylated form (MW$_{exp}$= 7986 Da), respectively. As discussed earlier, the Z-domain is subject to post-translational modifications that result in acetylation of its N-terminus and removal of its first Met residue (Wang *et al.*, 2003). Importantly, K$_7$Y-Z-domain, which would result from mischarging of mutRNA$_{CUA}$ with Y in place of NH$_2$Y, was not detected in the MALDI-TOF mass spectra. These results indicate that NH$_2$Y-RS is efficient and specific at suppressing the amber stop codon and inserting NH$_2$Y. In support of this finding, in the absence of NH$_2$Y, the amount of Z-domain produced was less than the lower limit of detection ($\approx$0.5 mg/L of culture). Therefore, in the absence of NH$_2$Y, the amber stop codon is not suppressed, resulting in the lack of expression or low levels of expression of the Z-domain.

6. Generation of Y$_{730}$NH$_2$Y-α2 and Y$_{731}$NH$_2$Y-α2

6.1. Overview

Having selected an NH$_2$Y-RS and demonstrated its ability to efficiently incorporate NH$_2$Y into the Z-domain, we next attempted to insert NH$_2$Y into the α subunit of RNR. As discussed previously, while the Z-domain allows for rapid assessment of the quality of the evolved RS, the conditions used for optimized expression and suppression of the amber stop codon in the target protein are usually different, as we and other research groups have experienced (Farrell *et al.*, 2005; Neumann *et al.*, 2008). In this section, we describe our initial unsuccessful attempts at incorporating NH$_2$Y into α2, as well as our detailed protocol for successful production of NH$_2$Y-α2s in 100-mg quantities with high specificity.

6.2. Unsuccessful attempts to incorporate NH$_2$Y into α2

Successful incorporation of NH$_2$Y into α2 first requires identification of a condition that yields high levels of expression of α2, and then suppression of the amber stop codon under the same growth condition. We first attempted to insert NH$_2$Y into α2 using the procedure successful with the Z-domain. pBAD-*nrdA* (or *nrdA* with a site-specifically encoded amber codon) carries the gene for α2 under the control of an L-Ara-inducible promoter with an *rrnB* terminator and the mutRNA$_{CUA}$ gene under the control of a *lpp* promoter with an *rrnC* terminator and a Tet^R marker (Seyedsayamdost *et al.*, 2007b; Zhang *et al.*, 2003). pBK-NH$_2$Y-RS contains the gene for

NH_2Y-RS described previously under the control of the constitutive *E. coli* Gln-RS promoter. Also, pBK-NH_2Y-RS and pBAD-*nrdA* contain compatible origins of replication, ColE1 and p15A, respectively.

Initially, we examined pBAD-*nrdA* alone to evaluate expression levels of α2 and α2 containing an amber stop codon (at residue 730) in GMML. In the former case, α2 was overexpressed to ≈10% of total protein; in the latter case, truncated α2 was overexpressed at similar levels. Ensuring expression in the absence of pBK-NH_2Y-RS is essential; however, it does not guarantee successful suppression of the amber stop codon.

Next, we examined the expression levels and amber stop codon suppression in cells cotransformed with pBAD-*nrdA* and pBK-NH_2Y-RS, grown in NH_2Y/DTT-supplemented GMML. Different conditions of induction and timing of addition of NH_2Y relative to induction were investigated. In all cases, levels of expression of α2 were low, as determined by SDS-PAGE. Western blot analysis indicated that the levels of α2 in the presence of NH_2Y were ≈2-fold that of endogenous α2 (α2 is an essential gene in *E. coli*, data not shown). We attempted growth under anaerobic conditions or aerobic growth in the presence of hydroxyurea. In the former case, the anaerobic class III RNR is operative in *E. coli*; therefore, wild-type class I β2 is not expressed. In the latter case, addition of hydroxyurea leads to reduction of the essential Y_{122} • in class I β2 so that it cannot react with NH_2Y-α2, if it is in fact expressed. In both cases, the reaction of wild-type β2 with NH_2Y-α2, which could lead to trapping of an NH_2Y • and subsequent inactivation of RNR, would be minimized. However, neither of the modifications in growth conditions resulted in significant overexpression of α2.

Continuous technological refinements from the Schultz lab have indicated that mutRNA$_{CUA}$ is often the limiting factor in production of sufficient levels of a recombinant target protein containing an unnatural amino acid (Ryu and Schultz, 2006). Failure to overexpress NH_2Y-α2, as described previously, might therefore have been due to limiting amounts of mutRNA$_{CUA}$ inside the cell. To increase the amounts of mutRNA$_{CUA}$, we next attempted expression of α2 with pAC-NH_2Y-RS, a vector developed by the Schultz lab to overcome this problem. pAC-NH_2Y-RS has six genetic copies of mutRNA$_{CUA}$ under the control of a *proK* promoter and terminator (Ryu and Schultz, 2006; Seyedsayamdost *et al.*, 2007b). It also carries the NH_2Y-RS gene under the control of a *glnS'* promoter, an *rrnB* terminator, and a Tet^R marker. While our analysis to understand the low levels of overexpression of NH_2Y-α2 was in progress, we found, as described subsequently, that pTrc-*nrdA* and pAC-NH_2Y-RS allowed overexpression of α2 and incorporation of NH_2Y; thus, this vector combination became the focus of our efforts. We have not yet reexamined expression NH_2Y-α2 with pBAD-*nrdA* and pAC-NH_2Y-RS and do not entirely understand the basis for the differences in the levels of expression of α2 with different vector combinations.

6.3. Successful incorporation of NH_2Y into $\alpha 2$

Reports of successful overexpression of *E. coli* nitroreductase using pTrc for production of the target protein (Jackson *et al.*, 2006) prompted us to investigate the pTrc-*nrdA*/pAC-NH_2Y-RS vector combination, simultaneously with efforts described in the preceding section. Thus, *nrdA* was cloned into the pTrc vector to generate pTrc-*nrdA*. This vector contains *nrdA* with an amber stop codon under the control of the highly active trp/lac (trc) promoter and an Amp^R marker (Amann *et al.*, 1988). Its p15a replicon is compatible with the ColE1 origin of pAC-NH_2Y-RS. Expression of $\alpha 2$ from pTrc-*nrdA* in DH10B cells was examined first. Expression and purification yielded 10 mg $\alpha 2$ per gram of wet cell paste, a $2\times$ to $4\times$ greater yield than with vector pMJ1-*nrdA*, which we have routinely used to generate wild-type and mutant $\alpha 2$s (data not shown) (Mao *et al.*, 1989; Seyedsayamdost *et al.*, 2007b). The activity of $\alpha 2$ produced from pTrc-*nrdA* was similar to that produced from pMJ1-*nrdA* (≈ 2500 nmol/min mg by the spectrophotometric assay) (Ge *et al.*, 2003). Therefore, pTrc-*nrdA* was suitable for production of $\alpha 2$, allowing expression of NH_2Y-$\alpha 2$ from pTrc-*nrdA* and pAC-NH_2Y-RS.

6.4. Protocol for successful expression of NH_2Y-$\alpha 2$s

Vectors pTrc-*nrdA*$_{730}$TAG and pTrc-*nrdA*$_{731}$TAG were generated from pTrc-*nrdA* using the site-directed mutagenesis kit (Stratagene), as already described (Seyedsayamdost *et al.*, 2007b). *E. coli* DH10B cells were transformed with pTrc-*nrdA*$_{730}$TAG and pAC-NH_2Y-RS and grown at 37 °C on LB/agar plates containing Amp (100 μg/mL) and Tet (25 μg/mL) for two days. All liquid culture growths contained Amp (100 μg/mL) and Tet (25 μg/mL) and were carried out in a shaker/incubator at 37 °C and 200 rpm. A single colony from the plate was inoculated into 5 mL of 2YT medium and grown to saturation (≈ 2 days). The 5 mL saturated culture was diluted into 180 mL of 2YT medium and grown to saturation (≈ 1 day). Twenty five mL of this culture were then inoculated into each of 6×6 L Erlenmeyer flasks, each containing 1 L of GMML medium supplemented with D-biotin (1 μg/mL, Sigma-Aldrich), thiamine (1 μg/mL, Sigma-Aldrich) and a $1\times$ heavy metal stock solution. A $1000\times$ heavy metal stock solution contains the following per liter, as previously described (Farrell *et al.*, 2005): 500 mg of $MoNa_2O_4 \bullet 2H_2O$, 250 mg of $CoCl_2$, 175 mg of $CuSO_4 \bullet 5H_2O$, 1 g of $MnSO_4 \bullet H_2O$, 8.75 g of $MgSO_4 \bullet 7H_2O$, 1.25 g of $ZnSO_4 \bullet 7H_2O$, 1.25 g of $FeCl_2 \bullet 4H_2O$, 2.5 g of $CaCl_2 \bullet 2H_2O$, and 1 g of H_3BO_3 in 1 M HCl. When $OD_{600\ nm}$ reached 0.6 (12 to 18 h), NH_2Y and DTT were added to final concentrations of 1 mM and 0.1 mM, respectively. After 15 min, IPTG was added to a final concentration of 1 mM and the growth continued for 4.5 h, at which point the cells were harvested by centrifugation, frozen in liquid N_2 and stored at $-80\,^{\circ}\mathrm{C}$. Expression of $Y_{731}NH_2Y\text{-}\alpha 2$ was carried out in identical fashion using $pTrc\text{-}nrdA_{731}TAG$ and $pAC\text{-}NH_2Y\text{-}RS$.

6.5. Results

Using the foregoing expression procedure, typically 1.5 g of wet cell paste was obtained per liter of culture. The results for the expression of $Y_{731}NH_2Y\text{-}\alpha 2$ in DH10B cells doubly transformed with $pAC\text{-}NH_2Y\text{-}RS$ and $pTrc\text{-}nrdA_{731}TAG$ are shown in Fig. 3.4A. In the presence of the IPTG inducer and NH_2Y/DTT, the amber stop codon is suppressed and $NH_2Y\text{-}\alpha 2$ is overexpressed. In the absence of NH_2Y, overproduction of only truncated $\alpha 2$ is observed. Finally, in the absence of inducer IPTG and NH_2Y, no expression of $\alpha 2$ occurs. A similar profile was obtained for the expression of $Y_{730}NH_2Y\text{-}\alpha 2$ (Fig. 3.4B).

6.6. Protocol for purification of $NH_2Y\text{-}\alpha 2s$

All purification steps were carried out at $4\,^{\circ}\mathrm{C}$. Typically, 10 g of wet cell paste (from ≈ 6 L of growth) were used for purification. Each gram of cell paste was resuspended in 5 mL of $\alpha 2$ buffer, which consists of 50 mM Tris, 1 mM EDTA, pH 7.6, supplemented with 1 mM PMSF (Sigma–Aldrich) and 5 mM DTT. PMSF and DTT were added from a 100 mM stock in isopropanol and a 1 M stock in water, respectively. The cells were lysed using a single passage through a French pressure cell operating at 16,000 psi. After lysis, cell debris was removed by centrifugation (15,000g, 35 min, $4\,^{\circ}\mathrm{C}$).

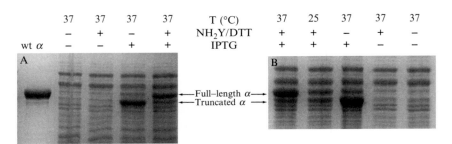

Figure 3.4 Incorporation of NH_2Y into $\alpha 2$. Expression of $Y_{731}NH_2Y\text{-}\alpha 2$ (A) and $Y_{730}NH_2Y\text{-}\alpha 2$ (B) are shown as a function of IPTG, NH_2Y/DTT and temperature. Cells were grown in the presence or absence of IPTG and NH_2Y/DTT as indicated and the level of expression assessed by SDS PAGE. The position of protein bands for full-length α and truncated α are denoted by arrows. For both constructs, no expression is observed in the absence of IPTG and expression of truncated $\alpha 2$ is observed in the absence of NH_2Y. Full-length $NH_2Y\text{-}\alpha 2$ is seen only in the presence of IPTG and NH_2Y/DTT.

The crude extract was transferred to a 100-mL beaker. DNA was precipitated by dropwise addition of 0.2 vol of α2 buffer containing 1 mM PMSF, 5 mM DTT, and 8 % (w/v) streptomycin sulfate (Sigma-Aldrich). The mixture was stirred for an additional 15 min, and the precipitated DNA was removed by centrifugation (15,000g, 35 min, 4 °C). The supernatant was transferred to a 100-mL beaker and 3.9 g of solid $(NH_4)_2SO_4$ were added per 10 mL of supernatant over 15 min, which corresponds to 66% $(NH_4)_2SO_4$ saturation. The mixture was stirred for an additional 30 min and the precipitated protein was isolated by centrifugation (15,000g, 45 min, 4 °C). The supernatant was discarded and the pellet redissolved in 3 to 4 mL of α2 buffer containing 1 mM PMSF and 5 mM DTT. This solution was desalted on a Sephadex G-25 column (1.5 cm × 25 cm, 45 mL, Sigma-Aldrich), which had been equilibrated in α2 buffer. The desalted protein was loaded directly onto a dATP affinity column (1.5 cm × 4 cm, 6 mL), which had been equilibrated in α2 buffer, at 0.5 mL/min (Berglund and Eckstein, 1972, 1974). The column was washed with 10 column-volumes of α2 buffer containing 1 mM PMSF and 5 mM DTT, followed by 2 volumes of α2 buffer containing 5 mM DTT. Then, NH₂Y-α2 was eluted in 3 to 4 column-volumes of α2 buffer containing 15 mM $MgSO_4$, 10 mM ATP (Sigma-Aldrich), and 10 mM DTT. Subsequently, ATP was removed using a Sephadex G-25 column as described earlier and NH₂Y-α2 concentrated to ≈40 μM using a Centriprep concentration device (Millipore). Concentrated NH₂Y-α2 was divided into 0.5- to 1-mL aliquots, flash frozen, and stored at −80 °C. Typically, 4 to 6 mg of pure NH₂Y-α2 were obtained per gram of wet cell paste.

6.7. Assessment of the fidelity and specificity of NH₂Y insertion into α2

The analytical methods available for assessing the level of NH₂Y incorporation into the Z-domain are not feasible with α2. For example, incorporation of NH₂Y results in a 0.017% mass increase (α_{MW} ≈ 86 kDa) in α2, as a result of the o-NH₂ moiety, which cannot be measured by ESI-MS methods. Further, analysis of trypsin digestion products of α2 is complicated given the large number of resulting peptides. Therefore, assessment of the efficiency of NH₂Y incorporation into α2 requires alternative methods. In the case of NH₂Y, two such methods have been considered. The first involves incorporation of [¹⁴C]-NH₂Y or [³H]-NH₂Y of known specific activity into α2 to determine the amount of NH₂Y by scintillation counting. A second method involves derivatization of NH₂Y with a known fluorophore and quantitation of the derivative by fluorescence spectroscopy (Hooker et $al.$, 2004). Both methods are currently being investigated.

The methods we have recently used to address the levels of contaminating wild-type α2 in our NH$_2$Y-α2 preparations consist of dCDP production assays in combination with the use of a stoichiometric mechanism-based inhibitor of RNRs, 2′-azido-2′-deoxyadenosine-5′-diphosphate (N$_3$ADP) (Thelander et al., 1976). NH$_2$Y-α2s were assayed for dCDP formation using the discontinuous radioactive assay with thioredoxin (TR), thioredoxin reductase (TRR), and NADPH as reductants. Our previous studies have shown that an activity {1/10^4} that of wild-type RNR is the lower limit of detection by this assay (Seyedsayamdost and Stubbe, 2006; Yee et al., 2003b). Observation of nucleotide reduction activity would suggest that either our NH$_2$Y-α2 preparations have a high level of Y incorporated at residues 730 (or 731) or that the NH$_2$Y-α2s are inherently active in nucleotide reduction.

To distinguish between the two options, N$_3$ADP was used to quantitate the amount of wild-type α2. Previous studies have revealed that wild-type RNR is inactivated with 1 equivalent of N$_3$NDP, resulting in loss of 50% Y$_{122}$ • (between 20 s and 2 min) and formation of 1 equivalent of a new substrate-derived nitrogen-centered radical (N •, Fig. 3.5A) (Fritscher et al., 2005; Salowe et al., 1993; Sjöberg et al., 1983; van der Donk et al., 1995). The amount of N • is directly proportional to the amount of active RNR in solution (Ekberg et al., 1998). Importantly, the EPR spectral features of N •, Y$_{122}$ •, and NH$_2$Y • do not overlap in the low-field region, thereby allowing spectral deconvolution and quantitation of these three species (Fig. 3.5B).

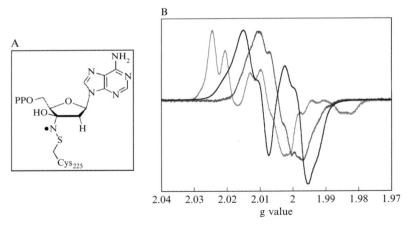

Figure 3.5 (A) Structure of the active site N-centered radical formed after reaction of RNR with N$_3$ADP. This radical is formed at the expense of Y$_{122}$ •. It is stable on the minute time scale and is abbreviated as N • in the text. (B) EPR spectra of N • (green), Y$_{122}$ • (blue), and NH$_2$Y$_{731}$ • (red). The distinct features in the low-field region between ≈2.02 and ≈2.03 were used to perform subtractions and quantitations of the three species in N$_3$ADP assays. (See Color Insert.)

6.8. Protocol for measurement of catalytic activities of NH₂Y-α2s

The assay contained in a final volume of 230 μL: 0.2, 1 or 3 μM NH$_2$Y-α2, a 5-fold molar excess of β2, 1 mM [5-^{3}H]-CDP (1190 cpm/nmol, Amersham Bioscience), 3 mM ATP, 30 μM TR, 0.5 μM TRR, and 1 mM NADPH (Sigma–Aldrich) in assay buffer (50 mM Hepes, 15 mM MgSO$_4$, 1 mM EDTA, pH 7.6) (Seyedsayamdost et al., 2006b). The purification of TR ($\approx$40 units/mg) (Chivers et al., 1997), TRR (1400 units/mg) (Russel and Model, 1985), and β2 (7200 nmol/min mg) (Salowe and Stubbe, 1986) have been described. For each reaction, nucleotides, TR, TRR, and α2 were incubated at 25 °C for 1.5 to 2 min. The reaction was then initiated by addition of wild-type β2 and NADPH. At defined time points, usually 0, 1, 2, 3, and 5 min, 40 μL were withdrawn and quenched in 25 μL of 2% (v/v) perchloric acid (Sigma–Aldrich). After the time course, the reactions were neutralized with 20 μL of 0.5 M KOH (Fisher Bioscience). The samples were incubated at -20 °C overnight to ensure complete precipitation of potassium perchlorate. The samples were then thawed on ice and spun down for 3 min at 21,000g in a tabletop centrifuge. Each supernatant was transferred to a 1.5-mL screw-top microfuge tube, to which was added 14 units of calf-intestine alkaline phosphatase (Roche), 120 nmol of carrier deoxycytidine (Sigma–Aldrich), and final concentrations of 75 mM Tris and 0.15 mM EDTA (pH 8.5). The samples were incubated at 37 °C for 2 h and the amount [^{3}H]-dC in each sample was quantitated using the method of Steeper and Steuart (1970).

6.9. Protocol for use of the mechanism-based RNR inhibitor, N₃ADP

The preparation of N$_3$ADP and other N$_3$NDPs, which may be used in this assay, has already been described (Hobbs and Eckstein, 1977; Robins et al., 1992; Salowe et al., 1987). The procedure for prereduction of NH$_2$Y-α2s has also been reported (Seyedsayamdost et al., 2007b). Prereduced NH$_2$Y-α2 (or wild-type α2) and dGTP (Sigma–Aldrich) were mixed with wild-type β2 and N$_3$ADP in assay buffer (50 mM Hepes, 15 mM MgSO$_4$, 1 mM EDTA, pH 7.6) to yield final concentrations of 20 μM, 1 mM, 20 μM, and 250 μM, respectively. The mixture was transferred to an EPR tube (Wilmad) and the reaction quenched after 20 s by inserting the tube into liquid N$_2$. EPR spectra were subsequently recorded at 77 K on a Bruker ESP-300 X-band spectrometer equipped with a quartz finger dewar filled with liquid N$_2$. The EPR parameters were as follows: microwave frequency = 9.34 GHz, power = 30 μW, modulation amplitude = 1.5 G, modulation frequency = 100 kHz, time constant = 5.12 ms, and scan time = 41.9 s. Analysis of the resulting spectra was carried out using an in-house written program in Excel and WinEPR (Bruker). These programs allow for accurate subtraction of the different spectral components in the reaction. In the experiments, deconvolution of the three signals observed $(Y_{122}\bullet$, $NH_2Y\bullet$, and $N\bullet$) was performed by first subtracting the $N\bullet$ spectrum, which is well established. Then, the well-characterized spectrum of unreacted $Y_{122}\bullet$ was subtracted, yielding the $NH_2Y\bullet$-$\alpha2$ spectrum. The ratio of the three signals was determined by comparing the double integral intensity of each trace. The EPR spin quantitation was carried out using a Cu^{II} standard as has been described elsewhere (Palmer, 1967).

6.10. Results of activity and N_3ADP assays

The activity assays for wild-type $\alpha2$, $Y_{730}NH_2Y$-$\alpha2$ and $Y_{731}NH_2Y$-$\alpha2$ are summarized in Table 3.1. The results show that the NH_2Y-$\alpha2$ variants retain significant nucleotide reduction activity. This activity may be inherent to NH_2Y-$\alpha2s$ or may be attributed to wild-type endogenous $\alpha2$ or wild-type $\alpha2$ generated by mischarging of $mutRNA_{CUA}$ with Y rather than NH_2Y. To distinguish among the three options, N_3ADP assays were carried out. As mentioned earlier, detailed characterization of wild-type $\alpha2$ with N_3NDPs by several groups has shown that after 20 to 120 s, 50 to 60% of the wild-type $\alpha2$ population will generate the well-characterized $N\bullet$ spectrum. Therefore, if the residual activity observed is due to wild-type $\alpha2$, we would expect that 3.5% and 2% of $\alpha2$ would contain the $N\bullet$ with $Y_{731}NH_2Y$-$\alpha2$ and $Y_{730}NH_2Y$-$\alpha2$, respectively.

The N_3ADP assays on NH_2Y-$\alpha2s$ showed that for both $Y_{730}NH_2Y$-$\alpha2$ and $Y_{731}NH_2Y$-$\alpha2$, 15% of the initial $Y_{122}\bullet$ generates a $N\bullet$ (Table 3.1). This amount is significantly greater than the amount of $N\bullet$ expected, if its formation was due to wild-type $\alpha2$. Therefore, the results indicate that the NH_2Y-$\alpha2s$ are inherently active and capable of generating a $C_{439}\bullet$, which gives rise to the $N\bullet$. Further, they show that wild-type $\alpha2$ levels in NH_2Y-$\alpha2$

Table 3.1 Activities of wild-type $\alpha2$ and NH_2Y-$\alpha2s$ using radioactive product formation and N_3ADP assays

$\alpha2$ variant	Radioactive RNR assay (% wt)[a]	N_3ADP assay (% $N\bullet$ at 20 s vs. initial $Y_{122}\bullet$)[b]
wt $\alpha2$	100[a]	52
$Y_{730}NH_2Y$-$\alpha2$	4 ± 0.5	15 ± 2[b]
$Y_{731}NH_2Y$-$\alpha2$	7 ± 0.5	15 ± 2[b]

[a] The activities are reported as percentage of wild-type $\alpha2$ activity, which was 2500 nmol/min mg. The error is the standard deviation from 2 independent measurements.
[b] Note that the amount of $N\bullet$ is reported relative to $[Y_{122}\bullet]$ at $t = 0$; the error is associated with EPR spin quantitation.

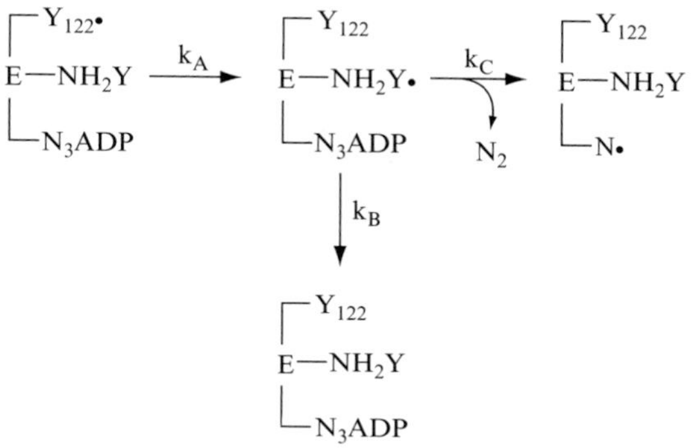

$$\begin{array}{ccc}
\left[\begin{array}{c} Y_{122}\bullet \\ E{-}NH_2Y \\ N_3ADP \end{array}\right] & \xrightarrow{k_A} & \left[\begin{array}{c} Y_{122} \\ E{-}NH_2Y\bullet \\ N_3ADP \end{array}\right] \xrightarrow[N_2]{k_C} \left[\begin{array}{c} Y_{122} \\ E{-}NH_2Y \\ N\bullet \end{array}\right] \\
& & \Big\downarrow k_B \\
& & \left[\begin{array}{c} Y_{122} \\ E{-}NH_2Y \\ N_3ADP \end{array}\right]
\end{array}$$

Figure 3.6 Working model for loss of spin during reaction of NH$_2$Y-α2s with N$_3$ADP/dGTP. After radical initiation and NH$_2$Y • formation (k$_A$), partitioning occurs between C$_{439}$• formation and subsequent N• formation (k$_C$) or quenching by yet unknown mechanisms (k$_B$).

preparations are low, indicating specific and efficient incorporation of NH$_2$Y. As we have discussed before, the activity of NH$_2$Y-α2s has major implications regarding the mechanism of radical propagation at residue Y$_{730}$ (Seyedsayamdost et al., 2007b). One problem with the N$_3$ADP assay is that NH$_2$Y • -α2 is unstable on the time scale of the experiment resulting in loss of 20% of total spin. The other two radicals in this reaction, Y$_{122}$ • and N • , have been shown to be stable over 20 s. Thus, the NH$_2$Y • may partition between two pathways (Fig. 3.6). It may give rise to a C$_{439}$ • , which generates a N • , or it may be reduced by components of the assay buffer. The latter pathway results in loss of spin.

Together, results from RNR activity and N$_3$ADP assays suggest that NH$_2$Y incorporation into α2 was efficient and specific; however, an accurate determination of the level of NH$_2$Y insertion will require one of the quantitative methods discussed in section 6.7.

7. CHARACTERIZATION OF NH$_2$Y-α2s

7.1. Overview

For NH$_2$Y to serve as a useful probe for enzymes that employ Y • s or modified Y • s in catalysis, its spectroscopic properties must be known. In this section, we describe the procedures we have used to characterize the EPR and UV–vis properties of NH$_2$Y • -α2. We also present the protocols for determining the rate constant for NH$_2$Y • -α2 formation using stopped flow (SF) UV–vis and rapid freeze-quench (RFQ) EPR spectroscopic methods.

7.2. Protocol for determining the EPR and UV-vis properties of $NH_2Y\bullet$-$\alpha 2s$

To determine the EPR spectrum of $NH_2Y\bullet$-$\alpha 2$, prereduced NH_2Y-α 2 and ATP were mixed with wild-type $\beta 2$ and CDP in assay buffer to give final concentrations of 20 to 24 μM, 3 mM, and 20 to 24 μM, 1 mM, respectively. The mixture was transferred to an EPR tube and quenched at defined time points by freezing in liquid N_2. The EPR spectra were recorded at 77 K and analyzed as described in section 6.8.

To determine the absorption features of $NH_2Y_{730}\bullet$ and $NH_2Y_{731}\bullet$, SF UV-vis kinetics were carried out on an Applied Photophysics DX 17MV instrument equipped with the Pro-Data upgrade using PMT and diode array detection at 25 °C. Prereduced $Y_{730}NH_2Y$-$\alpha 2$ (or $Y_{731}NH_2Y$-$\alpha 2$) and ATP in one syringe were rapidly mixed in a 1:1 ratio with wild-type $\beta 2$ and CDP from a second syringe to yield final concentrations of 8 to 10 μM, 3 mM, and 8 to 10 μM, 1 mM, respectively, in assay buffer. Initially, data was collected with a PDA.1 diode array detector by recording a spectrum between 300 and 700 nm every 100 ms after mixing. After identifying the spectral region at which $NH_2Y\bullet$ absorbs, the reaction was repeated using PMT detection to obtain a more accurate absorption profile of $NH_2Y\bullet$-$\alpha 2$. Two to four traces were recorded on a 1.5 to 2 s time scale and averaged between 305 and 365 nm in 5-nm intervals. The absorption change was corrected for the absorption of $Y_{122}\bullet$ in this region, given its well-characterized ε at these λs, (Bollinger *et al.*, 1995; Gräslund *et al.*, 1985; Nyholm *et al.*, 1993) and then plotted against λ. Calculation of the ε for $Y_{730}NH_2Y\bullet$-$\alpha 2$ (10,500 $M^{-1}cm^{-1}$) and $Y_{731}NH_2Y\bullet$-$\alpha 2$ (11,000 $M^{-1}cm^{-1}$) was performed using the ε of $Y_{122}\bullet$ ($\varepsilon_{410\ nm} = 3,700\ M^{-1}cm^{-1}$) (Bollinger *et al.*, 1995), assuming that consumption of each mole of $Y_{122}\bullet$ leads to formation of 1 mole of $NH_2Y\bullet$ in $\alpha 2$. Curve fitting was performed with OriginPro or KaleidaGraph Software.

7.3. Results

To monitor changes in the concentration of $NH_2Y\bullet$, its UV-vis and EPR spectral features must first be determined. These properties have facilitated our studies on NH_2Y-$\alpha 2s$ and will aid other research groups that intend to use NH_2Y as a probe. The reaction of NH_2Y-$\alpha 2/\beta 2$ with CDP/ATP, monitored by EPR spectroscopy, is shown in Fig. 3.7. The observed spectrum (Fig. 3.7A, black trace) is a mixture of two species: $Y_{122}\bullet$ and $NH_2Y\bullet$. The distinct low-field features of $Y_{122}\bullet$ were used to subtract its contribution from the observed spectrum to reveal the red trace in Fig. 3.7A, the spectrum of $NH_2Y_{731}\bullet$. It consists of a fairly isotropic signal with a g value of 2.0043 as expected for an organic radical. In the absence of substrate and effector, only the $Y_{122}\bullet$ is observed (Fig. 3.7B). This is an important control indicating pathway-dependent formation of $NH_2Y\bullet$

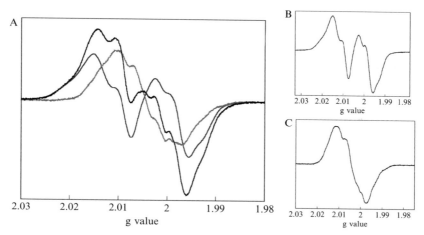

Figure 3.7 Reaction of NH_2Y-$\alpha2$/ATP with wild-type $\beta2$/CDP monitored by EPR spectroscopy. (A) The reaction components were mixed at 25 °C to yield final concentrations of 20 μM $Y_{731}NH_2Y$-$\alpha2$/$\beta2$ complex, 1 mM CDP, and 3 mM ATP in assay buffer. After 20 s, the reaction was quenched by hand-freezing in liquid N_2 and the EPR spectrum recorded at 77 K. Unreacted Y_{122}• (blue trace, 58% of total spin) was subtracted from the observed spectrum (black trace) to reveal the spectrum of NH_2Y_{731}• (red trace, 42 % of total spin). (B) Reaction of $Y_{731}NH_2Y$-$\alpha2$/$\beta2$ in the absence of CDP/ATP. Only the Y_{122}• spectrum is observed in this case. (C) Reaction of $Y_{730}NH_2Y$-$\alpha2$ with $\beta2$ and CDP/ATP, similar to that described in (A). The subtracted spectrum of NH_2Y_{730}• is shown. (See Color Insert.)

only when substrate and effector are bound to the NH_2Y-$\alpha2$/$\beta2$ complex. Interestingly, the spectra of NH_2Y_{730}• and NH_2Y_{731}• are similar but distinct, perhaps reflecting differences in the conformation(s) and/or environment around residue 730 vs. 731 (Fig. 3.7C).

Next, we wished to characterize the UV-vis features of NH_2Y•. We hypothesized that its absorption profile would be similar to that of DOPA • (Craw et al., 1984), given the structural similarity between the two species. SF diode array spectroscopy clearly showed changes in the region between 310 and 365 nm (data not shown). The extinction coefficients associated with Y_{122}• between 310 and 365 nm are known and small ($\varepsilon \approx 500$ to 1900 $M^{-1}cm^{-1}$) (Bollinger et al., 1995; Gräslund et al., 1985; Nyholm et al., 1993) and can be used in spectral deconvolution. To determine the features of NH_2Y• more accurately, SF UV-vis was carried out using PMT detection. The absorbance change at 1.5 s at each λ, corrected for the absorption by the Y_{122}•, was plotted against the λ. The results are shown in Fig. 3.8A and indicate that NH_2Y_{730}• and NH_2Y_{731}• have similar absorption profiles, which resemble the profile of DOPA• (Craw et al., 1984). The UV-vis spectrum of NH_2Y_{730}• consists of a broad feature with a λ_{max} at 325 nm ($\varepsilon \approx 10,500$ M^{-1} cm^{-1}). The NH_2Y_{731}• spectrum exhibits a λ_{max} at 320 nm ($\varepsilon \approx 11,000$ M^{-1} cm^{-1}) and a more defined shoulder at 350 nm.

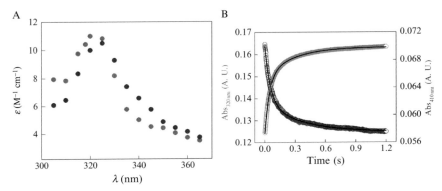

Figure 3.8 Characterization of NH_2Y-α2s by SF UV-vis spectroscopy. (A) Point-by-point reconstruction of the UV-vis spectrum of NH_2Y_{730} • (blue dots) and NH_2Y_{731} • (red dots). Prereduced $Y_{730}NH_2Y$-α2 and ATP in one syringe were mixed with wild-type β2 and CDP from another syringe, yielding final concentrations of 10 μM, 3 mM, 10 μM, and 1 mM, respectively. With $Y_{731}NH_2Y$-α2, the reaction was carried out at final concentrations of 9 μM $Y_{731}NH_2Y$-α2/β2, 1 mM CDP, and 3 mM ATP. The absorption change was monitored in 5-nm intervals; at each λ, 2 to 4 time courses were averaged and corrected for the absorption of Y_{122} • using previously determined ε in this spectral range. The corrected ΔOD was converted to ε, which was then plotted against λ. (B) Prereduced $Y_{731}NH_2Y$-α2 and CDP in one syringe were mixed in a 1:1 ratio with β2 and ATP from another syringe to yield the same concentrations as in (A). The concentration of Y_{122} • (blue trace) and NH_2Y_{731} • (red trace) were monitored at 410 nm and 320 nm, respectively. A total of 6 traces were averaged at each λ. See text for kinetic parameters. (See Color Insert.)

7.4. Protocol for determining the kinetics of NH_2Y•-α2 formation by SF UV-vis spectroscopy

To determine the kinetics of Y_{122} • consumption and NH_2Y • -α2 formation, a reaction identical to the one described previously was carried out. At least six individual traces were averaged at 410 nm (λ_{max} of Y_{122} •) and 325 nm (λ_{max} of NH_2Y_{730} •) or 320 nm (λ_{max} of NH_2Y_{731} •). Curve fitting was performed with OriginPro or KaleidaGraph Software. To obtain kinetic constants, iterative rounds of fitting and calculation of a residual plot and the R^2 correlation value were carried out until both were optimized.

7.5. Protocol for determining the kinetics of NH_2Y•-α2 formation by RFQ EPR spectroscopy

Rapid freeze-quench EPR samples were prepared using an Update Instruments 1019 Syringe Ram Unit, a Model 715 Syringe Ram Controller, and a quenching bath (which consists of a large outer stainless-steel pot that contains a smaller $\approx$7-L liquid isopentane bath). The temperature of the liquid isopentane bath was controlled with a Fluke 52 Dual Input Thermometer, equipped with an Anritsu Cu Thermocouple probe for the isopentane bath and the funnel. Stainless-steel packers were purchased from McMaster-Carr and were cut to a length of 40 cm and deburred at the MIT machine shop. The dead-time of the setup was determined to be 16 ± 2 ms with two independent measurements of the myoglobin/NaN$_3$ test reaction according to previously published procedures (Ballou and Palmer, 1974). A packing factor of 0.60 ± 0.05 was reproducibly obtained as tested with wild-type β2 samples. Routinely, a ram push velocity of 1.25 or 1.6 cm/s was used, and the displacement was adjusted to expel 300 μL of sample after the reaction.

Operation of the apparatus was similar to the procedures previously described (Ballou and Palmer, 1974; Bollinger et al., 1995). Briefly, the temperature of the isopentane bath was lowered to -135 to $-140\,^\circ$C with liquid N$_2$ in the outer bath. One syringe contained 38 μM prereduced Y$_{731}$NH$_2$Y-α2 and 6 mM ATP in assay buffer and was mixed with 38 μM wild-type β2 and 2 mM CDP in assay buffer from a second syringe. When the temperature of the EPR tube-funnel assembly had equilibrated to the bath temperature, the contents of each syringe were mixed rapidly and aged for a predetermined period by traversing the contents through a reaction loop. The sample was then sprayed into the EPR tube-funnel assembly, which was held at a distance of $\leq$ 1 cm from the spray nozzle. After the sample was sprayed into the funnel, the assembly was immediately returned to the bath and the crystals allowed to settle for 15 to 30 s. The sample was then packed into the EPR tube using the stainless-steel packers. After packing, excess isopentane was removed by flicking the EPR tube or by using a Pasteur pipette. The EPR tube was stored in liquid N$_2$ until the EPR spectrum was recorded. Acquisition of EPR spectra was carried out as described in section 6.8. Each time point was collected in duplicates. The quenching times, including the instrument dead time, are indicated in the figure legend.

7.6. Results of SF UV-vis and RFQ EPR spectroscopic studies

The SF UV-vis experiments were carried out to monitor the kinetics of Y$_{122}\bullet$ disappearance and NH$_2$Y$_{731}\bullet$ formation. The results are shown in Fig. 3.8B. At 410 nm, biexponential kinetics with rate constants of 17.3 ± 0.2 and 2.3 ± 0.1 s^{-1} were observed for loss of Y$_{122}\bullet$, similar to rate constants obtained for formation of NH$_2$Y$_{731}\bullet$ at 320 nm (21.0 ± 0.1 s^{-1} and 3.0 ± 0.1 s^{-1}). Analogous experiments carried out with Y$_{730}$NH$_2$Y-α2 showed that disappearance of Y$_{122}\bullet$ occurred biexponentially (12.0 ± 0.1 s^{-1} and 2.4 ± 0.1 s^{-1}), concomitant with formation of NH$_2$Y$_{730}\bullet$ (13.6 ± 0.1 s^{-1} and 2.5 ± 0.1 s^{-1}, data not shown). A control experiment was carried out in the absence of substrate and effector with $Y_{731}NH_2Y$-$\alpha 2$ (or $Y_{730}NH_2Y$-$\alpha 2$) and $\beta 2$. As indicated by EPR experiments, no loss of Y_{122} • or formation of NH_2Y • occurred under identical conditions (data not shown).

The slow and fast rate constants for NH_2Y • -$\alpha 2$ formation both occur in a fashion that is kinetically competent to be involved in dNDP formation. Thus, studies with the mutants provide the first direct evidence for their involvement in radical propagation. The slow rate constants, also observed in the DOPA-$\beta 2$ experiments, are similar to the steady-state rate constant for RNR turnover. See Seyedsayamdost et al. (2007b) for further discussion of the observed rate constants and amplitudes.

The RFQ-EPR data monitoring formation of NH_2Y_{731} • is shown in Fig. 3.9. A double exponential fit to the data is not warranted because fewer data points are accessible by this method, relative to SF UV-vis spectroscopy. We have therefore used the SF UV-vis data to fix the rate constant and amplitude of the slow phase in our double exponential fitting procedure of the RFQ-EPR data. The fit then yields the rate constant and amplitude of the fast phase, which are $19.2 \pm 2.4\ \text{s}^{-1}$ and $24 \pm 2\%$, respectively, in excellent agreement with the dominant kinetic phase in the SF UV-vis experiments. The total amplitude observed in this reaction, $35 \pm 4\%$

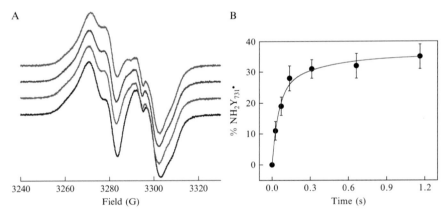

Figure 3.9 Analysis of NH_2Y_{731} • formation by rapid-freeze quench EPR spectroscopy. (A) $Y_{731}NH_2Y$-$\alpha 2$ and ATP were rapidly mixed with wild-type $\beta 2$ and CDP to yield final concentrations of 19 μM $\beta 2/Y_{731}NH_2Y$-$\alpha 2$, 3 mM ATP and 1 mM CDP. The reaction was quenched at various time points by spraying its contents into a liquid isopentane bath maintained at $\approx -140\ ^\circ C$. The EPR spectra were subsequently recorded at 77 K. Shown are the spectra at 0 ms (black), 72 ms (red), 312 ms (blue), and 1.16 s (green). (B) Kinetic trace obtained after subtracting the Y_{122} • component from each trace and plotting the percentage of NH_2Y_{731} • (relative to total spin) versus time. In this double exponential fit, the rate constant and amplitude for the slow phase have been held constant using the values from the SF UV-vis studies. Therefore, the fit yields the parameters for the fast kinetic phase. See text for kinetic parameters. (See Color Insert.)

$NH_2Y_{731}{}^\bullet$ relative to $[Y_{122}{}^\bullet]$ at $t = 0$, also agrees with the total amplitude for $NH_2Y_{731}{}^\bullet$ formation from the SF UV-vis studies ($35 \pm 1\%$ $NH_2Y_{731}{}^\bullet\%$).

8. SUMMARY

Studies using the evolved suppressor tRNA/RS pair to incorporate NH_2Y site-specifically into RNR show the power of this method to study mechanistic issues that are not accessible by any other methods currently available. Site-directed mutagenesis, a universally used method, took decades to evolve into the essential technology it is today. Likewise, incorporation of unnatural amino acids in a similarly robust fashion will require the efforts of many investigators and the sharing of methodological improvements. The radical propagation reaction catalyzed by RNR over 35 Å between two protein subunits is unprecedented. Only with the use of NH_2Y and other evolved suppressor tRNA/RS pairs (e.g. for fluorotyrosines (Seyedsayamdost et al., 2006a) and 3-nitrotyrosine (Neumann et al., 2008; Yee et al., 2003b)) will the mechanistic details of this remarkable reaction become unveiled.

ACKNOWLEDGEMENTS

We are grateful to Prof. Peter G. Schultz and Jianming Xie for sharing their technology in the development of NH_2Y-RS, Brenda N. Goguen for assistance with Fig. 3.2, and the National Institutes of Health Grant GM29595 for support.

REFERENCES

Alfonta, L., Zhang, Z., Uryu, S., Loo, J. A., and Schultz, P. G. (2003). Site-specific incorporation of a redox-active amino acid into proteins. *J. Am. Chem. Soc.* **125,** 14662–14663.

Amann, E., Ochs, B., and Abel, K. J. (1988). Tightly regulated tac promoter vectors useful for the expression of unfused and fused proteins in *Escherichia coli*. *Gene* **69,** 301–315.

Atkin, C. L., Thelander, L., Reichard, P., and Lang, G. (1973). Iron and free radical in ribonucleotide reductase. Exchange of iron and Mossbauer spectroscopy of the protein B2 subunit of the *Escherichia coli* enzyme. *J. Biol. Chem.* **248,** 7464–7472.

Ballou, D. P., and Palmer, G. (1974). Practical rapid quenching instrument for the study of reaction mechanisms by electron paramagnetic resonance spectroscopy. *Anal. Chem.* **46,** 1248–1253.

Bennati, M., Weber, A., Antonic, J., Perlstein, D. L., Robblee, J., and Stubbe, J. (2003). Pulsed ELDOR spectroscopy measures the distance between the two tyrosyl radicals in the R2 subunit of the *E. coli* ribonucleotide reductase. *J. Am. Chem. Soc.* **125,** 14988–14989.

Berglund, O., and Eckstein, F. (1972). Synthesis of ATP- and dATP-substituted sepharoses and their application in the purification of phage-T4-induced ribonucleotide reductase. *Eur. J. Biochem.* **28,** 492–496.

Berglund, O., and Eckstein, F. (1974). ATP-and dATP-substituted agaroses and the purification of ribonucleotide reductases. *Methods Enzymol.* **34,** 253–261.

Bollinger, J. M., Jr., Tong, W. H., Ravi, N., Huynh, B. H., Edmondson, D. E., and Stubbe, J. (1995). Use of rapid kinetics methods to study the assembly of the diferric-tyrosyl radical cofactor of *E. coli* ribonucleotide reductase. *Methods Enzymol.* **258,** 278–303.

Chin, J. W., Martin, A. B., King, D. S., Wang, L., and Schultz, P. G. (2002). Addition of a photocrosslinking amino acid to the genetic code of *Escherichia coli. Proc. Natl. Acad. Sci. USA* **99,** 11020–11024.

Chivers, P. T., Prehoda, K. E., Volkman, B. F., Kim, B. M., Markley, J. L., and Raines, R. T. (1997). Microscopic pKa values of *Escherichia coli* thioredoxin. *Biochemistry* **36,** 14985–14991.

Climent, I., Sjöberg, B.-M., and Huang, C. Y. (1992). Site-directed mutagenesis and deletion of the carboxyl terminus of *Escherichia coli* ribonucleotide reductase protein R2. Effects on catalytic activity and subunit interaction. *Biochemistry* **31,** 4801–4807.

Cotton, G. J., Ayers, B., Xu, R., and Muir, T. W. (1999). Insertion of a Synthetic Peptide into a Recombinant Protein Framework: A Protein Biosensor. *J. Am. Chem. Soc.* **121,** 1100–1101.

Craw, M., Chedekel, M. R., Truscott, T. G., and Land, E. J. (1984). The photochemical interaction between the triplet state of 8-methoxypsoralen and the melanin precursor L-3,4 dihydroxyphenylalanine. *Photochem. Photobiol.* **39,** 155–159.

Cukier, R. I., and Nocera, D. G. (1998). Proton-coupled electron transfer. *Annu. Rev. Phys. Chem.* **49,** 337–369.

Dawson, P. E., and Kent, S. B. (2000). Synthesis of native proteins by chemical ligation. *Annu. Rev. Biochem.* **69,** 923–960.

DeFelippis, M. R., Murthy, C. P., Broitman, F., Weinraub, D., Faraggi, M., and Klapper, M. H. (1991). Electrochemical properties of tyrosine phenoxy and tryptophan indolyl radicals in peptides and amino acid analogs. *J. Phys. Chem.* **95,** 3416–3419.

Ehrenberg, A., and Reichard, P. (1972). Electron spin resonance of the iron-containing protein B2 from ribonucleotide reductase. *J. Biol. Chem.* **247,** 3485–3488.

Ekberg, M., Pötsch, S., Sandin, E., Thunnissen, M., Nordlund, P., Sahlin, M., and Sjöberg, B.-M. (1998). Preserved catalytic activity in an engineered ribonucleotide reductase R2 protein with a nonphysiological radical transfer pathway. The importance of hydrogen bond connections between the participating residues. *J. Biol. Chem.* **273,** 21003–21008.

Ekberg, M., Sahlin, M., Eriksson, M., and Sjöberg, B.-M. (1996). Two conserved tyrosine residues in protein R1 participate in an intermolecular electron transfer in ribonucleotide reductase. *J. Biol. Chem.* **271,** 20655–20659.

Eklund, H., Uhlin, U., Färnegårdh, M., Logan, D. T., and Nordlund, P. (2001). Structure and function of the radical enzyme ribonucleotide reductase. *Prog. Biophys. Mol. Biol.* **77,** 177–268.

Farrell, I. S., Toroney, R., Hazen, J. L., Mehl, R. A., and Chin, J. W. (2005). Photo-cross-linking interacting proteins with a genetically encoded benzophenone. *Nat. Methods* **2,** 377–384.

Fritscher, J., Artin, E., Wnuk, S., Bar, G., Robblee, J. H., Kacprzak, S., Kaupp, M., Griffin, R. G., Bennati, M., and Stubbe, J. (2005). Structure of the nitrogen-centered radical formed during inactivation of *E. coli* ribonucleotide reductase by 2′-azido-2′-deoxyuridine-5′-diphosphate: Trapping of the 3′-ketonucleotide. *J. Am. Chem. Soc.* **127,** 7729–7738.

Ge, J., Yu, G., Ator, M. A., and Stubbe, J. (2003). Pre-steady-state and steady-state kinetic analysis of *E. coli* class I ribonucleotide reductase. *Biochemistry* **42**, 10071–10083.

Giriat, I., and Muir, T. W. (2003). Protein semi-synthesis in living cells. *J. Am. Chem. Soc.* **125**, 7180–7181.

Gräslund, A., Sahlin, M., and Sjöberg, B.-M. (1985). The tyrosine free radical in ribonucleotide reductase. *Environ. Health Perspect.* **64**, 139–149.

Gray, H. B., and Winkler, J. R. (1996). Electron transfer in proteins. *Annu. Rev. Biochem.* **65**, 537–561.

Hobbs, J. B., and Eckstein, F. (1977). A general method for the synthesis of 2′-azido-2′-deoxy- and 2′-amino-2′-deoxyribofuranosyl purines. *J. Org. Chem.* **42**, 714–719.

Hooker, J. M., Kovacs, E. W., and Francis, M. B. (2004). Interior surface modification of bacteriophage MS2. *J. Am. Chem. Soc.* **126**, 3718–3719.

Jackson, J. C., Duffy, S. P., Hess, K. R., and Mehl, R. A. (2006). Improving nature's enzyme active site with genetically encoded unnatural amino acids. *J. Am. Chem. Soc.* **128**, 11124–11127.

Jordan, A., and Reichard, P. (1998). Ribonucleotide reductases. *Annu. Rev. Biochem.* **67**, 71–98.

Jovanovic, S. J. S., Tosic, M., Marjanovic, B., and Simic, M. G. (1994). Flavonoids as antioxidants. *J. Am. Chem. Soc.* **116**, 4846–4851.

Kiga, D., Sakamoto, K., Kodama, K., Kigawa, T., Matsuda, T., Yabuki, T., Shirouzu, M., Harada, Y., Nakayama, H., Takio, K., Hasegawa, Y., Endo, Y., *et al.* (2002). An engineered *Escherichia coli* tyrosyl-tRNA synthetase for site-specific incorporation of an unnatural amino acid into proteins in eukaryotic translation and its application in a wheat germ cell-free system. *Proc. Natl. Acad. Sci. USA* **99**, 9715–9720.

Kobayashi, T., Nureki, O., Ishitani, R., Yaremchuk, A., Tukalo, M., Cusack, S., Sakamoto, K., and Yokoyama, S. (2003). Structural basis for orthogonal tRNA specificities of tyrosyl-tRNA synthetases for genetic code expansion. *Nat. Struct. Biol.* **10**, 425–432.

Kobayashi, T., Takimura, T., Sekine, R., Kelly, V. P., Kamata, K., Sakamoto, K., Nishimura, S., and Yokoyama, S. (2005). Structural snapshots of the KMSKS loop rearrangement for amino acid activation by bacterial tyrosyl-tRNA synthetase. *J. Mol. Biol.* **346**, 105–117.

Kovacs, E. W., Hooker, J. M., Romanini, D. W., Holder, P. G., Berry, K. E., and Francis, M. B. (2007). Dual-surface-modified bacteriophage MS2 as an ideal scaffold for a viral capsid-based drug delivery system. *Bioconjug. Chem.* **18**, 1140–1147.

Licht, S., and Stubbe, J. (1999). Mechanistic investigations of ribonucleotide reductases. *In* "Comprehensive Natural Products Chemistry." (S. D. Barton, K. Nakanishi, O. Meth-Cohn, and C. D. Poulter, eds.), pp. 163–203. Elsevier Science, New York.

Mao, S. S., Johnston, M. I., Bollinger, J. M., Jr., and Stubbe, J. (1989). Mechanism-based inhibition of a mutant *Escherichia coli* ribonucleotide reductase (cysteine-225→serine) by its substrate CDP. *Proc. Natl. Acad. Sci. USA* **86**, 1485–1489.

Marcus, R. A., and Sutin, N. (1985). Electron transfer in chemistry and biology. *Biochim. Biophys. Acta* **811**, 265–322.

Mootz, H. D., Blum, E. S., Tyszkiewicz, A. B., and Muir, T. W. (2003). Conditional protein splicing: A new tool to control protein structure and function in vitro and *in vivo*. *J. Am. Chem. Soc.* **125**, 10561–10569.

Moser, C. C., Keske, J. M., Warncke, K., Farid, R. S., and Dutton, P. L. (1992). Nature of biological electron transfer. *Nature* **355**, 796–802.

Muir, T. W. (2003). Semisynthesis of proteins by expressed protein ligation. *Annu. Rev. Biochem.* **72**, 249–289.

Muralidharan, V., and Muir, T. W. (2006). Protein ligation: An enabling technology for the biophysical analysis of proteins. *Nat. Methods* **3**, 429–438.

Neumann, H., Hazen, J. L., Weinstein, J., Mehl, R. A., and Chin, J. W. (2008). Genetically encoding protein oxidative damage. *J. Am. Chem. Soc.* **130,** 4028–4033.

Nilsson, B., Moks, T., Jansson, B., Abrahmsen, L., Elmblad, A., Holmgren, E., Henrichson, C., Jones, T. A., and Uhlen, M. (1987). A synthetic IgG-binding domain based on staphylococcal protein A. *Protein Eng.* **1,** 107–113.

Nordlund, P., Sjöberg, B.-M., and Eklund, H. (1990). Three-dimensional structure of the free radical protein of ribonucleotide reductase. *Nature* **345,** 593–598.

Nyholm, S., Thelander, L., and Gräslund, A. (1993). Reduction and loss of the iron center in the reaction of the small subunit of mouse ribonucleotide reductase with hydroxyurea. *Biochemistry* **32,** 11569–11574.

Page, C. C., Moser, C. C., Chen, X., and Dutton, P. L. (1999). Natural engineering principles of electron tunnelling in biological oxidation-reduction. *Nature* **402,** 47–52.

Page, C. C., Moser, C. C., and Dutton, P. L. (2003). Mechanism for electron transfer within and between proteins. *Curr. Opin. Chem. Biol.* **7,** 551–556.

Palmer, G. (1967). Electron paramagnetic resonance. *Methods Enzymol.* **10,** 595–610.

Pellois, J. P., and Muir, T. W. (2006). Semisynthetic proteins in mechanistic studies: Using chemistry to go where nature can't. *Curr. Opin. Chem. Biol.* **10,** 487–491.

Perler, F. B. (2005). Protein splicing mechanisms and applications. *IUBMB Life* **57,** 469–476.

Reece, S. Y., Hodgkiss, J. M., Stubbe, J., and Nocera, D. G. (2006). Proton-coupled electron transfer: The mechanistic underpinning for radical transport and catalysis in biology. *Philos. Trans. R. Soc. Lond. B Biol. Sci.* **361,** 1351–1364.

Reichard, P., and Ehrenberg, A. (1983). Ribonucleotide reductase: A radical enzyme. *Science* **221,** 514–519.

Robins, M. J., Hawrelak, S. D., Hernandez, A. E., and Wnuk, S. F. (1992). Nucleic acid related compounds. LXXXI. Efficient general synthesis of purine (amino, azido, and triflate)-sugar nucleosides. *Nucleosides Nucleotides* **11,** 821–834.

Russel, M., and Model, P. (1985). Direct cloning of the trxB gene that encodes thioredoxin reductase. *J. Bacteriol.* **163,** 238–242.

Ryu, Y., and Schultz, P. G. (2006). Efficient incorporation of unnatural amino acids into proteins in *Escherichia coli.* *Nat. Methods* **3,** 263–265.

Salowe, S., Bollinger, J. M., Jr., Ator, M., and Stubbe, J. (1993). Alternative model for mechanism-based inhibition of *Escherichia coli* ribonucleotide reductase by 2′-azido-2′-deoxyuridine 5′-diphosphate. *Biochemistry* **32,** 12749–12760.

Salowe, S. P., Ator, M. A., and Stubbe, J. (1987). Products of the inactivation of ribonucleoside diphosphate reductase from *Escherichia coli* with 2′-azido-2′-deoxyuridine 5′-diphosphate. *Biochemistry* **26,** 3408–3416.

Salowe, S. P., and Stubbe, J. (1986). Cloning, overproduction, and purification of the B2 subunit of ribonucleoside-diphosphate reductase. *J. Bacteriol.* **165,** 363–366.

Santoro, S. W., Wang, L., Herberich, B., King, D. S., and Schultz, P. G. (2002). An efficient system for the evolution of aminoacyl-tRNA synthetase specificity. *Nat. Biotechnol.* **20,** 1044–1048.

Seagle, R. L., and Cowgill, R. W. (1976). Fluorescence and the structure of proteins. XXI. Fluorescence of aminotyrosyl residues in peptides and helical proteins. *Biochim. Biophys. Acta* **439,** 461–469.

Seyedsayamdost, M. R., Chan, C. T., Mugnaini, V., Stubbe, J., and Bennati, M. (2007a). PELDOR spectroscopy with DOPA-$\beta2$ and NH_2Y-$\alpha2$s: Distance measurements between residues involved in the radical propagation pathway of *E. coli* ribonucleotide reductase. *J. Am. Chem. Soc.* **129,** 15748–15749.

Seyedsayamdost, M. R., Reece, S. Y., Nocera, D. G., and Stubbe, J. (2006a). Mono-, di-, tri-, and tetra-substituted fluorotyrosines: New probes for enzymes that use tyrosyl radicals in catalysis. *J. Am. Chem. Soc.* **128,** 1569–1579.

Seyedsayamdost, M. R., and Stubbe, J. (2006). Site-specific replacement of Y$_{356}$ with 3,4-dihydroxyphenylalanine in the β2 subunit of *E. coli* ribonucleotide reductase. *J. Am. Chem. Soc.* **128**, 2522–2523.

Seyedsayamdost, M. R., Xie, J., Chan, C. T., Schultz, P. G., and Stubbe, J. (2007b). Site-specific insertion of 3-aminotyrosine into subunit α2 of *E. coli* ribonucleotide reductase: Direct evidence for involvement of Y$_{730}$ and Y$_{731}$ in radical propagation. *J. Am. Chem. Soc.* **129**, 15060–15071.

Seyedsayamdost, M. R., Yee, C. S., Reece, S. Y., Nocera, D. G., and Stubbe, J. (2006b). pH rate profiles of F$_n$Y$_{356}$-R2s (n = 2, 3, 4) in *Escherichia coli* ribonucleotide reductase: Evidence that Y$_{356}$ is a redox-active amino acid along the radical propagation pathway. *J. Am. Chem. Soc.* **128**, 1562–1568.

Seyedsayamdost, M. R., Yee, C. S., and Stubbe, J. (2007c). Site-specific incorporation of fluorotyrosines into the R2 subunit of *E. coli* ribonucleotide reductase by expressed protein ligation. *Nat. Protoc.* **2**, 1225–1235.

Sjöberg, B.-M., Gräslund, A., and Eckstein, F. (1983). A substrate radical intermediate in the reaction between ribonucleotide reductase from *Escherichia coli* and 2′-azido-2′-deoxynucleoside diphosphates. *J. Biol. Chem.* **258**, 8060–8067.

Sjöberg, B.-M., Reichard, P., Gräslund, A., and Ehrenberg, A. (1978). The tyrosine free radical in ribonucleotide reductase from *Escherichia coli*. *J. Biol. Chem.* **253**, 6863–6865.

Steeper, J. R., and Steuart, C. D. (1970). A rapid assay for CDP reductase activity in mammalian cell extracts. *Anal. Biochem.* **34**, 123–130.

Stubbe, J. (1990). Ribonucleotide reductases: Amazing and confusing. *J. Biol. Chem.* **265**, 5329–5332.

Stubbe, J. (1998). Ribonucleotide reductases in the twenty-first century. *Proc. Natl. Acad. Sci. USA* **95**, 2723–2724.

Stubbe, J., Nocera, D. G., Yee, C. S., and Chang, M. C. Y. (2003). Radical initiation in the class I ribonucleotide reductase: long-range proton-coupled electron transfer? *Chem. Rev.* **103**, 2167–2201.

Stubbe, J., and Riggs-Gelasco, P. (1998). Harnessing free radicals: Formation and function of the tyrosyl radical in ribonucleotide reductase. *Trends Biochem. Sci.* **23**, 438–443.

Stubbe, J., and van der Donk, W. A. (1998). Protein radicals in enzyme catalysis. *Chem. Rev.* **98**, 705–762.

Thelander, L., Larsson, B., Hobbs, J., and Eckstein, F. (1976). Active site of ribonucleoside diphosphate reductase from *Escherichia coli*: Inactivation of the enzyme by 2′-substituted ribonucleoside diphosphates. *J. Biol. Chem.* **251**, 1398–1405.

Turner, J. M., Graziano, J., Spraggon, G., and Schultz, P. G. (2006). Structural plasticity of an aminoacyl-tRNA synthetase active site. *Proc. Natl. Acad. Sci. USA* **103**, 6483–6488.

Uhlin, U., and Eklund, H. (1994). Structure of ribonucleotide reductase protein R1. *Nature* **370**, 533–539.

van der Donk, W. A., Stubbe, J., Gerfen, G. G., Bellew, B. F., and Griffin, R. G. (1995). EPR investigations of the inactivation of *E. coli* ribonucleotide reductase with 2′-azido-2′-deoxyuridine 5′-diphosphate: Evidence for the involvement of the thiyl radical of C225-R1. *J. Am. Chem. Soc.* **117**, 8909–8916.

Wang, L., Brock, A., Herberich, B., and Schultz, P. G. (2001). Expanding the genetic code of *Escherichia coli*. *Science* **292**, 498–500.

Wang, L., and Schultz, P. G. (2001). A general approach for the generation of orthogonal tRNAs. *Chem. Biol.* **8**, 883–890.

Wang, L., and Schultz, P. G. (2004). Expanding the genetic code. *Angew. Chem. Int. Ed. Engl.* **44**, 34–66.

Wang, L., Xie, J., and Schultz, P. G. (2006). Expanding the genetic code. *Annu. Rev. Biophys. Biomol. Struct.* **35**, 225–249.

Wang, L., Zhang, Z., Brock, A., and Schultz, P. G. (2003). Addition of the keto functional group to the genetic code of *Escherichia coli*. *Proc. Natl. Acad. Sci. USA* **100**, 56–61.

Xie, J., and Schultz, P. G. (2005). An expanding genetic code. *Methods* **36**, 227–238.

Xie, J., and Schultz, P. G. (2006). A chemical toolkit for proteins: An expanded genetic code. *Nat. Rev. Mol. Cell Biol.* **7**, 775–782.

Yee, C. S., Chang, M. C. Y., Ge, J., Nocera, D. G., and Stubbe, J. (2003a). 2,3-Difluorotyrosine at position 356 of ribonucleotide reductase R2: A probe of long-range proton-coupled electron transfer. *J. Am. Chem. Soc.* **125**, 10506–10507.

Yee, C. S., Seyedsayamdost, M. R., Chang, M. C. Y., Nocera, D. G., and Stubbe, J. (2003b). Generation of the R2 subunit of ribonucleotide reductase by intein chemistry: Insertion of 3-nitrotyrosine at residue 356 as a probe of the radical initiation process. *Biochemistry* **42**, 14541–14552.

Zhang, Y., Wang, L., Schultz, P. G., and Wilson, I. A. (2005). Crystal structures of apo wild-type *M. jannaschii* tyrosyl-tRNA synthetase (TyrRS) and an engineered TyrRS specific for O-methyl-L-tyrosine. *Protein Sci.* **14**, 1340–1349.

Zhang, Z., Smith, B. A., Wang, L., Brock, A., Cho, C., and Schultz, P. G. (2003). A new strategy for the site-specific modification of proteins *in vivo*. *Biochemistry* **42**, 6735–6746.

Zhang, Z., Wang, L., Brock, A., and Schultz, P. G. (2002). The selective incorporation of alkenes into proteins in *Escherichia coli*. *Angew. Chem. Int. Ed. Engl.* **41**, 2840–2842.

CHAPTER FOUR

SEMISYNTHESIS OF PROTEINS USING SPLIT INTEINS

Christina Ludwig,* Dirk Schwarzer,*,† Joachim Zettler,*
Daniel Garbe,* Petra Janning,‡ Claus Czeslik,*
and Henning D. Mootz*,1

Contents

Abstract

Protein splicing is an autocatalytic reaction in which an internal protein domain, the intein, excises itself out of a precursor protein and concomitantly links the two flanking sequences, the exteins, with a native peptide bond. In split inteins, the intein domain is divided into two parts that undergo fragment association followed by protein splicing *in trans*. Thus, the extein sequences joined in the

* Fakultät Chemie – Chemische Biologie, Technische Universität Dortmund, Dortmund, Germany
† Current address: Leibniz-Institut für Molekulare Pharmakologie, Berlin, Germany
‡ Max-Planck-Institut für Molekulare Physiologie, Dortmund, Germany
1 Corresponding author: henning.mootz@uni-dortmund.de

Methods in Enzymology, Volume 462
ISSN 0076-6879, DOI: 10.1016/S0076-6879(09)62004-8

process originate from two separate molecules. The specificity and sequence promiscuity of split inteins make this approach a generally useful tool for the preparation of semisynthetic proteins. To this end, the recombinant part of the protein of interest is expressed as a fusion protein with one split intein fragment. The synthetic part is extended by the other, complementary fragment of the split intein. A recently introduced split intein, in which the N-terminal fragment consists of only 11 native amino acids, has greatly facilitated preparation of the synthetic part by solid-phase peptide synthesis. This intein enables the chemoenzymatic synthesis of N-terminally modified semisynthetic proteins. The reaction can be performed under native conditions and at protein and peptide concentrations in the low micromolar range. In contrast to chemical ligation procedures like native chemical ligation and expressed protein ligation, the incorporation of a thioester group and an aminoterminal cysteine into the two polypeptides to be linked is not necessary. We discuss properties of useful inteins, design rules for split inteins and intein insertion sites and we describe selected examples in detail.

1. INTRODUCTION

Semisynthetic proteins in the most general definition are hybrid compounds that consist of one part obtained by gene expression (i.e. a native or a recombinant protein) and a second part that is derived from chemical synthesis. The chemical moiety can be introduced in a variety of ways. Although chemically modified proteins have been prepared and investigated for decades, recently developed techniques have significantly expanded the protein chemists' toolbox (Durek and Becker, 2005; Muir, 2003; Nilsson *et al.*, 2005; Prescher and Bertozzi, 2005; Wang and Schultz, 2004). The new techniques, in principle, allow for manipulation of virtually any atom, bond, or substituent in a protein of interest with the precision of a synthetic organic chemist. Obviously, such semisynthetic proteins are valuable compounds for a wide array of biochemistry, cell biology, and protein-engineering investigations and applications, ranging from mechanistic studies via probing of pathways in living cells to therapeutic proteins and building blocks in bio- and nanotechnology.

Earlier approaches to prepare semisynthetic proteins mostly involved bioconjugation to reactive chemical groups in the proteinogenic amino acids and feeding of structural analogs that are taken up by the cell's biosynthetic machinery (Budisa, 2004; Hermanson, 1996). Thus, both approaches are generally limited in their scope and selectivity by the composition of the protein and by the substrate tolerance of the native biosynthetic pathways, respectively. From a variety of recently described approaches to achieve highly selective control on the chemical structure of semisynthetic proteins (Baskin *et al.*, 2007; Beckett *et al.*, 1999; Carrico *et al.*, 2007; de Araujo *et al.*, 2006; Dirksen *et al.*, 2006; Fernandez-Suarez *et al.*, 2007; Griffin *et al.*, 2000; Keppler *et al.*, 2003; Nguyen *et al.*, 2007; Popp *et al.*, 2007; Saxon *et al.*, 2000; Song *et al.*, 2008; Wehofsky *et al.*, 2003; Yin *et al.*, 2005), two techniques can be highlighted that arguably have had the greatest impact so far. The first example is the extension of site-directed mutagenesis to unnatural amino acids. The original concept of site-specific unnatural amino acid mutagenesis involved the preparation of chemically misacylated suppressor tRNAs for *in vitro* translation assays (Noren *et al.*, 1989). This process can further be adapted to larger-scale protein expression in *E. coli* or yeast cells by evolving and implanting an additional aminoacyl-tRNA synthetase/tRNA pair that is orthogonal to the endogenous machinery and specifically accepts the unnatural amino acid (Wang *et al.*, 2001; Xie and Schultz, 2006). Recently an orthogonal ribosome/mRNA pair could be established, which exhibits a significantly decreased affinity to release factor 1 and therefore a 3 to 20-fold increased incorporation of the unnatural amino acid (Wang *et al.*, 2007). The second major breakthrough was accomplished by the introduction of chemical ligation reactions (Dawson and Kent, 2000; Muir, 2003). In particular, native chemical ligation (NCL) is a chemoselective reaction between a C-terminal thioester and an N-terminal cysteine residue that generates a native peptide bond at the ligation site of two peptides (Dawson *et al.*, 1994). Thus, proteins can be assembled from smaller synthetic parts, which may contain the desired chemical modifications. Importantly, the reaction is also compatible with larger recombinant proteins, provided that they harbor either an N-terminal cysteine (Erlanson *et al.*, 1996) or a thioester at their C-terminus. In the latter case the reaction is also known as expressed protein ligation (EPL) (Evans *et al.*, 1998; Muir *et al.*, 1998).

Here we discuss a new approach to semisynthetic proteins by using an artificial form of protein splicing *in trans* performed by a split intein (see Fig. 4.1). Special emphasis is on a particular version of a split intein that enables the synthetic peptide to be linked to the N terminus of a recombinant protein by a native peptide bond. The product is thus similar to a semisynthetic protein prepared by NCL; however, because the C-terminal thioester and the N-terminal cysteine are replaced by fragments of an autocatalytic intein, the reaction can be regarded as chemo-enzymatic and brings about advantages for certain applications. Protein splicing *in trans* has been previously used to prepare C-terminally modified semisynthetic proteins (Evans *et al.*, 2000; Giriat and Muir, 2003; Kwon *et al.*, 2006; Lew *et al.*, 1998, 1999; Olschewski *et al.*, 2007). In the following, we will first give a short introduction on protein splicing and intein structure. Then, we will discuss different forms of split inteins and previous applications. Finally, we will explain experimental procedures of our new approach.

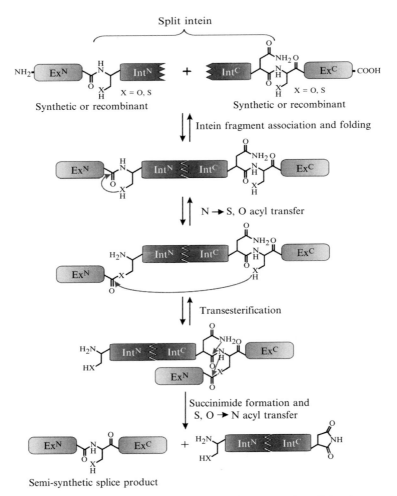

Figure 4.1 Principle and mechanism of semisynthetic protein splicing *in trans*. The semisynthetic protein is obtained from protein splicing *in trans*. One of the split intein fusion proteins is chemically synthesized, whereas the other is prepared recombinantly. Either the N-terminal or the C-terminal part can be the synthetic fragment. The connected polypeptides represent the extein sequences of the intein (Ex^N and Ex^C). In the first step, both intein fragments (Int^N and Int^C) associate and reconstitute to form the functional intein. The scissile bond between the N-extein and the intein is activated to a linear ester/thioester intermediate by an N→O or N→S acyl rearrangement. Subsequently, this ester/thioester is attacked by the hydroxyl/thiol group of the first amino acid of the C-extein, resulting in a branched intermediate. By a cyclization of the conserved C-terminal asparagine residue of the intein a succinimide derivative is formed and the liberated free α-amino group spontaneously rearranges by O→N or S→N acyl transfer to the native peptide bond between the N- and the C-exteins. Depending on the intein the atoms labeled with X may be O (= serine residue) or S (= cysteine residue) and the nucleophilic amino acid at the Int^C-Ex^C splice junction may also be a threonine. Some inteins also contain an alanine at the first position of the intein, in which case a direct attack of the peptide bond leading to the branched intermediate is assumed.

2. PROTEIN SPLICING *IN CIS* AND *IN TRANS* IS PERFORMED BY INTEINS

Protein splicing refers to the posttranslational excision of an internal polypeptide, the intein, out of a longer, gene-encoded precursor protein. Concomitantly, the flanking N- and C-terminal sequences of the protein, the so-called N- and C-exteins (Ex^N and Ex^C), are fused with a native peptide bond (Giriat *et al.*, 2001; Noren *et al.*, 2000; Paulus, 2000). This remarkable autocatalytic reaction performed by the intein proceeds through peptide backbone rearrangements involving thio- and/or oxoesters of amino acid side chains (see Fig. 4.1). All information for the reaction is stored within the intein. Other proteins, cofactors, or energy sources like ATP are not necessary for the splicing reaction to proceed. Importantly, the intein can in principle work in different sequence contexts, that is, it can be genetically introduced into or fused to other proteins of interest. These features provide the basis for the use of inteins as generic tools in protein chemical approaches. For example, the thioester intermediates of the splicing reaction are exploited in arrested intein mutants to generate recombinant protein thioesters, used for the previously mentioned EPL, from C-terminal intein fusion proteins (Chong *et al.*, 1997; Evans *et al.*, 1998; Muir *et al.*, 1998). However, the approach on which we wish to focus here directly uses the peptide-bond-forming ability of inteins for the preparation of semisynthetic proteins. For this purpose, polypeptides from two different molecules need to be spliced together. This is realized in the process of protein splicing *in trans* (or protein *trans*-splicing), a variation of regular protein splicing (or protein splicing *in cis*). In the former case, the intein domain is split into two parts, the Int^N and Int^C fragments, each connected to an extein sequence that represents one part of the protein of interest. For protein splicing *in trans* to occur, both fragments associate with each other upon mixing and fold into the active intein domain (see Fig. 4.1). Thus, the crucial prerequisite for protein semisynthesis applications is that the N- and C-terminal parts of a split intein can be produced separately (i.e. one synthetically and the other recombinantly) and that both pieces can be reconstituted into the active intein. Our ability to prepare either the Int^N or the Int^C fragment synthetically (i.e. by solid-phase peptide synthesis (SPPS)) will obviously depend on their sizes (see the subsequent paragraph). Split inteins are found in nature and can also be artificially engineered. Both native and artificially split inteins have been used for protein semisynthesis (see section 3). There are to date only two natively split inteins known, namely the *Ssp* DnaE intein from *Synechosystis sp.* PCC6803 (Wu *et al.*, 1998) and the *Neq* Pol intein from *Nanoarchaeum equitants* (Choi *et al.*, 2006). The *Ssp* DnaE intein has several homologs in other organisms

(Caspi *et al.*, 2003; Dassa *et al.*, 2007; Iwai *et al.*, 2006). The fragments of the two split inteins have a natural affinity for each other and spontaneously form the active intein under native conditions. In contrast, most artificially split inteins usually do not spontaneously reconstitute from their fragments when mixed under native conditions, and often the fusion constructs form inclusion bodies on expression in *E. coli*. Reconstitution to the active intein can in many cases be achieved by a denaturation step in urea or guanidinium hydrochloride, followed by removal of the denaturant through dialysis in the presence of the complementary fragment (Mills *et al.*, 1998; Southworth *et al.*, 1998; Yamazaki, 1998). Obviously, this procedure limits the applicability of the split intein to proteins that can also be refolded. Among the artificially split inteins, the *Ssp* DnaB intein as well as the *Mxe* GyrA intein can undergo spontaneous and efficient protein splicing *in trans* under native conditions (Brenzel *et al.*, 2006; Kurpiers and Mootz, 2007, 2008; Ludwig *et al.*, 2006).

3. DESIGN OF SPLIT INTEINS AND CONSIDERATIONS ON THE INTEIN INSERTION SITE

What is the size and sequence of the intein fragments that have to be fused to the extein sequences to perform semisynthetic protein splicing *in trans*? How can the choice of the fragments be rationalized from the intein structure? The overall sequence homology between different inteins is low and mostly concentrated in a few conserved sequence motifs (Noren *et al.*, 2000; Perler, 2002). Despite this finding, all inteins investigated so far share the same structural fold, termed the "HINT domain" (Hall *et al.*, 1997), which is about 140 to 150 aa in length, and can carry sequence insertions at various positions. The most prominent insertion found in many intein sequences is another protein, a homing endonuclease domain, which is functionally unrelated to protein splicing (Noren *et al.*, 2000). Inteins with a removed endonuclease domain are called mini–inteins. The separation of the intein into two regions of ≈ 100 and ≈ 40 aa in size, located N- and C-terminal to the endonuclease domain, respectively, has also defined the split site of most artificially split inteins (see Fig. 4.2A). This split position is also referred to as the S0 site (Sun *et al.*, 2004). Indeed, the natively split *Ssp* DnaE intein, though lacking an endonuclease domain, is made up of the corresponding Int^N and Int^C pieces (Sun *et al.*, 2005), and the same seems to be true for the *Neq* Pol intein (Choi *et al.*, 2006). Thus, only the Int^C fragment is in the realm of SPPS with its upper length limit of about 40 to 60 aa that can be routinely achieved using optimized protocols. So far, the Int^C fragments of the *Ssp* DnaE intein (36 aa) and the artificially split *Mtu* RecA intein (35 aa) have been synthesized with Ex^C-sequences of a few amino acids and shown

A

Removed endonuclease
and religated intein

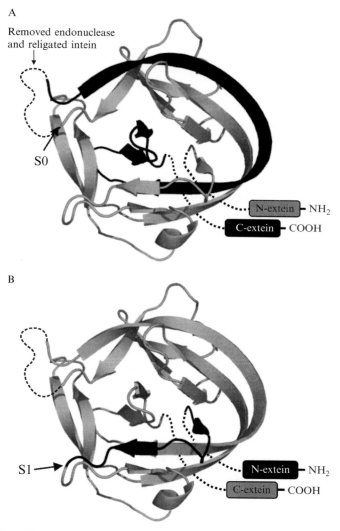

Figure 4.2   Different ways to generate artificially split *Ssp* DanB inteins. (A) A split site at the peptide bond after amino acid 104 (S0) results in the two intein fragments Int^N (104 aa, shown in black) and Int^C (47 aa, shown in grey). This position corresponds to the insertion site of the removed homing endonuclease (Wu *et al.*, 1998). (B) The split site S1 at the peptide bond behind amino acid 11 gives the small Int^N-fragment (11 aa, shown in black) and a large Int^C part (154 aa, shown in grey) (Sun *et al.*, 2004). This figure was created with PyMol from PDB file 1MI8 (Ding *et al.*, 2003).

to be active in subsequent *trans*-splicing reactions with recombinant counter-parts carrying the Int^N-fragment (Evans *et al.*, 2000; Lew *et al.*, 1998, 1999). Kwon and Camarero used the semisynthetic *Ssp* DnaE intein for the site-specific immobilization of proteins through their C-termini onto modified glass surfaces (Kwon *et al.*, 2006). Recently, Becker and co-workers exploited the same intein to access a double-lipidated prion protein, which represents a mimic of the native glycosylphosphatidylinositol (GPI) post-translational modification of prion proteins (Olschewski *et al.*, 2007). With the goal of producing semisynthetic proteins inside living cells, Giriat and Muir (2003) have reported protein splicing *in trans* in mammalian cells between a synthetic *Ssp* $DnaE^C$ part and a recombinant protein of interest fused to the *Ssp* $DnaE^N$ fragment. The synthetic intein fragment with a few amino acids of Ex^C-sequence were imported into the cells by a conjugated cell penetrating peptide. Furthermore, this example nicely highlights another advantage of split inteins over chemical ligation methods. Because of the natural affinity of the complementary intein fragments, they recognize only their partner fragments and the protein-splicing reaction occurs only from the reconstituted intein complex. Because of these features, the reaction is so selective that it can be performed even in the presence of the complex protein mixture inside a cell.

Although the split position S0 at the endonuclease insertion site (see Fig. 4.2A) gave rise to several artificially split inteins, this position also brings about limitations in terms of applicability of the resulting intein fragments for protein semisynthesis. First, the Int^N fragment is too large to be obtained synthetically with reasonable effort, therefore restricting the approach to semisynthetic products that carry the synthetic peptide at their C-terminal end. Second, the synthesis of the Int^C fragment with a size of $\approx35–45$ aa is already quite laborious and leaves room for only a few more amino acids to be included as the Ex^C-sequence. We have reported that the peptide bond following amino acid 11 in the *Ssp* DnaB intein is an alternative split site (Ludwig *et al.*, 2006, 2008), as already suggested in genetic studies by Liu and co-workers (Sun *et al.*, 2004). The active intein can be reconstituted from a short synthetic peptide including amino acids 1–11 of the intein as the Int^N component and a recombinant $Int^C(12–154)$ fragment with various POIs fused as Ex^C-segment (Fig. 4.3). Protein splicing *in trans* of this semisynthetic intein thus leads to products with N-terminal modifications. Figure 4.2B illustrates the position of the split site S1 in the crystal structure of the *Ssp* DnaB mini-intein (Ding *et al.*, 2003; Sun *et al.*, 2004). Although S1 is located within a turn region on the surface of the molecule, the interwoven arrangement of the resulting Int^N and Int^C fragments suggests that their spontaneous association and reconstitution into the active intein must be accompanied by a remarkable folding process. It also seems likely from the tertiary structure that the $Int^C(12–154)$ fragment will not be able to adopt a nativelike conformation on its own but rather will be at least

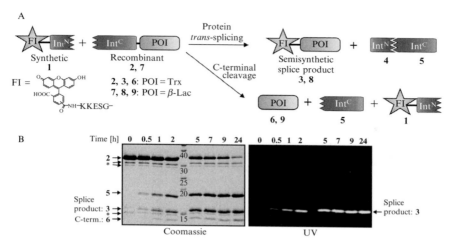

Figure. 4.3 Semisynthesis of N-terminally modified proteins using protein splicing *in trans*. (A) General scheme of protein semisynthesis using the *Ssp* DnaB intein split at position S1. Here, the peptide sequence 5(6)-carboxyfluorescein-KKESG (Fl) was linked to the two proteins of interest (POI) thioredoxin (Trx) and β-lactamase (βLac), which are expressed as recombinant fusion proteins with the Int^C fragment. The desired semisynthetic products (3 and 8) were formed in the protein *trans*-splicing reaction. C-terminal cleavage (6 and 9) occurred as a side-reaction due to premature asparagine cyclization. (B) Reaction progress of 40 μM peptide 1 (2.0 kDa) and 40 μM protein 2 (30.2 kDa) monitored on an SDS-polyacrylamide gel using Coomassie staining (left) or UV light illumination (right). The three new product bands visible by Coomassie staining represent the semisynthetic product Fl-Trx-His$_6$ (3, 14.6 kDa), the Int^C-fragment (5, 16.5 kDa) and the C-terminal cleavage product (6, 13.7 kDa). Under UV light only the semisynthetic Fl-Trx-His$_6$ (3) can be detected. (* = impurity protein bands).

partially unfolded or misfolded. The same is true for both intein fragments generated from splitting at the S0 site (compare Fig. 4.2A). The actual folding behavior of such split intein fragments and the consequences on fusion protein folding, solubility, and activity remains to be studied more systematically and may vary depending on the extein sequence. Misfolding of the individual fragments is certainly the likely explanation for the requirement of a renaturation procedure in the case of most artificially split inteins investigated so far (see end of section 2). The intein for which the largest number of active fusion proteins have been reported is the natively split *Ssp* DnaE intein, possibly because the fragments of this intein have undergone optimisation by natural evolution and are therefore more robust in a heterologous context. Among the artificially split inteins, the *Ssp* DnaB intein exhibits remarkable stability, as both split sites S0 and S1 have been shown to yield fragments that are active in protein splicing *in trans* under native conditions, upon incubation after separate expression and purification. However, fusion of artificially split intein fragments can decrease the solubility of the protein of interest and might result in reduced splicing yields. In this context we observed that the *Mxe* GyrA intein split at position S0 is superior over the similarly split *Ssp* DnaB intein (Kurpiers and Mootz, 2008).

Another point that often is important for the activity of a split intein within an heterologous sequence context is the nature of the directly flanking extein residues. Given their location in the immediate vicinity to the intein-extein junctions, they are likely to have electronic or steric effects that may destabilize the active site. There are a few studies that have addressed amino acid exchanges at one of these positions systematically (Chong *et al.*, 1998; Iwai *et al.*, 2006; Kinsella *et al.*, 2002; Mathys *et al.*, 1999; Southworth *et al.*, 1999; Xu *et al.*, 2000). From the current knowledge, it is clear only that an intein will not work with any given flanking residues, but, luckily, will work with many nonnative neighboring residues. Interestingly, protein splicing seems to be more delicate in terms of sequence context than the partial reactions of thioester formation and C-terminal cleavage through asparagine cyclization. This observation probably reflects the perfect alignment of the individual steps in the splicing mechanism required for successful product formation (see Fig. 4.1). As a rule of thumb, 3 to 5 residues each of the native Ex^N- and Ex^C-sequences can be added to the intein to minimize potential sequence restraints (Noren *et al.*, 2000; Ozawa *et al.*, 2003). However, this would result in a 6 to 10 amino acid stretch inserted in the final spliced product. When selecting the intein insertion site in the POI, we therefore recommend looking for a few consecutive amino acids that are most similar to those that flank the chosen intein in its native context. Note that the appearance of an extra amino acid at the carboxy-terminal side of the splice junction is an invariant consequence of the protein-splicing reaction. This is the first residue of the C-extein, referred to the +1 residue, which donates its side chain for the branched intermediate (see Fig. 4.1). Depending on the intein, it is a cysteine (as for the *Ssp* DnaE intein), a serine (as for the *Ssp* DnaB intein), or a threonine. In the ligation product, this residue corresponds to the cysteine required for NCL and EPL. In case of the *Ssp* DnaB intein we could further show that the first amino acid of Ex^N (glycine −1) is essential for efficient splice product formation (Ludwig *et al.*, 2008). This result is in good agreement with previously performed systematic studies on position −1 within the *Ssp* DnaB intein (Mathys *et al.*, 1999). Consequently, the splice junction for the *Ssp* DnaB intein must be located on the dipeptidyl unit Gly-Ser. Ideally this unit is already present in the target protein and can be chosen as ligation site, in order to minimize sequence changes in the splicing product. Alternatively, an unconserved region not important for the protein function can be chosen and the two residues can be introduced by site-directed mutagenesis.

4. MATERIALS AND METHODS

In the following sections, the approach for an N-terminal modification by semisynthetic protein *trans*-splicing is illustrated for two different proteins: (1) a thioredoxin domain (Trx) (Ludwig *et al.*, 2006, 2008), and (2) the protein β-lactamase (βLac) (Ludwig *et al.*, 2006). On the basis of these two constructs, we explain and discuss experimental designs and protocols of this technique.

If not otherwise specified standard protocols were used. All of the described oligonucleotides were purchased from Operon and the final plasmids were routinely verified by DNA sequencing. Kanamycin was used at 50 μg/mL concentration. Reagents were purchased from Acros, Novabiochem, Roth, or Sigma–Aldrich. Restriction enzymes and markers were obtained from Fermentas. All reactions and assays were performed at least in duplicate.

4.1. Solid-phase peptide synthesis of Ex^N-Int^N peptides

The synthetic peptide 5(6)-carboxyfluorescein-KKESGCISGDSLISLA-OH (Int^N underlined) includes the minimally required sequence of the N-terminal intein fragment (Int^N), which comprises the first 11 amino acids of the *Ssp* DnaB intein (Ludwig *et al.*, 2008). The used N-extein sequence (Ex^N) includes the first three native Ex^N amino acids of the *Ssp* DnaB Intein (ESG), two additional lysines, and a chemical modification in the form of 5,6-carboxyfluorescein, which is coupled to the N-terminal amino group via an amid bond. This peptide **1** was synthesized by standard Fmoc-synthesis and HPLC-purified to >95%. To prevent racemization, the Cys building block was coupled as pentafluorophenyl ester without the addition of base. The two additional lysine residues within the Ex^N-sequence were introduced to improve peptide solubility under acidic conditions applied for preparative HPLC (buffer A, H_2O + 0.05% TFA; buffer B, acetonitrile + 0.05% TFA). The introduction of additional functions for improved solubility can also be carried out at the C terminus of the Int^N-sequence (Ludwig *et al.*, 2008), a region which will not be inserted into the final splice product. Accordingly, two additional C-terminal lysine residues (data not shown) as well as a C-terminal PEG-linker and a trimethoprim ligand (Ando *et al.*, 2007) were tolerated. If improved peptide solubility under acidic conditions is not achievable, peptide solubilization and purification can also be performed using basic conditions (addition of 10 mM NH_3).

After purification peptide **1** was dissolved in 50 mM Tris, 300 mM NaCl, 1 mM EDTA, 2 mM TCEP, pH 7.0 at a stock concentration of 1 mM and finally flash-frozen and stored at $-80\,^{\circ}$C. The peptide concentration was determined photometrically at pH 9.0 and 492 nm using the molar extinction coefficient of fluorescein (ε_{492} = 65000 $M^{-1}cm^{-1}$).

4.2. Construction of Int^C-POI plasmids

In general the Int^C fragment started with amino acid Ser12 of the *Ssp* DnaB mini-intein (Wu *et al.*, 1998). Because of the cloning strategy, which incorporated the restriction sites *Nco*I and *Spe*I, four additional amino acids (MGTS) were added to the N terminus of the fragment. The C-terminus of the Int^C fragment ended with the sequence IIVHN<u>S</u>, of which the under-lined serine represents the first amino acid of the Ex^C (residue $+1$). Two additional amino acids (IE) were added to Ex^C, according to the native Ex^C-sequence of the *Ssp* DnaB intein, which thus comprised the primary structure H$_2$N-MGTS-Int^C(12-154)-**SIE**-*POI*-COOH (the required amino acids at the C-extein are shown in bold). Additionally, for more convenience in cloning of different POI coding gene fragments to the Int^C fragment, we introduced an *Eco*RV restriction site close to the C-terminus of the Int^C fragment. This was possible by a silent point mutation of the codon of Asp149 (GAC→GAT).

In this paragraph the detailed cloning procedures of the expression plasmids for the indicated proteins are given. **Int^C-Trx-His$_6$** (**2**): By using the oligonucleotides 5′-ATA<u>CCATGG</u>GCACTAGTAGCACAGG AAAAAGAGTTTC-3′ and 5′-ATA<u>GGTACC</u>GGATCCTTCAA TACTGTTATGGACAATGATG-3′ the Int^C fragment was amplified from pMST (Wu *et al.*, 1998). The *Sph*I-*Eco*RI fragment including the T7 promoter of the latter plasmid was then transferred to pSU38 (Bartolome *et al.*, 1991) (pCL11). A Trx-encoding fragment was amplified from pBAD202 (Invitrogen) using oligonucleotides 5′-ATA<u>GGATCC</u>GGAG GAGGATCTGATAAAATTATTCATC-3′ and 5′-TAT<u>GGATCC</u>AG AGCCGGCCAGGTTAGCG-3′ and ligated into the *Bam*HI site of pCL11 to give pCL20 (Ludwig *et al.*, 2006). **Int^C-βLac-His$_6$** (**7**): The *bla* gene was amplified with the oligonucleotides 5′-ATAGAATTC <u>GGATCC</u>CACCCAGAAACGCTGGTGAAAG-3′ and 5′-ATAAAG CTT<u>AGATCT</u>CCAATGCTTAATCAGTGAGG-3′ from pUC18 with-out the first 23 aa, which encode a signal sequence. After treatment with *Bam*HI and *Bgl*II the resulting fragment was ligated into the *Bam*HI site of pCL11 to give expression plasmid pHM137.

4.3. Expression and purification of Int^C-POI fusion proteins

For protein expression, *E. coli* BL21 (DE3) cells were transformed with the appropriate plasmids by using a heat shock at 42 °C. After growing in LB medium with the adequate antibotics at 37 °C to OD (600 nm) 0.5 to 0.7, the temperature was shifted to 30 °C. The expression was induced by adding isopropyl-β-thiogalactopyranoside (IPTG) with a final concentration of 0.4 mM. After 3 to 5 h, the cells were harvested by centrifugation, resus-pended in the appropriate buffer for the following purification step, and either stored at $-80\ °C$ or directly lysed by two passages through an emulsifier (Avestin EmulsiFlex C5). The insoluble material could be removed by centrifugation at 30,000g.

The fusion protein Int^C-Trx-His$_6$ (2) was purified from the soluble fraction on a Ni^{2+}-NTA superflow column (Qiagen) under native conditions. Int^C-βLac-His$_6$ (7) formed insoluble inclusion bodies and necessitated a Ni^{2+}-NTA purification under denaturing conditions (8 M urea). Pooled fractions of pure protein were dialyzed against splicing buffer (50 mM Tris, 300 mM NaCl, 1 mM EDTA, 2 mM DTT, 10% glycerine, pH 7.0) and flash frozen at $-80\ °C$. Refolded protein 7 was recovered in this one-step dialysis without further optimisation in about 50% yield. Overall yields were about 20 mg (2) and 5 mg (7) per liter of expression medium. Protein concentrations were determined by using a calculated molecular extinction coefficient at 280 nm.

4.4. Protein *trans*-splicing assays

A typical splicing reaction was performed by mixing the synthetic peptide and the recombinant protein in a ratio of 1:1 or up to 2:1 in splicing buffer at pH 7.0 and 25 °C. At various time points, aliquots were removed and analyzed by SDS-PAGE. For this purpose, the aliquots were mixed in a ratio of 3:1 with 4x SDS-PAGE loading buffer (containing 8% SDS and 20% β-mercaptoethanol) and heated for 10 min at 95 °C. For quantitative analysis and kinetic investigations the relative intensities of protein bands (stained with Coomassie brilliant blue) were densitometrically determined using the program Scion Image (http://www.scioncorp.com). All reactions were performed at least in duplicate.

Importantly, none of the intein fusion proteins showed any formation of splicing or cleavage product without the addition of the counterpart peptide 1 (data not shown), indicating that only the reconstituted intein is catalytically active (Ludwig et al., 2006). Incubation of protein 2 (Int^C-Trx-His$_6$, 40 μM) with synthetic peptide 1 (Fl-Int^N, 40 μM) yielded the desired semisynthetic protein 3 (Fl-Trx-His$_6$) in a time-dependent fashion, with a final yield of 75% after 24 h (see Fig. 4.3B). Interestingly, the formation of a second C-terminal cleavage by-product (6) was detected with 20% yield. Depending on the used protein of interest as well as the peptide the yields in splice product formation and the ratio between splicing and C-terminal cleavage may vary. For example, in the reaction peptide 1 with protein 7 (Int^C-βLac-His$_6$) yields of 35% splice product 8 and 35% C-terminal cleavage product 9 were determined (see Fig. 4.4B). The splicing reaction can also be performed at higher temperature; however, at 37 °C we observed an increase in the C-terminal cleavage at the expense of splice product formation (Ludwig et al., 2006). At lower temperatures, between 4 °C and 25 °C, the ratio remained constant; however, overall yields

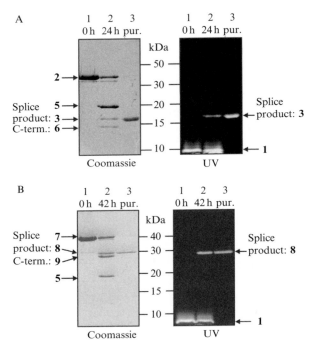

Figure 4.4 Preparation and purification of semisynthetic products. Shown is the SDS-PAGE analysis of the preparation and purification of the semisynthetic splice products Fl-Trx-His$_6$ (construct 3, panel A) and Fl-βLac-His$_6$ (construct 8, panel B). The gels were visualized by Coomassie staining (left) as well as UV light illumination (right). In all cases lane 1 and 2 show the reaction mixtures of peptide 1 and fusion protein (2 or 7) incubated at 25 °C for the indicated time periods. Lane 3 represents a sample of the purified products (3 and 8). See also Figure 4.3 for an illustration of the underlying reactions.

decreased. Changes in buffer conditions are tolerated, like varying salt concentrations (from 300 to 50 mM), reducing agents (2 mM DTT or 2 mM TCEP), glycerine addition (0–30%) or different buffer systems based on phosphate or HEPES. Consistent with previous studies on inteins the optimal pH value for protein *trans*-splicing was 7.0.

4.5. Interaction studies of the Int^N/Int^C association

An investigation of the association of the synthetic peptide and its recombinant counterpart was performed using a triple mutant of construct **2**, comprising three substitutions of catalytically essential amino acids (His$_{73}$Ala, Asn$_{154}$Ala and Ser$_{+1}$Ala). With this protein exclusively the intein fragment association with peptide **1** could be investigated, while the following steps of the protein-splicing mechanism were inhibited. Based on time-dependent changes in fluorescence anisotropy the kinetic parameters of the complex formation were determined as $K_d = 1.1 \pm 0.2\ \mu M$, $k_{on} = 16.8 \pm 1.0\ M^{-1}\ s^{-1}$ and $k_{off} = 1.8 \times 10^{-5}\ s^{-1}$ (Ludwig et al., 2008). From these data we could conclude that the semisynthetic protein trans-splicing reaction can be performed at concentrations in the low micromolar range, but significantly lower concentrations will not lead to sufficient product formation.

4.6. Purification and functional analysis of semisynthetic splice products

The preparative utility of the described approach could be demonstrated based on the semisynthetic products **3** (Fl-Trx-His$_6$) and **8** (Fl-βLac-His$_6$) (see Fig. 4.4). For the purification of both constructs the corresponding splicing reaction mixtures were dialyzed against 20 mM Tris-HCl, 2 mM DTT, pH 8.0, and applied to an anion-exchange chromatography, using a HiTrap Q anion-exchange column (1 mL of bed volume, Amersham Bioscience) equilibrated with 20 mM Tris-HCl, pH 8.0. After washing with 10 mL of the same buffer, retained material was eluted at a flow rate of 1 mL/min with a 20-mL linear gradient from 0 to 0.5 M NaCl. In a second purification step, the pooled fractions containing the semisynthetic products **3** and **8** were applied onto a Ni^{2+}-NTA superflow column (Qiagen) to remove remaining traces of peptide **1** (see Fig. 4.4). The overall yields after purification for both semisynthetic products were about 30% (Fl-Trx-His$_6$ (**3**) $\approx$0.4 mg and Fl-βLac- His$_6$ (**8**) $\approx$0.3 mg).

The activities of the semisynthetic products were analyzed using photometric enzyme assays and compared to those of fully recombinantly generated enzymes. Fl-Trx-His$_6$ (**3**) exhibited in the insulin-precipitation assay (Holmgren, 1979) even a slightly higher enzymatic activity than the recombinant thioredoxin positive control (Ludwig et al., 2006). The semi-synthetic Fl-βLac-His$_6$ (**8**) was indistinguishable in a nitrocefin hydrolysis assay (O'Callaghan et al., 1972) from the control construct βLac-His$_6$ (Ludwig et al., 2006).

In conclusion, the two semisynthetic proteins (**3**) and Fl-βLac-His$_6$ (**8**) were produced in pure forms and good yields, and their catalytic activity remained unaffected by expression as an intein fusion and the subsequent protein-splicing reaction.

4.7. Protein splicing in complex mixtures

To address the question of whether the semisynthetic splice reaction can be performed within complex protein mixtures, we carried out a splicing assay in an E. coli cell lysate. Fusion protein **2** (Int^C-Trx-His$_6$) was expressed in E. coli

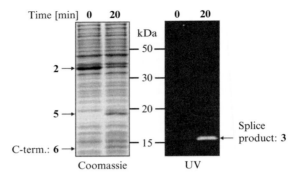

Figure 4.5 Semisynthetic protein *trans*-splicing within an *E. coli* cell lysate. SDS-PAGE analysis of the reaction performed by adding synthetic peptide **1** (Fl-Int^N, 90 μM) directly to a total lysate of *E. coli* BL21 (DE3) cells expressing protein **2** (Int^C-Trx-His$_6$, ~30 μM). After 20 min of incubation at 25 °C splice product **3** (Fl-Trx-His$_6$) could be detected, as shown in the UV-illuminated SDS-gel (right panel). The Coomassie-stained gel (left panel) suggested that also the C-terminal cleavage product **6** was formed.

BL21 (DE3) cells as described above. Subsequently the cells were resuspended in splicing buffer, lysed by two passages through a high pressure homogenizer, before peptide **1** was added directly to the soluble fraction in excess (90 μM; protein concentration of **2** was about 30 μM). Already after 20 minutes splice product formation could be detected (see Fig. 4.5). Furthermore, Nagamune and co-workers exploited the semisynthetic intein to modify proteins on cell surfaces. However, in this case the affinity of the two intein fragments needed to be increased by introducing an additional ligand/receptor pair (trimethoprim/dihydrofolat reductase) (Ando *et al.*, 2007). Taken together, these findings show that the splicing reaction can be performed efficiently and selectively in cell lysates or on living cells.

5. SUMMARY AND CONCLUSION

Protein splicing *in trans* is a new approach to prepare semisynthetic proteins consisting of a synthetic N-terminal and a recombinant C-terminal component. Herein, the synthetic part is extended by the first 11 amino acids of the *Ssp* DnaB mini-intein. This facilitates protein splicing *in trans* with the recombinant C-terminal component of the intein, which is expressed as a genetic fusion protein with the POI. Because inteins are promiscuous with respect to their fused extein sequences, this technique should allow one, in principle, to N-terminally modify any POI with any biophysical probe or chemical moiety. Importantly, this chemoenzymatic approach circumvents the C-terminal thioester group and N-terminal cysteine residue, which are essential for NCL and EPL. It should therefore also be useful to introduce chemical groups that are difficult or impossible to combine with a thioester group in a synthetic peptide. Furthermore, because of the affinity of the intein fragments, this reaction can be carried out at rather low protein and peptide concentrations (K_d value = 1.1 μM) as compared to chemical ligation, where high concentrations up to the millimolar range are favorable. Also the application in complex protein mixtures, like cell lysates or on cell surfaces, has already been demonstrated. Semisynthetic protein splicing in *trans* should thus provide a convenient and general route to semisynthetic proteins.

ACKNOWLEDGMENTS

This research was supported by the DFG (Emmy Noether grant MO1073/1 and FOR495 grant MO1073/2) and the Fonds der Chemischen Industrie. C.L. is a recipient of a Ph.D. fellowship from Fonds der Chemischen Industrie.

REFERENCES

Ando, T., *et al.* (2007). Construction of a small-molecule-integrated semisynthetic split intein for *in vivo* protein ligation. *Chem. Commun. (Camb)* 4995–4997.

Bartolome, B., *et al.* (1991). Construction and properties of a family of pACYC184-derived cloning vectors compatible with pBR322 and its derivatives. *Gene* **102,** 75–78.

Baskin, J. M., *et al.* (2007). Copper-free click chemistry for dynamic *in vivo* imaging. *Proc. Natl. Acad. Sci. USA* **104,** 16793–16797.

Beckett, D., *et al.* (1999). A minimal peptide substrate in biotin holoenzyme synthetase-catalyzed biotinylation. *Protein Sci.* **8,** 921–929.

Brenzel, S., Kurpiers, T., and Mootz, H. D. (2006). Engineering artificially split inteins for applications in protein chemistry: Biochemical characterization of the split Ssp DnaB intein and comparison to the split Sce VMA intein. *Biochemistry* **45,** 1571–1578.

Budisa, N. (2004). Prolegomena to future experimental efforts on genetic code engineering by expanding its amino acid repertoire. *Angew. Chem. Int. Ed. Engl.* **43,** 6426–6463.

Carrico, I. S., *et al.* (2007). Introducing genetically encoded aldehydes into proteins. *Nat. Chem. Biol.* **3,** 321–322.

Caspi, J., *et al.* (2003). Distribution of split DnaE inteins in cyanobacteria. *Mol. Microbiol.* **50,** 1569–1577.

Choi, J. J., *et al.* (2006). Protein trans-splicing and characterization of a split family B-type DNA polymerase from the hyperthermophilic archaeal parasite Nanoarchaeum equitans. *J. Mol. Biol.* **356,** 1093–1106.

Chong, S., *et al.* (1997). Single-column purification of free recombinant proteins using a self-cleavable affinity tag derived from a protein splicing element. *Gene* **192,** 271–281.

Chong, S., *et al.* (1998). Modulation of protein splicing of the Saccharomyces cerevisiae vacuolar membrane ATPase intein. *J. Biol. Chem.* **273,** 10567–10577.

Dassa, B., *et al.* (2007). Trans protein splicing of cyanobacterial split inteins in endogenous and exogenous combinations. *Biochemistry* **46,** 322–330.

Dawson, P. E., and Kent, S. B. (2000). Synthesis of native proteins by chemical ligation. *Annu. Rev. Biochem.* **69**, 923–960.

Dawson, P. E., *et al.* (1994). Synthesis of proteins by native chemical ligation. *Science* **266**, 776–779.

de Araujo, A. D., *et al.* (2006). Diels-Alder ligation of peptides and proteins. *Chemistry* **12**, 6095–6109.

Ding, Y., *et al.* (2003). Crystal structure of a mini-intein reveals a conserved catalytic module involved in side chain cyclization of asparagine during protein splicing. *J. Biol. Chem.* **278**, 39133–39142.

Dirksen, A., *et al.* (2006). Nucleophilic catalysis of oxime ligation. *Angew. Chem. Int. Ed. Engl.* **45**, 7581–7584.

Durek, T., and Becker, C. F. (2005). Protein semisynthesis: New proteins for functional and structural studies. *Biomol. Eng.* **22**, 153–172.

Erlanson, D. A., *et al.* (1996). The leucine zipper domain controls the orientation of AP-1 in the NFAT.AP-1.DNA complex. *Chem. Biol.* **3**, 981–991.

Evans, T. C., Jr., *et al.* (1998). Semisynthesis of cytotoxic proteins using a modified protein splicing element. *Protein Sci.* **7**, 2256–2264.

Evans, T. C., Jr., *et al.* (2000). Protein trans-splicing and cyclization by a naturally split intein from the dnaE gene of Synechocystis species PCC6803. *J. Biol. Chem.* **275**, 9091–9094.

Fernandez-Suarez, M., *et al.* (2007). Redirecting lipoic acid ligase for cell surface protein labeling with small-molecule probes. *Nat. Biotechnol.* **25**, 1483–1487.

Giriat, I., and Muir, T. W. (2003). Protein semisynthesis in living cells. *J. Am. Chem. Soc.* **125**, 7180–7181.

Giriat, I., *et al.* (2001). Protein splicing and its applications. *Genet. Eng. (NY)* **23**, 171–199.

Griffin, B. A., *et al.* (2000). Fluorescent labeling of recombinant proteins in living cells with FlAsH. *Methods Enzymol.* **327**, 565–578.

Hall, T. M., *et al.* (1997). Crystal structure of a Hedgehog autoprocessing domain: Homology between Hedgehog and self-splicing proteins. *Cell* **91**, 85–97.

Hermanson, G. T. (1996). Bioconjugate techniques. Academic Press, San Diego.

Holmgren, G. T. (1979). Thioredoxin Catalyzes the Reduction of Insulin Disulfides by Dithiothreitol and Dihydrolipoamide. *JBC* **254**, 9627–9632.

Iwai, H., *et al.* (2006). Highly efficient protein trans-splicing by a naturally split DnaE intein from Nostoc punctiforme. *FEBS Lett.* **580**, 1853–1858.

Keppler, A., *et al.* (2003). A general method for the covalent labeling of fusion proteins with small molecules *in vivo*. *Nat. Biotechnol.* **21**, 86–89.

Kinsella, T. M., *et al.* (2002). Retrovirally delivered random cyclic Peptide libraries yield inhibitors of interleukin-4 signaling in human B cells. *J. Biol. Chem.* **277**, 37512–37518.

Kurpiers, T., and Mootz, H. D. (2007). Regioselective cysteine bioconjugation by appending a labeled cystein tag to a protein by using protein splicing in trans. *Angew. Chem. Int. Ed. Engl.* **46**, 5234–5237.

Kurpiers, T., and Mootz, H. D. (2008). Site-Specific Chemical Modification of Proteins with a Prelabelled Cysteine Tag Using the Artificially Split Mxe GyrA Intein. *Chembiochem* **9**, 2317–2325.

Kwon, Y., *et al.* (2006). Selective immobilization of proteins onto solid supports through split-intein-mediated protein trans-splicing. *Angew. Chem. Int. Ed. Engl.* **45**, 1726–1729.

Lew, B. M., *et al.* (1998). Protein splicing *in vitro* with a semisynthetic two-component minimal intein. *J. Biol. Chem.* **273**, 15887–15890.

Lew, B. M., *et al.* (1999). Characteristics of protein splicing in trans mediated by a semisynthetic split intein. *Biopolymers* **51**, 355–362.

Ludwig, C., *et al.* (2006). Ligation of a synthetic peptide to the *N* terminus of a recombinant protein using semisynthetic protein trans-splicing. *Angew. Chem. Int. Ed. Engl.* **45**, 5218–5221.

Ludwig, C., *et al.* (2008). Interaction studies and alanine scanning analysis of a semi-synthetic split intein reveal thiazoline ring formation from an intermediate of the protein splicing reaction. *J. Biol. Chem.* **283**, 25264–25272.

Mathys, S., *et al.* (1999). Characterization of a self-splicing mini-intein and its conversion into autocatalytic N- and C-terminal cleavage elements: Facile production of protein building blocks for protein ligation. *Gene* **231**, 1–13.

Mills, K. V., *et al.* (1998). Protein splicing in trans by purified N- and C-terminal fragments of the Mycobacterium tuberculosis RecA intein. *Proc. Natl. Acad. Sci. USA* **95**, 3543–3548.

Muir, T. W. (2003). Semisynthesis of proteins by expressed protein ligation. *Annu. Rev. Biochem.* **72**, 249–289.

Muir, T. W., *et al.* (1998). Expressed protein ligation: A general method for protein engineering. *Proc. Natl. Acad. Sci. USA* **95**, 6705–6710.

Nguyen, U. T., *et al.* (2007). Exploiting the substrate tolerance of farnesyltransferase for site-selective protein derivatization. *Chembiochem* **8**, 408–423.

Nilsson, B. L., *et al.* (2005). Chemical synthesis of proteins. *Annu. Rev. Biophys. Biomol. Struct.* **34**, 91–118.

Noren, C. J., *et al.* (1989). A general method for site-specific incorporation of unnatural amino acids into proteins. *Science* **244**, 182–188.

Noren, C. J., *et al.* (2000). Dissecting the Chemistry of Protein Splicing and Its Applications. *Angew. Chem. Int. Ed. Engl.* **39**, 450–466.

O'Callaghan, C. H., *et al.* (1972). Novel method for detection of beta-lactamases by using a chromogenic cephalosporin substrate. *Antimicrob. Agents Chemother.* **1**, 283–288.

Olschewski, D., *et al.* (2007). Semisynthetic murine prion protein equipped with a GPI anchor mimic incorporates into cellular membranes. *Chem. Biol.* **14**, 994–1006.

Ozawa, T., *et al.* (2003). A genetic approach to identifying mitochondrial proteins. *Nat. Biotechnol.* **21**, 287–293.

Paulus, H. (2000). Protein splicing and related forms of protein autoprocessing. *Annu. Rev. Biochem.* **69**, 447–496.

Perler, F. B. (2002). InBase: The Intein Database. *Nucleic Acids Res.* **30**, 383–384.

Popp, M. W., *et al.* (2007). Sortagging: A versatile method for protein labeling. *Nat. Chem. Biol.* **3**, 707–708.

Prescher, J. A., and Bertozzi, C. R. (2005). Chemistry in living systems. *Nat. Chem. Biol.* **1**, 13–21.

Saxon, E., *et al.* (2000). A "traceless" Staudinger ligation for the chemoselective synthesis of amide bonds. *Org. Lett.* **2**, 2141–2143.

Song, W., *et al.* (2008). A photoinducible 1,3-dipolar cycloaddition reaction for rapid, selective modification of tetrazole-containing proteins. *Angew. Chem. Int. Ed. Engl.* **47**, 2832–2835.

Southworth, M. W., *et al.* (1998). Control of protein splicing by intein fragment reassembly. *EMBO J.* **17**, 918–926.

Southworth, M. W., *et al.* (1999). Purification of proteins fused to either the amino or carboxy terminus of the Mycobacterium xenopi gyrase A intein. *Biotechniques* **27**, 110–114, 116, 118–120.

Sun, P., *et al.* (2005). Crystal structures of an intein from the split dnaE gene of Synechocystis sp. PCC6803 reveal the catalytic model without the penultimate histidine and the mechanism of zinc ion inhibition of protein splicing. *J. Mol. Biol.* **353**, 1093–1105.

Sun, W., *et al.* (2004). Synthetic Two-piece and Three-piece Split Inteins for Protein trans-Splicing. *J. Biological Chemistry* **279**, 35281–35286.

Wang, K., *et al.* (2007). Evolved orthogonal ribosomes enhance the efficiency of synthetic genetic code expansion. *Nat. Biotechnol.* **25**, 770–777.

Wang, L., *et al.* (2001). Expanding the genetic code of Escherichia coli. *Science* **292,** 498–500.

Wang, L., and Schultz, P. G. (2004). Expanding the genetic code. *Angew. Chem. Int. Ed. Engl.* **44,** 34–66.

Wehofsky, N., *et al.* (2003). Reverse proteolysis promoted by in situ generated peptide ester fragments. *J. Am. Chem. Soc.* **125,** 6126–6133.

Wu, H., *et al.* (1998). Protein trans-splicing by a split intein encoded in a split DnaE gene of Synechocystis sp. *PCC6803. Proc. Natl. Acad. Sci. USA* **95,** 9226–9231.

Xie, J., and Schultz, P. G. (2006). A chemical toolkit for proteins--an expanded genetic code. *Nat. Rev. Mol. Cell. Biol.* **7,** 775–782.

Xu, M. Q., *et al.* (2000). Fusions to self-splicing inteins for protein purification. *Methods Enzymol.* **326,** 376–418.

Yamazaki, T. (1998). Segmental Isotope Labeling for Protein NMR Using Peptide Splicing. *J. Am. Chem. Soc.* **120,** 5591–5592.

Yin, J., *et al.* (2005). Genetically encoded short peptide tag for versatile protein labeling by Sfp phosphopantetheinyl transferase. *Proc. Natl. Acad. Sci. USA* **102,** 15815–15820.

CHAPTER FIVE

Expressed Protein Ligation for Metalloprotein Design and Engineering

Kevin M. Clark,* Wilfred A. van der Donk,*,† *and* Yi Lu*,†

Contents

Abstract

Metalloproteins contain highly specialized metal-binding sites that are designed to accept specific metal ions to maintain correct function. Although many of the

* Department of Biochemistry, University of Illinois at Urbana-Champaign, Urbana, Illinois, USA
† Department of Chemistry, University of Illinois at Urbana-Champaign, Urbana, Illinois, USA

Methods in Enzymology, Volume 462
ISSN 0076-6879, DOI: 10.1016/S0076-6879(09)62005-X

sites have been modified with success, the relative paucity of functional group availability within proteinogenic amino acids can sometimes leave open questions about specific functions of the metal binding ligands. Attaining a more thorough analysis of individual amino acid function within metalloproteins has been realized using expressed protein ligation (EPL). Here we describe our recent efforts using EPL to incorporate nonproteinogenic cysteine and methionine analogues into the type 1 copper site found in *Pseudomonas aeruginosa* azurin.

1. INTRODUCTION

Protein design and engineering have contributed greatly to our understanding of protein structure and function. A common practice in the field is to use site-directed mutagenesis to replace a specific amino acid in proteins with other natural amino acids. Although quite powerful, it has become increasingly evident that site-directed mutagenesis cannot address the precise role of a specific amino acid because, with a limitation of 20 natural amino acids, it is often difficult to change one factor, such as an electronic effect, without simultaneously changing other factors, such as steric effects. This limitation is even more evident in metalloprotein design and engineering, as even a smaller number of natural amino acids are capable of coordinating to metal ions (Lu *et al.*, 2001; Lu, 2006b). More importantly, the metal–binding sites are intricately designed to accept a specific metal ion among a large number of other metal ions for binding and activity. Because geometry and ligand donor sets of metal-binding sites are much more varied than those of metal-free active sites, a subtle change of one factor can have a profound effect on the structure and activity. Therefore, the need to introduce unnatural amino acids into metalloproteins for studies such as isostructural substitutions is even greater. Such an endeavor can not only define the role of amino acids more precisely but also produce novel metalloproteins with predicted and unprecedented properties (Lu, 2005, 2006a). As described herein, this need has recently been met by utilizing the expressed protein ligation (EPL) technique.

An excellent test case for engineering of metalloprotein is the type I (T1) blue copper proteins, a class of common electron transfer (ET) proteins involved in many important biological processes, such as photosynthesis and respiration (Lu, 2003; Vila and Fernandez, 2001). The blue copper center displays much higher reduction potentials and much more efficient electron transfer properties than those of the same copper complexes found outside of the protein environment. A substantial amount of research has been applied to understanding the origin of these exceptional functional properties. *Pseudomonas aeruginosa* azurin is a member of T1 or blue copper proteins. In its resting state, the azurin blue copper center contains a

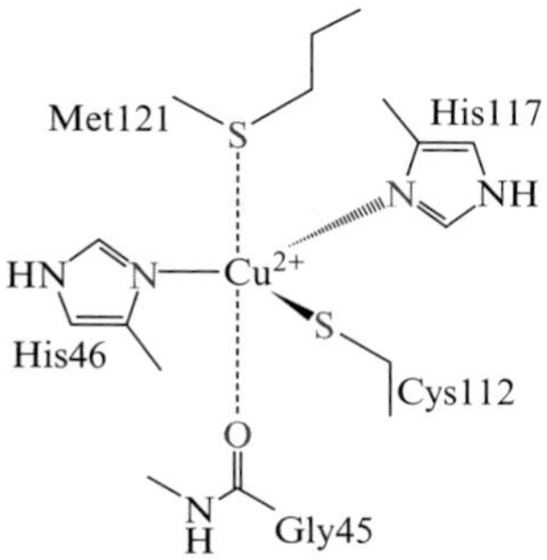

Figure 5.1 The T1 Copper site of *Pseudomonas aeruginosa* azurin.

Cu(II) ion bound in a trigonal bipyramidal plane by two equatorial histidines (His46, His117) and one equatorial cysteine (Cys112) (Fig. 5.1). Similar to the majority of other blue copper proteins, the azurin T1 copper site has a methionine sulfur thioether (Met121) as an axial ligand. Unique to azurin, however, the backbone carbonyl from a glycine residue (Gly45) is present opposite the methionine. The T1 copper site is defined by an unusually short copper to cysteine bond (2.1 Å), and this shortened bond, together with its unique metal-binding site geometry, is responsible for an intense ligand-to-metal charge transfer (LMCT) band resulting from the S(Cys112) pπ $\rightarrow$ Cu(II) d$_{x^2-y^2}$ charge transfer transition, giving azurin its strong UV/Vis absorption at 625 nm and thus its intense blue color (Malmström, 1994; Pierloot *et al.*, 1997; Solomon *et al.*, 1992). The highly covalent nature of the bond between the copper and Cys ligand is also responsible for the narrow hyperfine splittings in the parallel region of the EPR spectra (A$_{\parallel}$ $\approx 60 \times 10^{-4}$ cm^{-1}), smaller than those of "normal" copper coordination compounds (A$_{\parallel}$ $\approx 160 \times 10^{-4}$ cm^{-1}) (Solomon *et al.*, 1992).

In azurin, both the equatorial Cys112 ligand and the axial Met121 ligand have been implicated in the origin of the large increase in redox potential and more efficient ET properties over "normal" copper complexes. Serving as a model protein for the study of this class of metalloproteins, the T1 copper site in azurin has been exhaustively modified using site-directed mutagenesis, and its variants have been extensively studied using a broad range of spectroscopic and crystallographic techniques. These thorough investigations have shown that Cys112 is essential for protein function. Early studies established the loss of the azurin spectral properties when Cys112 was substituted with other natural amino acids differing greatly in size and electronics, whereas substitution of the remaining three ligands, His46, His117, and Met121, still yielded proteins dominated by the strong LMCT and blue color (Chang *et al.*, 1991; Den Blaauwen *et al.*, 1991; Germanas *et al.*, 1993; Karlsson *et al.*, 1989). However, the presence of the Cys112 thiol is not a requisite for copper binding. A single Cys112 mutant,

Cys112Asp, retained the ability to bind copper (Mizoguchi *et al.*, 1992), but all of the spectral features, including the distinct LMCT band at 625 nm, disappeared, and the ET properties of azurin were greatly reduced (Faham *et al.*, 1997; Mizoguchi *et al.*, 1992). Further spectroscopic evidence demonstrated that the original T1 copper site had been transformed to a type 2 (T2) "normal" copper site with a bidentate bound Asp (DeBeer *et al.*, 1999; Mizoguchi *et al.*, 1992). Although these site-directed mutagenesis studies suggest that the Cys ligand is essential in maintaining blue copper structural character and efficient ET properties, the structural features of Cys that are responsible for the character and properties of the site remain unclear, as mutation of Cys to other natural amino acids changes more than one factor at the time. A new methodology was needed to more precisely study the effects of Cys112 in the T1 copper site.

To achieve this goal, we chose to incorporate selenocysteine (Sec or U) as an isosteric cysteine analogue. Because Se and S share similar atomic radii, electronegativity, and oxidation states, the replacement by Sec is much more conservative than replacement of Cys with other proteinogenic amino acids (Fig. 5.2). The similarity also yields comparable oxidized analogues (e.g. disulfides and diselenides) and allows for mixed S–Se compounds (Gieselman *et al.*, 2002; Moroder, 2005), a feature that the natural amino acid Ser cannot mimic (Wessjohann *et al.*, 2007). However, although both Se and S are Group VIA elements with similar physical characteristics, they exhibit different reactivities. Most notably, the pK_a of Sec (5.2) is much lower than that of Cys (8.6), which suggests that, at physiological pH, Sec exists in its anionic selenolate form, which renders it more nucleophilic than Cys (Johansson *et al.*, 2005; Moroder, 2005). Unfortunately, the replacement of Cys with Sec is not easily accomplished by site-directed mutagenesis and requires the use of alternative techniques for its incorporation.

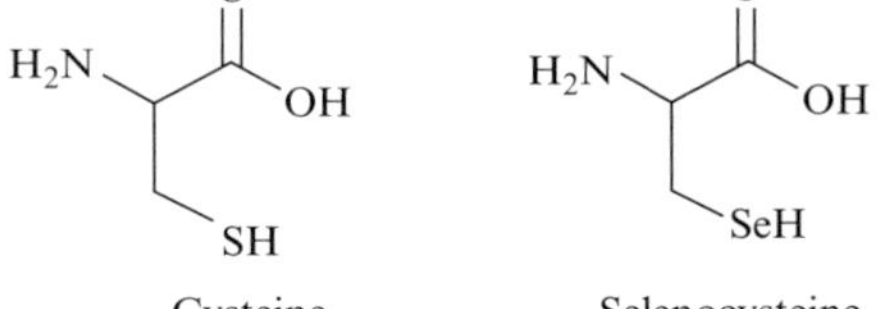

	Cysteine *Sulfur*	Selenocysteine *Selenium*
Electron configuration	[Ne] $3s^2.3p^4$	[Ar] $3d^{10}.4s^2.4p^4$
Atomic radius (pm)	100	115
Covalent radius (pm)	102	116
Bond length of dichalcogen (pm)	205 (S–S)	232 (Se–Se)
Electronegativity	2.58	2.55
Redox potential at pH <7 (mV)	231	276
Common oxidation states	−2, 0, +4, +6	−2, 0, +4, +6
pK_a	8.3	5.2

Figure 5.2   Comparison of the physiochemical properties of cysteine and selenocysteine.

2. METHODS OF SELENOCYSTEINE INCORPORATION

Several methods have been developed to incorporate selenocysteine, including three biosynthetic techniques (auxotrophic substitution, use of Sec insertion sequences (SECIS), and employment of tRNA/synthetase pairs), one chemical synthetic technique (native chemical ligation) (Casi *et al.*, 2007), and one semisynthetic technique (expressed protein ligation). The incorporation of seleno–amino acid analogues such as selenomethionine using *E. coli* auxotrophs has long been performed to solve phasing problems within protein crystallography. This methodology has also been applied to the use of cysteine auxotrophic bacteria to substitute cysteine residues with selenocysteine (Cheong *et al.*, 2007; Neuwald *et al.*, 1992). Unfortunately, the auxotrophic method for replacement of sulfur containing amino acids with their seleno-derivatives suffers from two key limitations. First, the method replaces any cysteine found in the protein with selenocysteine and thus alters more than one single site of interest without control over which positions are being substituted. Second, and more important, complete replacement of cysteine with selenocysteine in proteins using this technique has yet to be accomplished, with typical replacement percentages between 60 and 80% (Boschi-Muller *et al.*, 1998; Mueller *et al.*, 1994). Azurin has three Cys residues in the protein; thus, full replacement of Cys112 (the copper binding ligand) cannot be 100% efficient, thereby complicating characterization of the Sec mutant protein.

To overcome the inability to control both the location and the level of Sec replacement using cysteine *E. coli* auxotrophs, the incorporation of Sec into proteins may be accomplished by the use of SECIS elements. To correctly recognize the UGA codon for Sec insertion and not as a stop codon, the SECIS element is used. Incorporation of selenocysteine during protein translation parallels proteinogenic amino acid translation, using a specific, tRNA^{Sec}. The Sec tRNA is initially charged with a serine residue that is converted to selenocysteine by Sec synthase using selenomonophosphate as the selenium donor (Fischer *et al.*, 2007; Su *et al.*, 2005). The charged tRNA^{Sec} is then guided into the ribosome by the elongation factor EF–Tu homologue SelB, where it undergoes GTP hydrolysis to result in Sec incorporation at the UGA codon (Fischer *et al.*, 2007; Hoffmann and Berry, 2005).

Recombinant selenoprotein production is relatively difficult given not only the number of required elements needed for proper recognition of the UGA stop codon and subsequent Sec incorporation but also incompatibilities between the prokaryotic and eukaryotic SECIS sequences and the inability to incorporate multiple Sec residues in a single protein. The primary incompatibility between prokaryotes and eukaryotic SECIS elements is location of and distance between the SECIS loop and the

UGA codon. In prokaryotic systems, the SECIS loop is located in the coding region at a distance of approximately 20 nucleotides from the UGA codon. Thus, there are amino acids following the UGA codon that are needed for complete protein translation (Su *et al.*, 2005). However, the eukaryotic and archaea SECIS loops are found in the 3′ untranslated region and therefore have variable distances from the UGA codon (Kryukov *et al.*, 2003). This incompatibility hinders the usage of common prokaryotic expressions systems for recombinant expression of mammalian selenoproteins. In a recent development, however, a SECIS has been engineered for Sec incorporation in the middle of a protein in bacteria, making it possible for a wider range of applications (Aldag *et al.*, 2009).

The last and most current technique for specific biological incorporation of selenocysteine (and other unnatural amino acids) is the nonsense suppression technique, which uses orthogonal tRNA/synthetase pairs. By using rounds of positive and negative selection, a tyrosyl amber suppressor tRNA/tRNA synthetase pair from *Methanococcus jannaschii* was created that allows for translational incorporation of phenylselenocysteine into proteins at the desired location (Wang *et al.*, 2007). Phenylselenocysteine, however, cannot be used as a ligand because the phenyl group is not readily removable. Currently, no orthogonal tRNA/tRNA synthetase pair has been reported to accept and incorporate the free selenol form of selenocysteine using this methodology.

3. INCORPORATION OF SELENOCYSTEINE INTO THE TYPE 1 COPPER SITE OF AZURIN

Because EPL permits direct control over the location of amino acid substitutions with unnatural amino acids, this technique met our needs very well. Azurin is well suited for EPL, as the protein is relatively small, at 128 amino acids in length (Fig. 5.3). Furthermore, three of the four copper binding ligands are found within the C-terminal 17 amino acids (H₂N-CTFPGHSALMKGTLTLK-COOH) of the protein, making substitutions at these positions conveniently accessible. Therefore, azurin was prepared semisynthetically by using a recombinantly expressed truncated protein containing the first 111 amino acids with a thioester at its C-terminus and a synthetic peptide containing the last 17 amino acids, including the

```
AECSVDIQGN  DQMQFNTNAI  TVDKSCKQFT
VNLSHPGNLP  KNVMGHNWVL  STAADMQGVV
TDGMASGLDK  DYLKPDDSRV  IAHTKLIGSG
EKDSVTFDVS  KLKEGEQYMF  FCTFPGHSALMKGTLTLK
```

Figure 5.3 Amino acid sequence of *Pseudomonas aeruginosa* azurin. Bold residues are those involved in copper binding. The final 17 amino acids used for SPPS are italicized.

replacement of Cys with Sec to afford a full-length selenocysteine-substituted azurin variant (C112U azurin).

Substitution of Cys112 with selenocysteine requires a synthetic peptide with an N-terminal Sec residue that can participate directly in the ligation reaction. The use of Sec residues for EPL has been reported by several groups. Hondal and Raines (2002) showed that replacement of Cys110 in RNase A with Sec could be accomplished using a synthetic 15 amino acid peptide containing an N-terminal Sec residue. The EPL product protein was indistinguishable from the wild-type protein, with similar rates of catalysis indicating that replacement of Cys with Sec did not dramatically perturb either function or protein folding. More recently, Hondal has demonstrated Sec incorporation into High M_r thioredoxins without disruption of protein folding (Eckenroth et al., 2006, 2007). Concurrently, we developed and used similar methods for azurin as detailed herein (Berry et al., 2002; Gieselman et al., 2001; Ralle et al., 2004).

3.1. Synthesis of Fmoc-Sec(PMB)-OH

To accomplish substitution of Cys112 with selenocysteine, a derivative of selenocysteine was needed that was suitable for standard Fmoc-based solid – phase peptide synthesis (SPPS). The most common derivative of selenocysteine used for SPPS has been Fmoc-Sec(PMB)-OH (PMB: p-methoxybenzyl), primarily because of incompatibilities with the trityl (Trt) protecting group (Theodoropoulos et al., 1967a,b; Walter and Chan, 1967). Many procedures have been developed over the years to synthesize Fmoc-Sec(PMB)-OH from various starting materials. In recent years, we utilized a slightly modified and optimized scheme of our original methodology (Gieselman et al., 2001). In this route (Fig. 5.4), the final allyl deprotection using Pd(PPh$_3$)$_4$ was performed in the presence of sodium 2-ethylhexanoate as the allyl transfer reagent rather than morpholine to prevent the partial removal of the Fmoc group that was observed with the latter reagent.

Starting from inexpensive Fmoc-Ser-OH, the precursor Fmoc-Se (PMB)-Sec-OAll compound was prepared in 3 steps (Scheme 5.1). The final optimized step was carried out as described by McCombie with slight modifications (Jeffrey and McCombie, 1982). Pd(PPh$_3$)$_4$ (556 mg, 0.48 mmol) and PPh$_3$ (126 mg, 0.48 mmol) were combined in a dried round-bottom flask equipped with a septum and purged with Ar. The flask was covered with aluminum foil and dry CH_2Cl_2 (20 mL) was added. To a second round-bottom flask Fmoc-Se(PMB)-Sec-OAll (5.3 g, 9.63 mmol) and sodium 2-ethylhexanoate (1.62 g, 9.73 mmol) were added, dissolved in dry CH_2Cl_2 (20 mL), and the contents of the flask were transferred to the Pd catalyst via cannulation under Ar pressure. The reaction was stirred at 25 °C for 20 min under Ar. After completion,

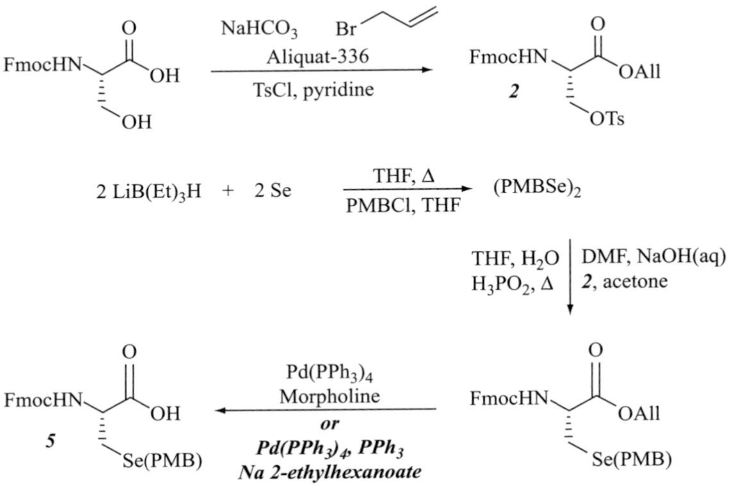

FmocHN—CH(—OH)—C(=O)—OH → [NaHCO₃, Br⁀, Aliquat-336, TsCl, pyridine] → FmocHN—CH(—OTs)—C(=O)—OАll   **2**

$$2\ LiB(Et)_3H\ +\ 2\ Se\ \xrightarrow[\text{PMBCl, THF}]{\text{THF, }\Delta}\ (PMBSe)_2$$

THF, H₂O | DMF, NaOH(aq)
H₃PO₂, Δ | **2, acetone**

FmocHN—CH(—Se(PMB))—C(=O)—OH **5** ← [Pd(PPh₃)₄, Morpholine *or* **Pd(PPh₃)₄, PPh₃ Na 2-ethylhexanoate**] ← FmocHN—CH(—Se(PMB))—C(=O)—OАll

Figure 5.4 Synthetic route for Fmoc-Sec(PMB)-OH from Fmoc-Ser-OH (Gieselman *et al.*, 2001). Conditions in **bold** are modified and optimized conditions used for allyl deprotection.

the reaction was quenched with EtOAc (50 mL) and washed with 2 × 50 mL each of 2 *M*, 4 *M*, and 6 *M* HCl. The organic layer was then dried over MgSO₄, filtered, and the solvent removed using a rotary evaporator to obtain a yellow oil. The crude product was purified by flash chromatography using 35:1 CH₂Cl₂/MeOH with 0.1% AcOH (v/v) to obtain the final product as a white solid (R_f = 0.5, 9:1 CH₂Cl₂/MeOH, 4.7 g, 95%). After successful synthesis of Fmoc-Sec(PMB)-OH, it was then incorporated into the C-terminal 17 amino acid sequence of azurin using standard Fmoc-based chemistry.

3.2. Solid-phase peptide synthesis (SPPS) of C-terminal 17-mer peptide

The C-terminal 17-mer peptide containing selenocysteine (H₂N-UTFPGHSALMKGTLTLK-COOH) was synthesized on a 0.1 mmol scale with a Rainin PS3 automated synthesizer using standard Fmoc chemistry. Preloaded Fmoc-Lys(Boc)-Wang resin was first swelled in DMF (3 × 8 min, 6 mL). All Fmoc deprotections were accomplished using 20% piperidine/DMF (v/v) (3 × 3 min, 6 mL). For amino acid couplings, a 5-fold excess of Fmoc-amino acid and O-benzotriazole-N,N,N′,N′,-tetramethyluronium hexafluorophosphate (HBTU) were combined, and 0.4 *M* N-methylmorpholine (1 × 1.5 min) was used as the activating reagent. The activated amino acids were then transferred to the reaction vessel and coupled under N₂ for 45 min.

The coupling of Fmoc-Sec(PMB)-OH was performed using 2 equivalents of 1-hydroxybenzotriazole (HOBT, 6 equiv) and N,N'-diisopropylcarbodiimide (DIC, 6 equiv) in DMF to reduce the possibility of racemization (Han et al., 1997). The N-terminal Fmoc group was then removed using 20% piperidine/DMF (v/v, 3 × 2 min, 6 mL), and the resin was washed extensively with CH_2Cl_2 and dried in a dessicator for 10 min. Because of the sensitivity of selenocysteine to both racemization and beta elimination, special selenol deprotection conditions were needed after SPPS.

3.3. Deprotection of selenocysteine-containing 17-mer peptide

Several studies have been performed using a multitude of reagents to successfully and fully deblock selenols and thiols protected with PMB, but many of the reagents are harsh, do not result in full deblocking of the thiol/selenol, and can promote deselenization (Hunter and Komives, 1995; Otaka et al., 1991; Yajima et al., 1988). A review of such techniques was published explaining the advantages and disadvantages of each (Hondal, 2005). Recently, a new, milder technique for deblocking selenols protected with PMB was developed using 2,2'-dithiobis(5-nitropyridine) (Harris et al., 2007). A modified version of this procedure was used in our work to afford fully deblocked peptide during the cleavage from the resin.

For a 0.1-mmol scale peptide synthesis, 2,2'-dithiobis(5-nitropyridine) (1.3 mol equiv, 41 mg) was added to the dried resin-bound peptide, and 2% (v/v) triisopropylsilane (100 μL) and ddH$_2$O (100 μL) were added followed by the addition of 2.5% (v/v) thioanisole (125 μL). To the mixture, 5 mL of neat TFA was added and the reaction was stirred vigorously at 25 °C for 2 h. The reaction was filtered through a coarse-grain Büchner funnel and the resin was washed with neat TFA (3 × 5 mL). The TFA was then removed via rotary evaporation to obtain a yellow oil. The peptide was precipitated from the oil by the addition of cold Et$_2$O and pelleted from the solution by centrifugation at 4000 rpm for 5 min. The pellet was then washed with Et$_2$O (3 × 30 mL) with centrifugation steps between each washing. The precipitated peptide was then dissolved in a minimal amount of 1:1 MeCN/H$_2$O and lyophilized overnight. The crude lyophilized product was purified by preparative RP-HPLC on a Waters C18 column (130 mm × 25 mm) with a water-acetonitrile solvent system using a gradient of 20 to 70% of solvent B over 20 min (A: H$_2$O, 0.1% TFA; B: CH$_3$CN 80% in H$_2$O v/v, 0.1% TFA). Peptides were lyophilized after purification to obtain the final product as a pale yellow solid and analyzed using MALDI–TOF-MS; Calculated: 1852.06 (SeH), 2006.21 (Se(pys)), 3704.1 (dimer); observed: 1852.08 (SeH), 3703.6 (dimer). The typical yield for a 0.1-mmol scale synthesis was 70 to 80 mg (≈95% coupling efficiency). The mass spectrum displayed the typical isotopic distribution pattern expected for a peptide containing one selenium atom (Fig. 5.5).

To accomplish incorporation of selenocysteine into azurin using EPL, the first 111 amino acids of azurin were recombinantly expressed as an

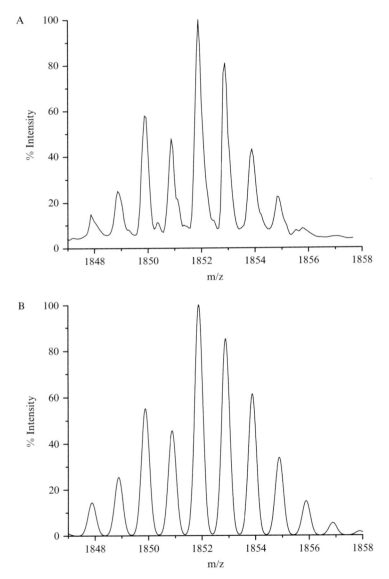

Figure 5.5 MALDI-TOF-MS for (A) Deprotected C112U Azurin 17mer (B) Simulated data.

intein–chitin binding domain (CBD) fusion using NEB vector PtxB1 intein splicing technology. After expression and intein-mediated cleavage with various thiol reagents, the desired recombinant thioester could not be isolated given its insolubility and facile thioester hydrolysis. Therefore, ligations were performed on the column with the final 17 amino acid peptide in the presence of N-methylmercaptoacetamide (NMA).

3.4. General procedure for EPL of Azurin$^{(1-111)}$- Intein-CBD and the C-terminal 17–mer peptide

The azurin$^{(1-111)}$-intein-CBD fusion was overexpressed in *E. coli* BL21 (DE3) in LB media for 8 h at 37 °C. The cells from 5 mL of starter culture were used to inoculate 2 L of LB media, and the cells were grown at 37 °C for 16 h at 210 rpm. Protein expression was induced at ≈16 h with 0.3 mM IPTG, and induction was continued for 4 h at 37 °C. The cells were then harvested at 9000 rpm, frozen at −20 °C, and used when needed.

For ligations, the frozen cell stock was resuspended in a lysis buffer (20 mM HEPES, pH 8.0, 250 mM NaCl, 1 mM EDTA, 1 mM PMSF, 0.1% Triton-x-100) and was lysed twice with a French press. The cell lysate was then treated with 2 M urea for 10 min, and the lysate was centrifuged at 13,000 rpm for 30 min. The resulting supernatant was added to chitin resin (80 mL) preequilibrated in 20 mM HEPES, pH 8.0, 250 mM NaCl, 1 mM EDTA (buffer 1). The mixture was shaken gently at 4 °C for 1 to 2 h and poured into a column. The column was sealed with a septum and the headspace of the column was purged with Ar. The column was then washed with 3 column-volumes of buffer 1, containing 2 M urea under Ar pressure, followed by 5 to 7 column-volumes of buffer 1.

The 17-mer peptide (0.675 mM, 80 mg) and 2 equiv tris-(2-carboxyethyl)phosphine (TCEP, 1.35 mM, 27 mg) were dissolved in 40 mL of buffer 1, containing 75 mM NMA under Ar, and transferred to the column under Ar. The column was then agitated gently at 4 °C for 66 h to promote mixing and reaction. After ligation, the column was mounted, eluted and washed with 1 column-volume of buffer 1. The eluent was centrifuged at 14,000 rpm for 30 min, and the supernatants were combined and concentrated using 10,000 MWCO Centricon concentration spin tubes at 5000 rpm and exchanged into 50 mM MOPS buffer, pH 6.6, via PD10 (G25 Sephadex) columns. The desalted sample was then titrated with incremental additions of freshly made 20 mM CuSO$_4$ solution in ddH$_2$O that had been pretreated with Chelex. The titration was performed at 4 °C until maximal absorbance at 675 nm was reached. Samples were then purified using a Pharmacia Q HiTrap HP column (5 mL) that had been washed with 10 column-volumes of buffer B (1 M NaCl in H$_2$O) and then equilibrated with 10 column-volumes of buffer A (50 mM MOPS, pH 6.6).

The sample was then concentrated and analyzed using MALDI–TOF–MS and ESI–MS (C112U Apo-azurin calculated: 13,999; observed: 14,000. C112U azurin + Cu calculated: 14,055; observed: 14,054).

3.5. Characterization of selenocysteine azurin

This procedure resulted in a C112U azurin mutant that was identical to that previously reported but with slightly better yield ($\approx$0.4 mg/L cell growth) than the previously reported yield ($\approx$0.25 mg/L cell growth) (Berry *et al.*, 2002; Ralle *et al.*, 2004). The C112U mutant protein demonstrated distinct differences from the wild-type protein. The first characteristic change was a red shift in the UV-Vis spectrum from 625 nm for wild type to 677 nm for the C112U mutant, likely resulting from the lower ionization energy of Se as compared to S. More dramatic are the changes noted in the EPR spectra between the two proteins (Fig. 5.6). A marked increase of the parallel hyperfine splitting (100 G) of the C112U mutant compared to wild-type azurin (56 G). Furthermore, the observed increased rhombic character of the site with a larger increase in g_y than in the g_z tensor is consistent with increased axial interaction or loss of Cu(II) interaction with the selenolate. This finding was expected as Se is softer than S and thus would prefer coordination with Cu(I) rather than Cu(II). Surprisingly, however, no dramatic change was noted in the reduction potential of the site when the ligand was changed from Cys to Sec (Berry *et al.*, 2002). A more detailed spectroscopic studies using magnetic circular dichroism and resonance Raman spectroscopies, together with density functional theory calculations suggest that SCys and SeCys have similar covalent interactions with Cu at their respective bond lengths of 2.1 and 2.3 Å (Sarangi *et al.*, 2008).

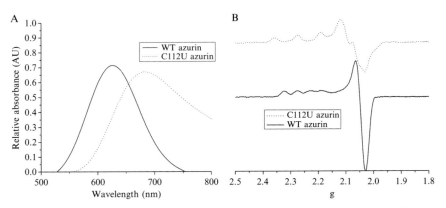

Figure 5.6 Spectroscopic characterization of C112U azurin compared to wild-type azurin (A) UV-Vis spectrum (B) EPR spectrum.

4. Tuning the Type 1 Copper Reduction Potential Using Isostructural Methionine Analogues

In addition to substitution of the Cys ligand, several studies have reported replacement of the axial methionine ligand (see Fig. 5.1). A broad range of amino acid substitutions have been used, and the variants typically still retained the native structure and function. Although a full set of saturated mutational data of the axial methionine (M121) to all other natural amino acids in azurin has been reported (Chang *et al.*, 1991; Karlsson *et al.*, 1991), precise elucidation of the role of this ligand remains challenging because mutations to other proteinogenic amino acids changed more than one factor at a time. To circumvent the limitations of proteinogenic amino acids, an early attempt to incorporate unnatural amino acids at position 121 has been reported (Frank *et al.*, 1985). Using a *P. aeruginosa* strain exposed to rounds of selenomethionine (SeMet) selection, an expression system was created to substitute Met with SeMet. This methodology is an effective means for incorporating SeMet. However, absolute control over the location of incorporation could not be completed, resulting in fully substituted SeMet azurin. Although useful, the inability to individually replace Met121 with SeMet could not be achieved, demonstrating the need for techniques such as EPL to more precisely control the nature of the axial ligand.

Recently, our laboratory has reported the substitution of the axial methionine with nonproteinogenic amino acids using EPL (Berry *et al.*, 2003; Garner *et al.*, 2006). By incorporating a broad range of methionine analogues at the Met121 position, new insights have been obtained that were not possible by proteinogenic amino acid substitution.

4.1. Use of EPL for incorporation of methionine analogues

As the final 17 amino acid sequence contains all of the copper binding ligands except His46, incorporation of methionine analogues used similar methodology as the incorporation of cysteine analogues into the azurin T1 copper site. Furthermore, the same protein thioester truncate of azurin containing the N-terminal 111 amino acids was utilized.

4.2. General procedure for the SPPS of methionine analogue 17-mer peptides

The C-terminal 17-mer peptides containing methionine analogues (H2N-CTFPGHSALxKGTLTLK-COOH, x = Met, SeMet, OxoMethionine (OxM), Leu, Nle, DifluoroMethionine (DFM), or TrifluoroMethionine (TFM)) were synthesized on a 0.1-mmol scale with an Advanced Chemtech

Model 90 peptide synthesizer using standard Fmoc chemistry. Preloaded Fmoc-Lys(Boc)-Wang resin was swelled in DMF, and all Fmoc deprotections were accomplished using 20% piperidine/DMF (v/v). For the standard amino acid couplings, a 4-fold excess of Fmoc-amino acid and O-benzotriazole-N,N,N′,N′,-tetramethyl-uroniumhexafluoro-phosphate (HBTU) were combined, and 0.4 M N-methylmorpholine (1 × 1.5 min) was used as the activating reagent. The activated amino acids were then transferred to the reaction vessel and coupled under N_2. The coupling of the methionine analogues was performed using 1.5 equivalents of 2-(7-aza-1H-benzotriazole-1-yl)-1,1,3,3-tetramethyluronium hexafluorophosphate (HATU). The N-terminal Fmoc protection was removed from the peptide using 20% piperidine/DMF (v/v) and the resin was dried overnight prior to cleavage.

4.3. General procedure for the peptide cleavage from the resin

To the product of a 0.1-mmol synthesis of peptide 5% (v/v) ddH$_2$O, thioanisole, and anisole (500 μL each) were added. To the mixture, 10 mL of neat TFA was added, and the reaction was stirred at RT for 2 h. The reaction was filtered through a coarse-grain Büchner funnel, and the resin was washed with neat TFA (3 × 5 mL). The TFA was then removed via rotary evaporation and the peptide was precipitated by the addition of cold Et$_2$O. The crude peptide was purified by preparative RP-HPLC on a Waters C18 column (130 mm × 25 mm) with a water–acetonitrile solvent system using a gradient of 20 to 70% of solvent B over 20 min (A: H$_2$O, 0.1% TFA; B: CH$_3$CN 80% in H$_2$O v/v, 0.08% TFA). Peptides were lyophilized after purification to obtain final products as a white solid and were analyzed using MALDI–TOF–MS.

4.4. General procedure for EPL of azurin$^{(1-111)}$-intein-CBD and the C-terminal peptides containing methionine analogues

The azurin$^{(1-111)}$-intein-CBD fusion was overexpressed as described previously, but in *E. coli* BLR1 at 30 °C and without IPTG induction. The ligation procedure was performed as previously described but with the following differences. A buffer, 20 mM HEPES, pH 8.0, 250 mM NaCl, 1 mM EDTA (buffer 2) was used. Although the stringent anaerobic environment needed for selenocysteine ligation was not needed for ligation of the Met analogue peptides, the column headspace was still purged with Ar before ligation. The column was equilibrated with buffer 2 before adding the 17-mer peptide (10 mg) predissolved in 4 mL of buffer 1 containing 50 mM NMA. The peptide mixture was then transferred to the column and shaken gently at 4 °C for 60 h. After ligation, the column was eluted, the eluent was treated and concentrated exactly as before but exchanged into 50 mM ammonium acetate buffer, pH 5.1, and titrated with incremental additions of freshly made CuSO$_4$ solution (10 mM) in ddH$_2$O. The titration was performed at 4 °C until maximal absorbance at 625 nm was reached. Samples were then purified via FPLC using anion exchange chromatography.

4.5. Characterization of methionine analogues of azurin

The incorporation of methionine analogues into the azurin T1 copper site demonstrated that the use of isosteric structures reduces the perturbation of the site. UV-vis spectroscopy of all the variants resulted in <6-nm shifts in the S(Cys)-Cu charge transfer (CT) band in the electronic absorption spectra and <8-G changes in the copper hyperfine coupling constants (A$_{\parallel}$) in the X-band EPR spectra.

In contrast to the minimal spectral feature changes that are visible from the substitution of the axial Met with isostructural amino acids, substantial deviations from the wild-type reduction potential are evident. It is well known that Met plays an important role in fine-tuning the redox potential of the blue copper center, but the molecular origin remains unclear because conclusions drawn from site-directed mutagenesis depend heavily on the choice of natural amino acids used to replace Met. The incorporation of isostructural unnatural amino acids using EPL has made it possible to deconvolute different factors affecting the reduction potentials of the blue copper center. A careful analysis, including a plot of observed reduction potentials of the wild-type azurin and its variants and the corresponding hydrophobicity of the axial ligand side chain defined by the log of the partition coefficient between octanol and water (logP) (Fig. 5.7) revealed hydrophobicity as the dominant factor in tuning the reduction potentials of blue copper centers by axial ligands (Berry *et al.*, 2003). Extension of such a finding in azurin to other T1 blue copper proteins established a linear correlation between hydrophobicity of the axial ligand and the reduction potential of all T1 copper centers, independent of the protein scaffold, experimental conditions, measurement techniques, and steric modifications (Garner *et al.*, 2006).

5. Conclusion

In this work, the specific functionalities of the cysteine ligand and the axial methionine ligand in azurin have been extensively probed using unnatural amino acid substitutions via EPL. The incorporation of Sec to replace Cys created a variant in which the overall function and redox activity of azurin remained relatively unperturbed and for which noticeable

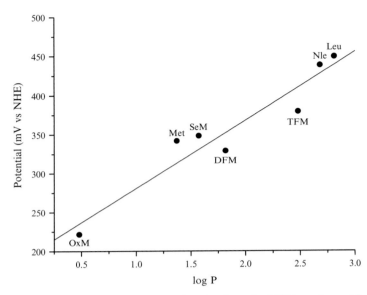

Figure 5.7 Reduction potential versus log P of azurin M121 variants. Unnatural amino acid labels: OxM = oxomethionine, SeM = selenomethionine, DFM = difluoromethionine, TFM = trifluoromethionine, and Nle = norleucine.

spectroscopic changes can be attributed to the differences in the electronic characteristics between sulfur and selenium. A thorough investigation of the axial methionine by replacement with a broad range of electronic and hydrophobic mimics of Met have definitively shown that the axial methionine is a key determinant in the magnitude of the reduction potential of the T1 copper site found in azurin. Variants with more hydrophobic side chains demonstrated large increases in reduction potential when compared to the wild-type protein. We have shown that engineering of metalloproteins can be accomplished using EPL to gain a better understanding of protein function and to alter the coordination sphere of the metal ion to fine-tune the desired properties of the engineered variant.

ACKNOWLEDGMENTS

We wish to thank Dr. Mark Nilges for help with EPR experiments and Mr. Furong Sun for technical assistance with mass spectrometry analysis. The materials described here are based on work supported by the National Science Foundation under Award No. CHE 05-52008 and Special Creativity Extension (to Y.L.), and the National Institutes of Health under Award No. GM58822 (to W.A.V.).

REFERENCES

Aldag, C., Gromov, I. A., García-Rubio, I., von Koenig, K., Schlichting, I., Jaun, B., and Hilvert, D. (2009). Probing the role of the proximal heme ligand in cytochrome P450cam by recombinant incorporation of selenocysteine. *Proc. Natl. Acad. Sci. USA* **106,** 5481–5486.

Berry, S. M., Gieselman, M. D., Nilges, M. J., van der Donk, W. A., and Lu, Y. (2002). An Engineered Azurin Variant Containing a Selenocysteine Copper Ligand. *J. Am. Chem. Soc.* **124,** 2084–2085.

Berry, S. M., Ralle, M., Low, D. W., Blackburn, N. J., and Lu, Y. (2003). Probing the Role of Axial Methionine in the Blue Copper Center of Azurin with Unnatural Amino Acids. *J. Am. Chem. Soc.* **125,** 8760–8768.

Boschi-Muller, S., Muller, S., Van Dorsselaer, A., Bock, A., and Branlant, G. (1998). Substituting selenocysteine for active site cysteine 149 of phosphorylating glyceraldehyde 3-phosphate dehydrogenase reveals a peroxidase activity. *FEBS Lett.* **439,** 241–245.

Casi, G., Roelfes, G., and Hilvert, D. (2007). Selenoglutaredoxin as a Glutathione Peroxidase Mimic. *ChemBioChem.* **9,** 1623–1631.

Chang, T. K., Iverson, S. A., Rodrigues, C. G., Kiser, C. N., Lew, A. Y. C., Germanas, J. P., and Richards, J. H. (1991). Gene Synthesis, Expression, and Mutagenesis of the Blue Copper Proteins Azurin and Plastocyanin. *Proc. Natl. Acad. Sci. USA* **88,** 1325–1329.

Cheong, J.-J., Hwang, I., Rhee, S., Moon, T., Choi, Y., and Kwon, H.-B. (2007). Complementation of an *E. coli* cysteine auxotrophic mutant for the structural modification study of 3'(2'),5'-bisphosphate nucleotidase. *Biotechnol. Lett.* **29,** 913–918.

DeBeer, S., Kiser, C. N., Mines, G. A., Richards, J. H., Gray, H. B., Solomon, E. I., Hedman, B., and Hodgson, K. O. (1999). X-ray Absorption Spectra of the Oxidized and Reduced Forms of C112D Azurin from *Pseudomonas aeruginosa. Inorg. Chem.* **38,** 433–438.

Den Blaauwen, T., Van de Kamp, M., and Canters, G. W. (1991). Type I and II copper sites obtained by external addition of ligands to a His117Gly azurin mutant. *J. Am. Chem. Soc.* **113,** 5050–5052.

Eckenroth, B., Harris, K., Turanov, A. A., Gladyshev, V. N., Raines, R. T., and Hondal, R. J. (2006). Semisynthesis and Characterization of Mammalian Thioredoxin Reductase. *Biochemistry* **45,** 5158–5170.

Eckenroth, B. E., Lacey, B. M., Lothrop, A. P., Harris, K. M., and Hondal, R. J. (2007). Investigation of the C-Terminal Redox Center of High-Mr Thioredoxin Reductase by Protein Engineering and Semisynthesis. *Biochemistry* **46,** 9472–9483.

Faham, S., Mizoguchi, T. J., Adman, E. T., Gray, H. B., Richards, J. H., and Rees, D. C. (1997). Role of the active-site cysteine of *Pseudomonas aeruginosa* azurin. Crystal structure analysis of the CuII(Cys112Asp) protein. *J. Biol. Inorg. Chem.* **2,** 464–469.

Fischer, N., Paleskava, A., Gromadski, K. B., Konevega, A. L., Wahl, M. C., Stark, H., and Rodnina, M. V. (2007). Towards Understanding Selenocysteine Incorporation into Bacterial Proteins. *Biol. Chem.* **388,** 1061–1067.

Frank, P., Licht, A., Tullius, T. D., Hodgson, K. O., and Pecht, I. (1985). A selenomethionine-containing azurin from an auxotroph of Pseudomonas aeruginosa. *J. Biol. Chem.* **260,** 5518–5525.

Garner, D. K., Vaughan, M. D., Hwang, H. J., Savelieff, M. G., Berry, S. M., Honek, J. F., and Lu, Y. (2006). Reduction Potential Tuning of the Blue Copper Center in *Pseudomonas aeruginosa* Azurin by the Axial Methionine as Probed by Unnatural Amino Acids. *J. Am. Chem. Soc.* **128,** 15608–15617.

Germanas, J. P., Di Bilio, A. J., Gray, H. B., and Richards, J. H. (1993). Site saturation of the histidine-46 position in *Pseudomonas aeruginosa* azurin: Characterization of the His46Asp copper and cobalt proteins. *Biochemistry* **32**, 7698–7702.

Gieselman, M. D., Xie, L., and van der Donk, W. A. (2001). Synthesis of a Selenocysteine-Containing Peptide by Native Chemical Ligation. *Org. Lett.* **3**, 1331–1334.

Gieselman, M. D., Zhu, Y., Zhou, H., Galonic, D., and van der Donk, W. A. (2002). Selenocysteine Derivatives for Chemoselective Ligations. *ChemBioChem.* **3**, 709–716.

Han, Y., Albericio, F., and Barany, G. (1997). Occurrence and Minimization of Cysteine Racemization during Stepwise Solid-Phase Peptide Synthesis. *J. Org. Chem.* **62**, 4307–4312.

Harris, K. M., Flemer, S., and Hondal, R. J. (2007). Studies on deprotection of cysteine and selenocysteine side-chain protecting groups. *J. Pept. Sci.* **13**, 81–93.

Hoffmann, P. R., and Berry, M. J. (2005). Selenoprotein Synthesis: A Unique Translational Mechanism Used by a Diverse Family of Proteins. *Thyroid* **15**, 769–775.

Hondal, R. J. (2005). Incorporation of Selenocysteine into Proteins Using Peptide Ligation. *Protein Pept. Let.* **12**, 757–764.

Hondal, R. J., and Raines, R. T. (2002). Semisynthesis of proteins containing selenocysteine. *Methods Enzymol.* **347**, 70–83.

Hunter, M. J., and Komives, E. A. (1995). Deprotection of S-Acetamidomethyl Cysteine-Containing Peptides by Silver Trifluoromethanesulfonate Avoids the Oxidization of Methionines. *Anal. Biochem.* **228**, 173–177.

Jeffrey, P. D., and McCombie, S. W. (1982). Homogeneous, palladium(0)-catalyzed exchange deprotection of allylic esters, carbonates and carbamates. *J. Org. Chem.* **47**, 587–590.

Johansson, L., Gafvelin, G., and Arner, E. S. J. (2005). Selenocysteine in proteins--properties and biotechnological use. *Biochim. Biophys. Acta* **1726**, 1–13.

Karlsson, B. G., Aasa, R., Malmström, B. G., and Lundberg, L. G. (1989). Rack-induced bonding in blue copper proteins: Spectroscopic properties and reduction potential of the azurin mutant Met-121 --> Leu. *FEBS Lett.* **253**, 99–102.

Karlsson, B. G., Nordling, M., Pascher, T., Tsai, L.-C., Sjolin, L., and Lundberg, L. G. (1991). Cassette mutagenesis of Met121 in azurin from *Pseudomonas aeruginosa*. *Protein Eng.* **4**, 343–349.

Kryukov, G. V., Castellano, S., Novoselov, S. V., Lobanov, A. V., Zehtab, O., Guigo, R., and Gladyshev, V. N. (2003). Characterization of Mammalian Selenoproteomes. *Science* **300**, 1439–1443.

Lu, Y. (2003). "Cupredoxins", In *Comprehensive Coordination Chemistry II: From Biology to Nanotechnology* (Jon McCleverty and Tom J. Meyer, Eds.), Vol 8, *Biocoordination Chemistry*, (Lawrence Que, Jr. and William B. Tolman, Eds.), Elsevier: Oxford, pp. 91–122.

Lu, Y. (2005). Design and engineering of metalloproteins containing unnatural amino acids or non-native metal-containing cofactors. *Curr. Opin. Chem. Biol.* **9**, 118–126.

Lu, Y. (2006a). Biosynthetic Inorganic Chemistry. *Angew. Chem., Int. Ed.* **45**, 5588–5601.

Lu, Y. (2006b). Metalloprotein and Metallo-DNA/RNAzyme Design: Current Approaches, Success Measures, and Future Challenges. *Inorg. Chem.* **45**, 9930–9940.

Lu, Y., Berry, S. M., and Pfister, T. D. (2001). Engineering Novel Metalloproteins: Design of Metal-Binding Sites into Native Protein Scaffolds. *Chem. Rev.* **101**, 3047–3080.

Malmström, B. G. (1994). Rack-Induced Bonding In Blue-Copper Proteins. *Eur. J. Biochem.* **223**, 711–718.

Mizoguchi, T. J., Di Bilio, A. J., Gray, H. B., and Richards, J. H. (1992). Blue to type 2 binding. Copper(II) and cobalt(II) derivatives of a Cys112Asp mutant of *Pseudomonas aeruginosa* azurin. *J. Am. Chem. Soc.* **114**, 10076–10078.

Moroder, L. (2005). Isosteric Replacement of Sulfur with Other Chalcogens in Peptides and Proteins. *J. Pept. Sci.* **11**, 187–214.

Mueller, S., Senn, H., Gsell, B., Vetter, W., Baron, C., and Boeck, A. (1994). The Formation of Diselenide Bridges in Proteins by Incorporation of Selenocysteine Residues: Biosynthesis and Characterization of (Se)2-Thioredoxin. *Biochemistry* **33**, 3404–3412.

Neuwald, A. F., Krishnan, B. R., Brikun, I., Kulakauskas, S., Suziedelis, K., Tomcsanyi, T., Leyh, T. S., and Berg, D. E. (1992). cysQ, a gene needed for cysteine synthesis in *Escherichia coli* K-12 only during aerobic growth. *J. Bacteriol.* **174**, 415–425.

Otaka, A., Koide, T., Shide, A., and Fujii, N. (1991). Application of dimethylsulphoxide (DMSO) / trifluoroacetic acid(TFA) oxidation to the synthesis of cystine-containing peptide. *Tetrahedron Lett.* **32**, 1223–1226.

Pierloot, K., De Kerpel, J. O. A., Ryde, U., and Roos, B. O. (1997). Theoretical Study of the Electronic Spectrum of Plastocyanin. *J. Am. Chem. Soc.* **119**, 218–226.

Ralle, M., Berry, S. M., Nilges, M. J., Gieselman, M. D., van der Donk, W. A., Lu, Y., and Blackburn, N. J. (2004). The Selenocysteine-Substituted Blue Copper Center: Spectroscopic Investigations of Cys112SeCys *Pseudomonas aeruginosa* Azurin. *J. Am. Chem. Soc.* **126**, 7244–7256.

Sarangi, R., Gorelsky, S. I., Basumallick, L., Hwang, H. J., Pratt, R. C., Stack, T. D. P., Lu, Y., Hodgson, K. O., Hedman, B., and Solomon, E. I. (2008) Spectroscopic and Density Functional Theory Studies of the Blue-Copper Site in M121SeM and C112SeC Azurin: Cu-Se Versus Cu-S Bonding. *J. Am. Chem. Soc.* **130**, 3866–3877

Solomon, E. I., Baldwin, M. J., and Lowery, M. D. (1992). Electronic Structures of Active Sites in Copper Proteins: Contributions to Reactivity. *Chem. Rev.* **92**, 521–542.

Su, D., Li, Y., and Gladyshev, V. N. (2005). Selenocysteine insertion directed by the 3'-UTR SECIS element in *Escherichia coli*. *Nucleic Acids Res.* **33**, 2486–2492.

Theodoropoulos, D., Schwartz, I. L., and Walter, R. (1967a). New synthesis of L-selenocysteine derivatives and peptides. *Tetrahedron Lett.* **8**, 2411–2414.

Theodoropoulos, D., Schwartz, I. L., and Walter, R. (1967b). Synthesis of Selenium-Containing Peptides. *Biochemistry* **6**, 3927–3932.

Vila, A. J., and Fernandez, C. O. (2001). Copper in Electron transfer Proteins, In Handbook on Metalloproteins. Wiley. New York, NY, pp. 813–856.

Walter, R., and Chan, W.-Y. (1967). Syntheses and pharmacological properties of selenium isologs of oxytocin and deaminooxytocin. *J. Am. Chem. Soc.* **89**, 3892–3898.

Wang, J., Schiller, S. M., and Schultz, P. G. (2007). A Biosynthetic Route to Dehydroalanine-Containing Proteins. *Angew. Chem. Int. Ed.* **46**, 6849–6851.

Wessjohann, L. A., Schneider, A., Abbas, M., and Brandt, W. (2007). Selenium in chemistry and biochemistry in comparison to sulfur. *Biol. Chem.* **388**, 997–1006.

Yajima, H., Fujii, N., Funakoshi, S., Watanabe, T., Murayama, E., and Otaka, A. (1988). New strategy for the chemical synthesis of proteins. *Tetrahedron* **44**, 805–819.

CHAPTER SIX

USING EXPRESSED PROTEIN LIGATION TO PROBE THE SUBSTRATE SPECIFICITY OF LANTIBIOTIC SYNTHETASES

Xingang Zhang* *and* Wilfred A. van der Donk[†]

Contents

Abstract

The lantibiotics are a class of ribosomally synthesized and posttranslationally modified peptide antibiotics containing the thioether cross-links lanthionine (Lan) and 3-methyllanthionine (MeLan) and typically also the dehydroamino acids dehydroalanine (Dha) and (Z)-dehydrobutyrine (Dhb). The modifications are formed by dehydration of Ser/Thr residues to produce the Dha and Dhb

* Department of Chemistry, University of Illinois at Urbana-Champaign, Urbana, Illinois, USA
† Departments of Chemistry and Biochemistry, University of Illinois at Urbana-Champaign, Urbana, Illinois, USA

Methods in Enzymology, Volume 462
ISSN 0076-6879, DOI: 10.1016/S0076-6879(09)62006-1

structures, and subsequent conjugate additions of Cys residues onto the unsaturated amino acids to form thioether rings (Lan and MeLan). Because of their ribosomal origin, investigations of the substrate specificity of lantibiotic synthetases have typically focused on site-directed mutagenesis. With the *in vitro* reconstitution of lacticin 481 synthetase, its substrate specificity was explored in much greater detail by the incorporation of a series of nonproteinogenic Ser, Thr, and Cys analogues into a truncated prelacticin peptide LctA using a combination of solid-phase peptide synthesis (SPPS) and expressed protein ligation (EPL). The strategy described can be used for the growing number of ribosomally synthesized and posttranslationally modified natural products such as the microcins and patellamides.

1. INTRODUCTION

The lantibiotics are a class of ribosomally synthesized and posttranslationally modified peptide antibiotics (Chatterjee *et al.*, 2005b; Cotter *et al.*, 2005; Willey and van der Donk, 2007). The modifications involve dehydrations of Ser and Thr residues followed by intramolecular Michael-type additions of Cys residues to the dehydroamino acids in a stereoselective fashion (Fig. 6.1). This sequence of events produces lanthionines (Lan) from Ser and methyllanthionines (MeLan) from Thr, the thioether cross-links from which lantibiotics derive their name. A typical lantibiotic contains 3 to 6 (methyl)lanthionines in addition to several dehydroalanines (Dha) and dehydrobutyrines (Dhb) (Chatterjee *et al.*, 2005b). The enzymatic systems that carry out the modifications during the biosynthesis of lacticin 481 (Xie *et al.*, 2004), nisin (Li *et al.*, 2006), and haloduracin (McClerren *et al.*, 2006) have been reconstituted *in vitro* providing a platform for investigating their catalytic mechanisms and their potential for applications in bioengineering. During the maturation process for lacticin 481, the synthetase LctM executes the dehydration of four Ser and Thr residues in the propeptide region of LctA, followed by the intramolecular regio- and stereoselective addition of three cysteines to a subset of the Dha and Dhb residues to generate three cyclic thioethers (Fig. 6.2). Using site-directed mutagenesis, the synthetase was shown to display a high degree of substrate promiscuity (Xie *et al.*, 2004), which suggests the intriguing possibility of introducing nonproteinogenic amino acids into its peptide substrate.

The ribosomal origin of the lantibiotics currently limits *in vivo* engineering in the producing strains to the 20 to 22 proteinogenic amino acids when using mutagenesis strategies (Chatterjee *et al.*, 2005b; Cotter *et al.*, 2005; Kuipers *et al.*, 1996; Lubelski *et al.*, 2007). The potential use of substrate analogues that go beyond this restriction would greatly expand the functional and structural diversity that can be incorporated into lantibiotics via the biosynthetic enzymes. Various protein-engineering approaches have

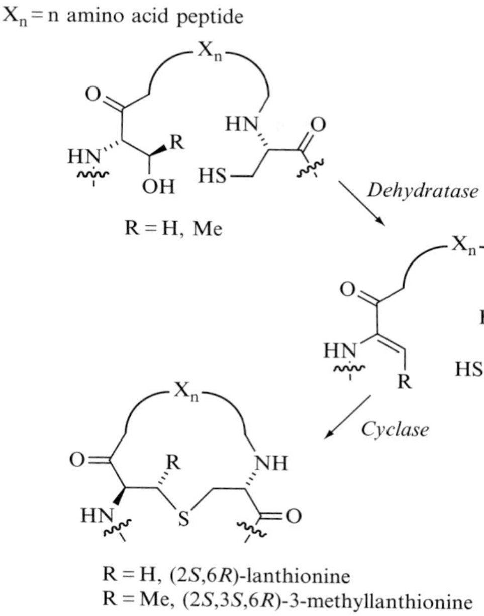

X_n = n amino acid peptide

Figure 6.1 The posttranslational modifications by lantibiotic synthetases involve dehydration of Ser and Thr residues followed by intramolecular Michael-type additions of Cys residues to the dehydroamino acids.

been developed to site-specifically incorporate nonnative structures into proteins. The techniques range from classical chemical-labeling methods to more recent technologies highlighted in this volume: nonsense suppression mutagenesis (Wang et al., 2001; Wang and Schultz, 2002) and expressed protein ligation (EPL) (Evans et al., 1998; Muir et al., 1998). Given the moderate size of the substrate peptides for lantibiotics ($\approx$50 to 60 amino acids) and the fact that the modifications all take place in the C-terminal half of the peptides, we initially chose to use EPL technology to investigate the substrate scope of LctM. Herein, we describe a general approach to access semisynthetic analogues of the LctA substrate based on the native chemical ligation reaction (NCL) (Dawson and Kent, 2000; Dawson et al., 1994) between bacterially expressed N-terminal peptides and synthetic C-terminal peptides. Using this approach, we show that LctM displays relaxed substrate specificity, which may be used for the design of tailor-made lantibiotics or for the installation of thioether rings into nonlantibiotic therapeutic peptides (Kluskens et al., 2005). A similar strategy should be applicable to other natural products that are generated by posttranslational modification of ribosomally synthesized peptides, such as the microcins (Duquesne et al., 2007; Kazakov et al., 2007; Roy et al., 1999b;

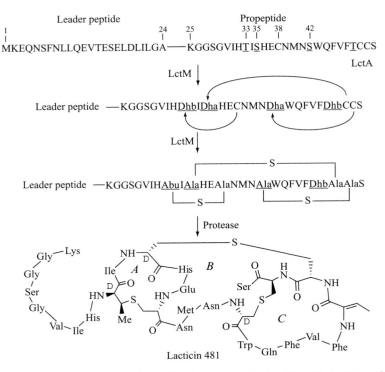

Figure 6.2 The posttranslational modification process for lacticin 481. LctM catalyzes both the dehydration of Ser and Thr residues in the propeptide region of the substrate LctA and the subsequent regio- and stereoselective conjugate addition of cysteines to the Dha and Dhb residues. The bifunctional enzyme LctT then proteolytically removes the leader peptide and transports the mature product across the cell membrane.

Severinov *et al.*, 2007; Vassiliadis *et al.*, 2007) and the patellamides (Donia *et al.*, 2006; Milne *et al.*, 2006). Indeed, one such study has been reported on microcin B17 (Roy *et al.*, 1999a).

2. USE OF EXPRESSED PROTEIN LIGATION TO PREPARE SUBSTRATE ANALOGUES

2.1. Overview

A series of amino acids were designed as possible Ser, Thr, and Cys analogues that could serve to investigate the substrate specificity of the dehydration reaction and the subsequent intramolecular conjugate addition (Fig. 6.3). The analogues were synthesized with appropriate protecting

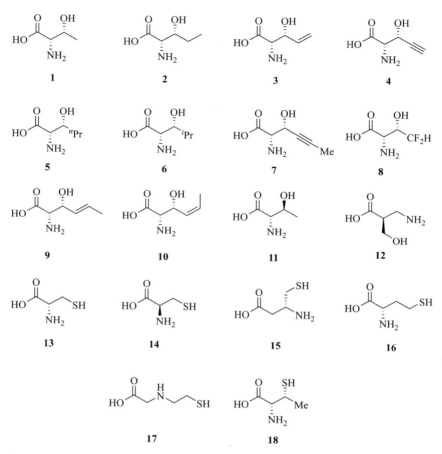

Figure 6.3 After incorporation into the LctA substrate peptide using semisynthesis as described in the text, amino acids 1 to 18 were used to probe the substrate tolerance of the lacticin 481 synthetase LctM.

groups for Fmoc-based solid-phase peptide synthesis (Li and van der Donk, 2007; Zhang et al., 2005; Zhang and van der Donk, 2007).

For the construction of LctA mutants incorporating the amino acids in Fig. 6.3, LctA was divided into two fragments. The N-terminal segment consisted of LctA(1–37) (Fig. 6.4) expressed in E. coli as a fusion to the intein domain from the gyrA gene (Evans et al., 1998; Telenti et al., 1997). The intein in turn was fused to a chitin binding domain (CBD) for purification purposes. Plasmids containing the gene for the intein-CBD fusion and a multiple cloning region for insertion of a gene of interest are available from New England Biolabs. Purification of the LctA(1–37)-intein-CBD fusion

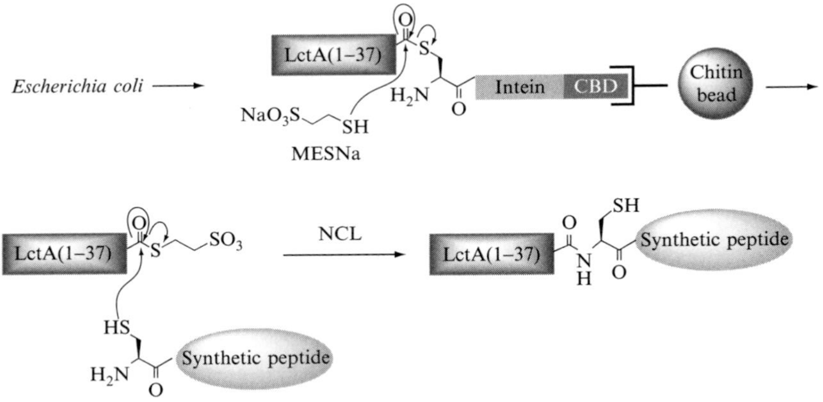

Figure 6.4 Semisynthetic strategy toward LctA analogues.

protein using chitin affinity chromatography and elution with the sodium salt of mercaptoethanesulfonic acid (MESNA) resulted in the desired thioester for NCL (see Fig. 6.4). The C-terminal partners for the ligation reactions were short synthetic peptides corresponding to amino acids 38 to 43 of LctA. For peptides designed to investigate the dehydration activity of the bifunctional lacticin 481 synthetase, the Ser/Thr analogues **2** to **12** were incorporated at position 42 of the synthetic peptides normally occupied by Ser (see Figs. 6.2 and 6.3). In these peptides, a Cys occupied the N-terminus and the ligation reactions represented standard native chemical ligations (NCLs, entries 1–12, Table 6.1).

Peptides containing amino acids **14** to **18** were designed to probe the tolerance of LctM with respect to cyclization of Cys analogues. These analogues were incorporated at the N-terminus of the synthetic peptides and subsequently successfully ligated to the LctA(1-37)-MES thioester. Notably, the native chemical ligations with β^3-homocysteine (**15**), the peptoid analogue of Cys (**17**), and (R)-3-methylcysteine (**18**) had not been reported before to the best of our knowledge. Once ligated to the LctA(1-37)-MES thioester, the Cys analogues occupied position 38 of LctA and were used in a cyclization assay that reports not only whether cyclization occurs but also its regio- and chemoselectivity (see section 2.8).

The synthesis of LctA(38-43) was based on standard Fmoc SPPS chemistry during which the hydroxyl groups of the Thr and Ser analogues were protected with tert-butyl groups, and the thiol of the Cys analogues were protected as disulfides (Li and van der Donk, 2007; Zhang et al., 2005). The overall yields for the syntheses of the peptides ranged from 60 to 80% after purification by reverse phase (C18) preparative high-performance liquid chromatography (RP-HPLC).

Table 6.1 Use of expressed protein ligation to prepare LctA analogues LctA(1-43)

Entry	Amino Acids	Ligation products
1	**1**	LctA(1-37)CNMN**X1**WA
2	**2**	LctA(1-37)CNMN**X2**WA
3	**3**	LctA(1-37)CNMN**X3**WA
4	**4**	LctA(1-37)CNMN**X4**WA
5	**5**	LctA(1-37)CNMN**X5**WA
6	**6**	LctA(1-37)CNMN**X6**WA
7	**7**	LctA(1-37)CNMN**X7**WA
8	**8**	LctA(1-37)CNMN**X8**WA
9	**9**	LctA(1-37)CNMN**X9**WA
10	**10**	LctA(1-37)CNMN**X10**WA
11	**11**	LctA(1-37)CNMN**X11**WA
12	**12**	LctA(1-37)CNMN**X12**WA
13	**13**	LctA(1-37)**X13**NMNSWA-Ser35Ala[a]
14	**14**	LctA(1-37)**X14**NMNSWA-Ser35Ala[a]
15	**15**	LctA(1-37)**X15**NMNSWA-Ser35Ala[a]
16	**16**	LctA(1-37)**X16**NMNSWA-Ser35Ala[a]
17	**17**	LctA(1-37)**X17**NMNSWA-Ser35Ala[a]
18	**18**	LctA(1-37)**X18**NMNSWA-Ser35Ala[a]

[a] For these peptides, Ser35 was replaced by Ala to test for the regioselectivity of cyclization (Zhang and van der Donk, 2007). The C-terminal Ala is not found in LctA at position 44 but was used for convenience of synthesis.

2.2. General procedure for solid-phase peptide synthesis

The C-terminal peptides were synthesized on an automated peptide synthesizer by standard Fmoc chemistry. Preloaded Wang-resin (Fmoc-Ala-Wang resin, 0.1 mmol) and Fmoc-amino acids (0.4 mmol) were used. O-Benzotriazole-N,N,N′,N′,-tetramethyluronium hexafluorophosphate (HBTU) was used as coupling reagent. Both Fmoc-amino acids and HBTU were used in 3-fold excess relative to the resin. For the coupling of nonproteinogenic Fmoc-amino acids a 2-fold excess was used to preserve the synthetic compounds. Piperidine (20% in DMF) was used as the deprotecting reagent, and N-methylmorpholine (NMM, 4.4% in DMF) was employed with HBTU as the activating reagent. Resins were first swollen in DMF (6 mL) for 30 min. The Fmoc group was then removed with 20% piperidine in DMF (5 × 5 min, 6 mL). After deprotection, the resins were washed with DMF (9 × 30 s, 6 mL). Amino acids were activated and transferred into the reaction vessel and the coupling was performed for 2 h with agitation by sparging with nitrogen (Levengood and van der Donk, 2006). To reduce the possibility of racemization (Han et al., 1997), the coupling of Fmoc-L-Cys-OH as well as Cys analogues was manually conducted with 1-hydroxybenzotriazole (HOBt, 4 eq.) and N,N′-diisopropylcarbodiimide

(DIC, 4 eq.) in DMF (6 mL) in the absence of external bases for 2 h. This same set of reagents was also used to couple nonproteinogenic amino acids. The completion of couplings was monitored by modified ninhydrin test (Levengood and van der Donk, 2006). After completion of the peptide synthesis the resins were rinsed with DMF (9 × 30 s, 6 mL), EtOH (5 × 10 mL), and CH$_2$Cl$_2$ (5 × 10 mL), and dried by storage in a vacuum dessicator for >4 h. Peptide cleavage from the resin was achieved with a mixture of TFA (6 mL), water (25 μL), and ethanedithiol (EDT, 25 μL) with the dropwise addition of triisopropylsilane (TIPS) until a colorless suspension was obtained, indicating consumption of the trityl cation. After the suspension was stirred for 1 h at room temperature, the resin was filtered. The filtrate was concentrated to give crude peptides that were washed with cold diethyl ether and lyophilized from 10% aqueous acetic acid after filtration through a 0.45 μm filter. The lyophilized crude peptides were purified by preparative RP-HPLC on a Vydac C18 column (2.2 cm × 25 cm) employing a water–acetonitrile solvent system. The standard HPLC gradient was 2 to 100% of solvent B over 45 min (A: H$_2$O, 0.1% TFA; B: CH$_3$CN 80% in H$_2$O v/v, 0.086% TFA). Peptides were detected by their absorbance at 220 nm, and analyzed by ESI-MS.

2.3. General procedure for purification of the peptide thioester His-LctA (1-37)-MES

BL21(DE3) cells carrying a pET15b-derived plasmid encoding the His-LctA(1-37)-intein-CBD fusion protein (Chatterjee *et al.*, 2005a, 2006) were plated on Luria–Bertani (LB) agar containing 100 μg/mL ampicillin and grown at 37 °C for 12 h. A single colony was used to inoculate 5 mL of LB medium (NaCl 10 mg/mL, tryptone 10 mg/mL, yeast extract 5 mg/mL), and a filter-sterilized solution of ampicillin (5 μL, 100 mg/mL) was added. The cells were shaken at 37 °C for 12 h. The cells were harvested by centrifugation, and the cell pellet was resuspended in 5 mL of LB. From the suspension, 0.5 mL was used to inocculate 50 mL of LB and a sterile ampicillin solution (50 μL, 100 mg/mL) was added. The cells were shaken at 37 °C for 12 h, harvested by centrifugation, and the cell pellet was resuspended in 50 mL of LB, and added to 3 L of LB. Filter-sterilized ampicillin solution (3 mL, 100 mg/mL) was added, and the cells were shaken at 37 °C until the OD$_{600\ nm}$ was 0.6 to 0.7.

Protein expression was induced with IPTG (1 M, 1.95 mL), and the cells were grown for an additional 6 h at 25 °C. The cells were harvested by centrifugation at 18.5 kg for 15 min at 4 °C, and the cell pellet ($\approx$10 g) was resuspended in 20 mL of cell lysis buffer (20 mM Tris, 500 mM NaCl, 0.1% Triton X-100, 1 mM EDTA, pH 7.5 at 4 °C). The cells were lysed on ice by sonication at Amp 75, 5 s on, 9.9 s off for 15 to 20 min. The lysate was centrifuged at 27,000g for 30 min. The supernatant containing the fusion protein was bound to a pre-equilibrated capped column containing chitin resin (20 mL) at 4 °C by gentle shaking for 2 h. The column was subsequently mounted and washed with column buffer (20 mM HEPES, 500 mM NaCl, 1 mM EDTA, pH 7.2 at 4 °C) until the OD_{280} of the flow through was less than 0.01. Then the column was rapidly washed with three column-volumes of cleavage buffer (100 mM HEPES, 200 mM NaCl, 50 mM MESNA, 1 mM EDTA, pH 7.75 at 4 °C).

For overnight intein-mediated cleavage of the truncated His–LctA mutants from the resin and generation of the MES thioester, two column-volumes of cleavage buffer were added to the column before it was capped and gently shaken for 12 h at 4 °C. After mounting the column, it was drained and washed with one column-volume of buffer. The eluate containing the peptide thioester was concentrated by ultrafiltration using an Amicon YM1 membrane at 4 °C and lyophilized after acidification with 0.1% TFA. The lyophilized peptide thioester was purified by preparative RP-HPLC using a C4 column employing a water-acetonitrile solvent system (the standard HPLC gradient was 2 to 100% of solvent B over 45 min. A: H_2O, 0.1% TFA; B: CH_3CN 80% in H_2O, v/v 0.086% TFA; peptides were detected by their absorbance at 220 nm). The peptides were analyzed by MALDI-TOF MS. His–LctA(1–37)MES eluted on a preparative RP-HPLC C4 column (2.2 cm × 25 cm) at 24.8 to 27.0 min (MALDI-TOF MS calculated: 6181 (M); found: 6180). Note: The MES thioesters LctA(1–37)MES hydrolyze readily. Therefore, after the elution of the truncated His–LctA mutants with cleavage buffer, the eluate was acidified with 0.1% TFA to pH of approximately 6.5 and concentrated by Amicon immediately.

2.4. General procedure for ligation of LctA(1-37)MES with short peptides

Synthetic peptides (3 to 5 mg) that were previously purified by C18 RP-HPLC were dissolved in ligation buffer (160 μL) containing 100 mM HEPES, 200 mM NaCl, 50 mM MESNA, pH 7.75 at 4 °C. For hydrophobic peptides, precipitation was sometimes observed. Lowering the peptide concentration or addition of DMSO lead to greater solubility. In cases where the cysteine and cysteine analogues were present as tert-butyl disulfides, three equivalents of TCEP per equivalent of the amino acid were added and the mixture incubated at 25 °C for 30 min. Then a solution of His_6–LctA(1–37)MES thioester (≈0.5 mg) in ligation buffer (60 μL) was added to the peptide solution to obtain a final concentration of 0.5 ≈ 1.0 mM of the thioester. The pH of the resulting mixture was adjusted to 7.6 ≈ 7.8 with 1 N NaOH, and the ligation reaction was allowed to proceed for 12 to 16 h at 4 °C with gentle shaking. For the ligation of homocysteine (16), the peptoid analogue of Cys (17), and (R)-3-methylcysteine (18) containing peptides, 24 h were needed to complete the ligation.

The crude ligation products were analyzed by MALDI-TOF MS for completeness of the reaction before acidification with 5% TFA. For MS analysis, 5 μL of the ligation buffer mixture was mixed with 9 μL of sinapic acid matrix (prepared in 30% CH_3CN, containing 0.1 TFA), and 5 μL of the mixture was spotted on a MALDI target plate. The acidified ligation mixture was incubated with TCEP ($\approx$5 mg) for 30 min at 25 °C to reduce any intermolecular disulfides before purification by C4 analytical RP-HPLC.

2.5. Investigation of the dehydration reaction

The LctA analogues corresponding to LctA1–44 were incubated with LctM in the presence of ATP and Mg^{2+}, which the enzyme uses for phosphorylation of the hydroxyl groups of Ser and Thr that undergo elimination (Chatterjee et al., 2005a). The assays with the LctA analogues were then analyzed by MALDI-TOF mass spectrometry. L-Threonine (**1**) incorporated at position 42 of the truncated LctA peptide as a control (LctA1–43-S42T) was cleanly converted to a 3-fold dehydrated product, resulting from dehydration of Thr33, Ser35, and Thr42. Substitution of Ser42 with (R)-3-ethylserine (**2**), (R)-3-vinylserine (**3**), and (R)-3-ethynylserine (**4**) also led to clean 3-fold dehydration, demonstrating that LctM can tolerate substituents larger than methyl at the β-carbon. However, both propyl and isopropyl groups proved too large for the enzyme, as the LctA analogues with (R)-3-propylserine (**5**) and (R)-3-isopropylserine (**6**) at position 42 resulted in just two dehydrations after incubation with LctM. However, a three-carbon substituent that is less sterically demanding, such as the propynyl group in **7**, was accepted by the enzyme. Similarly, a difluoromethyl group, which is larger than methyl but has smaller dimensions than isopropyl, was tolerated by the enzyme, as analogue **8** was dehydrated when incorporated at position 42. The dehydration of **8** resulted in a difluorodehydrobutyrine (Fig. 6.5) that, as anticipated, proved very reactive, ultimately resulting in nonenzymatic release of both fluorides and cleavage of the peptide. The restrictions of the active side pocket with respect to the substituent at the β-carbon of Thr analogues is also shown with amino acids **9** and **10**. Whereas the substrate peptide incorporating the E-alkene **9** was dehydrated, the Z-alkene **10** was not.

Dehydration of Thr results in Z-dehydrobutyrine in all lantibiotics investigated to date indicating an *anti* elimination mechanism (Chatterjee et al., 2005b). Substitution of Ser42 with *allo*-Thr (**11**), if tolerated by the enzyme, would result in formation of an E-dehydrobutyrine. However, *allo*-Thr was not dehydrated when incorporated at position 42, demonstrating that (R)-stereochemistry at the β-carbon is essential. Another change in substrate structure that was not tolerated is the use of (S)-β^2-homoserine (**12**), which still resulted in phosphorylation by LctM, but the enzyme-catalyzed elimination was prohibited.

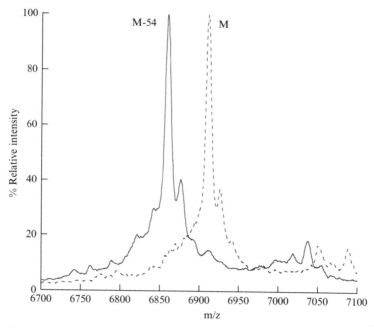

Figure 6.5 MALDI-MS spectrum of the assay product formed upon incubation of an LctA analogue containing difluoromethylthreonine 8 (dashed line) with LctM for 7 min at pH 7.0. A 3-fold dehydrated product was produced (solid line), resulting from dehydration of Thr33, Ser35, and the difluorothreonine analogue at position 42.

2.6. General procedure for LctM dehydration assays with truncated LctA analogues

The truncated substrate (5 to 10 μg) was dissolved in sterile Millipore water (20 μL) and 5 μL of solution was used for the assay. The peptide solution was incubated with buffer (2 μL, 500 mM Tris.HCl, 100 mM MgCl$_2$, 250 μg/mL, pH 7.5, 25 °C), 100 mM ATP (1 μL), and sterile Millipore water (8 μL). Heterologously expressed and purified His–tagged LctM (Xie et al., 2004) (1.0 mg/mL, 4 μL) was added to the buffered peptide solution. The assay mixture was incubated at 25 °C for 5 to 12 h. Assay products were analyzed by MALDI-TOF MS in one of two ways. In the first method, 1 μL of the assay mixture was mixed with 9 μL of sinapic acid matrix (prepared in 30% CH$_3$CN, containing 0.1% TFA) and 5 μL was spotted directly on the target plate. A second procedure involved acidification of the assay mixture with 5 μL of 5% TFA (final pH $\approx$2). The acidified solution was then loaded onto a C18 zip tip, and the assay product eluted with 8 μL of α-hydroxycinnamic acid matrix (prepared in 50% CH$_3$CN, containing 0.1% TFA) of which 5 μL was spotted on the MALDI plate.

2.7. Procedure for LctM-catalyzed dehydration of LctA(1–43) with difluoromethylthreonine (8) at position 42

The truncated substrate (5 to 10 μg) was dissolved in sterile Millipore water (20 μL) and 5 μL of solution was used for the assay. Buffer (2 μL, 500 mM Tris.HCl, 100 mM MgCl$_2$, 250 μg/mL, pH 7.5, 25 °C), 100 mM ATP (1 μL), and sterile Millipore water (8 μL) were added to the peptide solution. The pH of the resulting mixture was adjusted to 7.0 with 1 N NaOH. Heterologously expressed and purified His-tagged LctM (Xie *et al.*, 2004) (1.0 mg/mL, 20 μL) was added to the buffered peptide solution. The assay mixture was incubated at 25 °C for 7 min and quenched immediately with 5% TFA to pH $\approx$2. The acidified solution (5 μL) was then loaded onto a C18 zip tip, and the assay product eluted with 8 μL of α-hydroxycinnamic acid matrix (prepared in 50% CH$_3$CN, containing 0.1% TFA) of which 5 μL was spotted on the MALDI plate.

2.8. Investigation of the cyclization reaction

The relaxed substrate specificity discussed for the dehydration reaction was also observed in an early study on the cyclization activity of LctM, as several Cys analogues were converted to the corresponding thioether rings (Chatterjee *et al.*, 2006). However, those experiments analyzed for the formation of a single ring and could not distinguish between enzymatic and nonenzymatic cyclization. Furthermore, a more stringent test of the substrate specificity would not only focus on the ability to achieve cyclization but also to control the regio- and chemoselectivity of thioether ring formation. A more recent assay addresses these issues and relies on a truncated analogue **19** of the LctA substrate for lacticin 481 synthetase (Fig. 6.6). After LctM dehydrates this substrate, it contains two dehydro amino acids, a Dhb at position 33 and a Dha at position 42, as well as one Cys (or Cys analogue) at residue 38.[1] On the basis of model studies, we anticipated that after dehydration of this substrate, any nonenzymatic cyclization would occur between Cys38 and Dha42 to give product **21** given the much higher reactivity of Dha compared to Dhb (Okeley *et al.*, 2000; Zhou and van der Donk, 2002; Zhu *et al.*, 2003). In contrast, the enzyme was expected to overcome the inherent kinetic bias against formation of a methyllanthionine and catalyze the addition of Cys38 to Dhb33 to provide peptide **20**, containing the A-ring of lacticin 481 (see Fig. 6.2). The two reaction products were distinguished by treatment with cyanogen bromide.

---

[1] Note that the Ser present at position 35 in wild-type LctA (see Fig. 6.2) was mutated to Ala in this substrate, such that the peptide after dehydration would contain just one Dhb and one Dha resulting from dehydration of Thr33 and Ser42, respectively. The removal of Ser35 prevents the formation of another Dha, which would further complicate analysis of the cyclization reaction. See also Zhang and van der Donk (2007) and Paul *et al.* (2007).

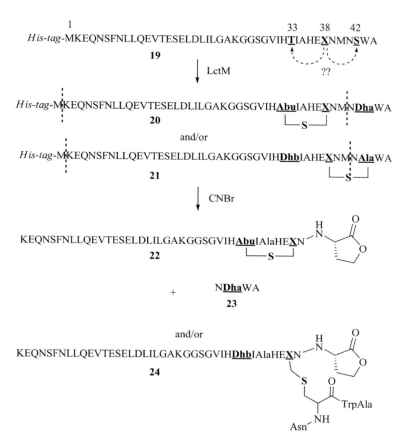

Figure 6.6   LctM and LctM–C781A assays with truncated LctA analogues and analysis of the assay products with CNBr. After incubation of truncated LctA analogue **19** with LctM, the enzymatic and nonenzymatic products **20** and **21** were distinguished by the treatment with cyanogen bromide.

This reagent fragments peptides C-terminal to Met residues, converting the Met into a homoserine (Hse) lactone. In the case of peptides **20** and **21**, CNBr treatment resulted in cleavage of the amide bonds between Met1 and Lys2 as well as between Met40 and Asn41 (dashed lines, Fig. 6.6). For product **20**, in which Cys38 is engaged in a methyllanthionine linkage to residue 33, the cleavage resulted in two peptide products, **22** and **23**, in addition to a fragment originating from the His-tag. In contrast, in peptide **21**, in which Cys38 is linked to residue 42 through a lanthionine linkage, CNBr treatment would result in peptide **24** in addition to the His-tag peptide fragment. The sulfur atom of the (methyl)lanthionines can also react with CNBr (Mckendrick *et al.*, 1998), but products expected from this reaction were not observed.

When this assay was employed with Cys at position 38, indeed the major observed product was peptide **22**, whereas use of a LctM mutant that retains dehydration activity but has no cyclization activity above background (Cys781Ala) (Paul *et al.*, 2007) predominantly resulted in peptide **24** (Zhang *et al.*, 2007). Similar results were obtained with peptides containing D-Cys (**14**) and β^3-homocysteine (**15**) at position 38 of LctA (Fig. 6.7A), indicating that they are enzymatically cyclized by LctM. Homocysteine (**16**) (Fig. 6.7B) and the peptoid analogue of Cys (**17**), however, resulted in a mixture of enzymatically and nonenzymatically cyclized product, and (*R*)-3-methylcysteine (**18**) produced only nonenzymatically cyclized products (Fig. 6.7C). In all cases, control reactions with the cyclization deficient mutant LctM-C781A resulted in the formation of nonenzymatically cyclized product **24**.

The data demonstrate that the enzyme can be used to produce the analogues shown in Fig. 6.8 of the naturally occurring isomer of (methyl) lanthionine. Although the stereochemistry at carbons 2 and 3 of the resulting thioethers could not be rigorously determined in this work, the fact that cyclization is enzyme catalyzed combined with the known stereoselectivity of the Michael-type addition catalyzed by LctM strongly suggests that the stereochemistry of the products is (2*S*,3*S*).

2.9. General procedure for LctM and LctM-C781A assays with truncated LctA analogues and analysis of the assay products with CNBr

The truncated substrate (5 to 10 μg) was dissolved in sterile Millipore water (20 μL) and 5 μL of solution was used for the assay. The peptide solution was incubated with buffer (2 μL, 500 mM Tris.HCl, 100 mM MgCl$_2$, 250 μg/mL, pH 7.5, 25 °C), 100 mM ATP (1 μL), and sterile Millipore water (8 μL). His$_6$-LctM or His$_6$-LctM-C781A (1.0 mg/mL, 4 μL) was added to the buffered peptide solution. The assay mixture was incubated at 25 °C for 2 to 20 h. Assay products were analyzed by MALDI TOF MS in one of two ways. In the first method, 1 μL of the assay mixture was mixed with 9 μL of sinapic acid matrix (prepared in 30% CH$_3$CN, containing 0.1% TFA) and 5 μL was spotted directly on the target plate. A second procedure involved acidification of the assay mixture with 5 μL of 5% TFA. The acidified solution was then loaded onto a C18 zip tip, and the assay product was eluted with 8 μL of α-hydroxycinnamic acid matrix (prepared in 50% CH$_3$CN, containing 3% TFA), of which 5 μL was spotted on the MALDI plate.

The product (15 μL) resulting from the assay of truncated His-LctA analogues with LctM or LctM-C781A was purified using a zip tip eluting

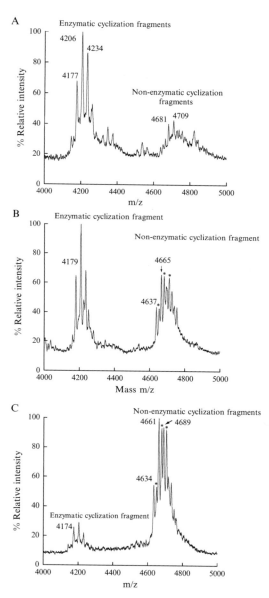

Figure 6.7   MALDI mass spectra of peptides 22 and 24 following the protocol depicted in Fig. 6.6. (A) X = β^3-Hcys (**15**), (B) X = Hcys (**16**), (C) X = β-methylCys (**18**). In addition to the expected masses, peptides corresponding to formylated products (M + 28 and M + 56) are also observed because of the incubation in formic acid during the CNBr treatment (Zhang *et al.*, 2007). Peaks labeled with asterisks are oxidation products (M + 16).

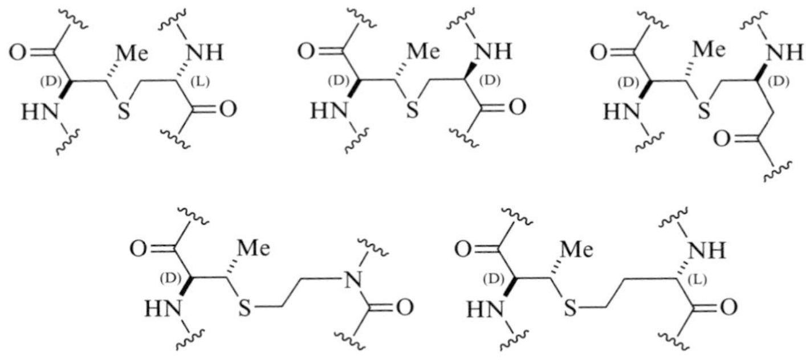

Figure 6.8 The naturally occurring (2*S*,3*S*)-(methyl)lanthionine and analogues produced from thiol containing amino acids **14** to **17**.

with 10-15 μL of 50% CH_3CN, 40% H_2O, 10% aqueous (3%) TFA. After the concentration using a speed-vac, the resulting peptide was treated with CNBr (10 μL, 50 mg/mL in 70% HCOOH) in the dark overnight at room temperature. The reaction mixture was concentrated again using a speed-vac and 10 μL of water was added. Then 5 μL of the solution was placed into an Eppendorf tube, followed by 5 μL of matrix (α-cyano-4-hydroxy-cinnamic acid supersolution in 50% CH_3CN, 40% H_2O, 10% aqueous (3%) TFA). Then the sample was monitored by MALDI-TOF MS.

3. CONCLUSION

A series of modified LctA substrates containing the amino acids **1** to **18** were prepared using a combination of SPPS and expressed protein ligation technology. Several Cys analogues proved efficient partners in the ligation reaction. Use of the semisynthetic peptides in dehydration assays with lacticin 481 synthetase demonstrated that analogues **2** to **4** and **7** to **9** were dehydrated by the enzyme whereas residues **5**, **6**, and **10** through **12** were not. Furthermore, cyclization assays showed that the enzyme regioselectively catalyzed the cyclization of peptides containing the Cys analogues **14** and **15**, but that analogues **16** and **17** resulted in only partial enzyme catalyzed cyclization; Cys analogue **18** was not cyclized by the enzyme. Thus, the ability to alter the structure of the substrate peptide by expressed protein ligation beyond the limited set of substitutions accessible through standard site-directed mutagenesis allowed a more in depth analysis of the substrate scope of the posttranslational modification catalyzed by lacticin 481 synthetase. These techniques are applicable to reengineering the structure of lantibiotics and should also be applicable to other posttranslational modification processes that result in natural products with interesting biological activities such as the microcins and patellamides.

ACKNOWLEDGMENTS

This work was supported by the National Institutes of Health (GM58822).

REFERENCES

Chatterjee, C., Miller, L. M., Leung, Y. L., Xie, L., Yi, M., Kelleher, N. L., and van der Donk, W. A. (2005a). Lacticin 481 Synthetase Phosphorylates its Substrate during Lantibiotic Production. *J. Am. Chem. Soc.* **127,** 15332–15333.

Chatterjee, C., Patton, G. C., Cooper, L., Paul, M., and van der Donk, W. A. (2006). Engineering Dehydro Amino Acids and Thioethers into Peptides using Lacticin 481 Synthetase. *Chem. Biol.* **13,** 1109–1117.

Chatterjee, C., Paul, M., Xie, L., and van der Donk, W. A. (2005b). Biosynthesis and Mode of Action of Lantibiotics. *Chem. Rev.* **105,** 633–684.

Cotter, P. D., Hill, C., and Ross, R. P. (2005). Bacterial lantibiotics: Strategies to improve therapeutic potential. *Curr. Protein Pept. Sci.* **6,** 61–75.

Dawson, P. E., and Kent, S. B. H. (2000). Synthesis of native proteins by chemical ligation. *Ann. Rev. Biochem.* **69,** 923–960.

Dawson, P. E., Muir, T. W., Clark-Lewis, I., and Kent, S. B. (1994). Synthesis of proteins by native chemical ligation. *Science* **266,** 776–779.

Donia, M. S., Hathaway, B. J., Sudek, S., Haygood, M. G., Rosovitz, M. J., Jacques Ravel, J., and Schmidt, E. W. (2006). Natural combinatorial peptide libraries in cyanobacterial symbionts of marine ascidians. *Nat. Chem. Biol.* **2,** 729–735.

Duquesne, S., Destoumieux-Garzon, D., Zirah, S., Goulard, C., Peduzzi, J., and Rebuffat, S. (2007). Two Enzymes Catalyze the Maturation of a Lasso Peptide in *Escherichia coli*. *Chem. Biol.* **14,** 793–803.

Evans, T. C., Benner, J., and Xu, M. Q. (1998). Semisynthesis of Cytotoxic Proteins Using a Modified Protein Splicing Element. *Protein Science* **7,** 2256–2264.

Han, Y., Albericio, F., and Barany, G. (1997). Occurrence and Minimization of Cysteine Racemization during Stepwise Solid-Phase Peptide Synthesis. *J. Org. Chem.* **62,** 4307–4312.

Kazakov, T., Metlitskaya, A., and Severinov, K. (2007). Amino acid residues required for maturation, cell uptake, and processing of translation inhibitor microcin C. *J. Bacteriol.* **189,** 2114–2118.

Kluskens, L. D., Kuipers, A., Rink, R., de Boef, E., Fekken, S., Driessen, A. J. M., Kuipers, O. P., and Moll, G. N. (2005). Post-translational Modification of Therapeutic Peptides By NisB, the Dehydratase of the Lantibiotic Nisin. *Biochemistry* **44,** 12827–12834.

Kuipers, O. P., Bierbaum, G., Ottenwälder, B., Dodd, H. M., Horn, N., Metzger, J., Kupke, T., Gnau, V., Bongers, R., van den Bogaard, P., Kosters, H., Rollema, H. S., *et al.* (1996). Protein engineering of lantibiotics. *Antonie van Leeuwenhoek* **69,** 161–170.

Levengood, M. R., and van der Donk, W. A. (2006). Dehydroalanine-containing peptides: Preparation from phenylselenocysteine and utility in convergent ligation strategies. *Nat. Protoc.* **1,** 3001–3010.

Li, B., Yu, J.-P. J., Brunzelle, J. S., Moll, G. N., van der Donk, W. A., and Nair, S. K. (2006). Structure and Mechanism of the Lantibiotic Cyclase Involved in Nisin Biosynthesis. *Science* **311,** 1464–1467.

Li, G., and van der Donk, W. A. (2007). Efficient synthesis of suitably protected beta-difluoroalanine and gamma-difluorothreonine from L-ascorbic acid. *Org. Lett.* **9,** 41–44.

Lubelski, J., Rink, R., Khusainov, R., Moll, G. N., and Kuipers, O. P. (2008). Biosynthesis, immunity, regulation, mode of action and engineering of the model lantibiotic nisin. *Cell Mol. Life Sci.* **65,** 455–476.

McClerren, A., Cooper, L. E., Quan, C., Thomas, P. M., Kelleher, N. L., and van der Donk, W. A. (2006). Discovery and *in vitro* biosynthesis of haloduracin, a new two-component lantibiotic. *Proc. Nat. Acad. Sci. USA* **103,** 17243–17248.

Mckendrick, J. E., Frormann, S., Luo, C., Semchuck, P., Vederas, J. C., and Malcolm, B. A. (1998). Rapid mass spectrometric determination of preferred irreversible proteinase inhibitors in combinatorial libraries. *Int. J. Mass. Spectrom.* **176,** 113.

Milne, B. F., Long, P. F., Starcevic, A., Hranueli, D., and Jaspars, M. (2006). Spontaneity in the patellamide biosynthetic pathway. *Org. Biomol. Chem.* **4,** 631–638.

Muir, T. W., Sondhi, D., and Cole, P. A. (1998). Expressed protein ligation: A general method for protein engineering. *Proc. Natl. Acad. Sci. USA* **95,** 6705–6710.

Okeley, N. M., Zhu, Y., and van der Donk, W. A. (2000). Facile Chemoselective Synthesis of Dehydroalanine-Containing Peptides. *Org. Lett.* **2,** 3603–3606.

Paul, M., Patton, G. C., and van der Donk, W. A. (2007). Mutants of the Zinc Ligands of Lacticin 481 Synthetase Retain Dehydration Activity but Have Impaired Cyclization Activity. *Biochemistry* **46,** 6268–6276.

Roy, R. S., Allen, O., and Walsh, C. T. (1999a). Expressed protein ligation to probe regiospecificity of heterocyclization in the peptide antibiotic microcin B17. *Chem. Biol.* **6,** 789–799.

Roy, R. S., Gehring, A. M., Milne, J. C., Belshaw, P. J., and Walsh, C. T. (1999b). Thiazole and oxazole peptides: biosynthesis and molecular machinery. *Nat. Prod. Rep.* **16,** 249–263.

Severinov, K., Semenova, E., Kazakov, A., Kazakov, T., and Gelfand, M. S. (2007). Low-molecular-weight post-translationally modified microcins. *Mol. Microbiol.* **65,** 1380–1394.

Telenti, A., Southworth, M., Alcaide, F., Daugelat, S., Jacobs, W. R. Jr., and Perler, F. B. (1997). The Mycobacterium xenopi GyrA protein splicing element: Characterization of a minimal intein. *J. Bacteriol.* **179,** 6378–6382.

Vassiliadis, G., Peduzzi, J., Zirah, S., Thomas, X., Rebuffat, S., and Destoumieux-Garzon, D. (2007). Insight into siderophore-peptide biosynthesis: Enterobactin is a precursor for microcin E492 post-translational modification. *Antimicrob. Agents Chemother.* **51,** 3546–3553.

Wang, L., Brock, A., Herberich, B., and Schultz, P. G. (2001). Expanding the genetic code of Escherichia coli. *Science* **292,** 498–500.

Wang, L., and Schultz, P. G. (2002). Expanding the genetic code. *Chem. Commun.* 1–11.

Willey, J. M., and van der Donk, W. A. (2007). Lantibiotics: Peptides of Diverse Structure and Function. *Annu. Rev. Microbiol.* **61,** 477–501.

Xie, L., Miller, L. M., Chatterjee, C., Averin, O., Kelleher, N. L., and van der Donk, W. A. (2004). Lacticin 481: *In vitro* reconstitution of lantibiotic synthetase activity. *Science* **303,** 679–681.

Zhang, X., Ni, W., and van der Donk, W. A. (2005). Synthesis of nonproteinogenic amino acids to probe lantibiotic biosynthesis. *J. Org. Chem.* **70,** 6685–6692.

Zhang, X., Ni, W., and van der Donk, W. A. (2007). On the Regioselectivity of Thioether Formation by Lacticin 481 Synthetase. *Org. Lett.* **9,** 3343–3346.

Zhang, X., and van der Donk, W. A. (2007). On the substrate specificity of the dehydratase activity of lacticin 481 synthetase. *J. Am. Chem. Soc.* **129,** 2212–2213.

Zhou, H., and van der Donk, W. A. (2002). Biomimetic Stereoselective Formation of Methyllanthionine. *Org. Lett.* **4,** 1335–1338.

Zhu, Y., Gieselman, M., Zhou, H., Averin, O., and van der Donk, W. A. (2003). Biomimetic studies on the mechanism of stereoselective lanthionine formation. *Org. Biomol. Chem.* **1,** 3304–3315.

CHAPTER SEVEN

SEMISYNTHESIS OF K⁺ CHANNELS

Alexander G. Komarov,* Kellie M. Linn,* Jordan J. Devereaux,*
and Francis I. Valiyaveetil*,[1]

Contents

Abstract

The ability to selectively conduct K⁺ ions is central to the function of K⁺ channels. Selection for K⁺ and rejection of Na⁺ takes place in a conserved structural element referred to as the selectivity filter. The selectivity filter consists of four K⁺-specific ion binding sites that are created using predominantly the backbone carbonyl oxygen atoms. Due to the involvement of the protein backbone, experimental manipulation of the ion binding sites in the selectivity filter is not possible using traditional site directed mutagenesis. The limited suitability of the site-directed mutagenesis for studies on the selectivity filter

* Program in Chemical Biology, Department of Physiology and Pharmacology, Oregon Health and Sciences University, Portland, Oregon, USA
[1] Corresponding author: valiyave@ohsu.edu

Methods in Enzymology, Volume 462
ISSN 0076-6879, DOI: 10.1016/S0076-6879(09)62007-3

has motivated the development of a semisynthesis approach, which enables the use of chemical synthesis to manipulate the selectivity filter. In this chapter, we describe the protocols that are presently used in our laboratory for the semi-synthesis of the bacterial K^+ channel, KcsA. We show the introduction of a spectroscopic probe into the KcsA channel using semisynthesis. We also review previous applications of semisynthesis in investigations of K^+ channels. While the protocols described in this chapter are for the KcsA K^+ channel, we antici-pate that similar protocols will also be applicable for the semisynthesis of other integral membrane proteins.

1. INTRODUCTION

K^+ channels are ubiquitous, tetrameric, integral membrane proteins that catalyze the selective permeation of K^+ ions across biological mem-branes (Hille, 2001). Key aspects of ion permeation through a K^+ channel are high selectivity for K^+ over Na^+ and ion-permeation rates that are close to the diffusion limit (Hille, 2001). The ion conduction pathway through a K^+ channel lies within the pore domain. The structure of the K^+ channel pore was first determined when the crystal structure of the bacterial K^+ channel KcsA was solved (Fig. 7.1A) (Doyle *et al.*, 1998; Zhou *et al.*, 2001). The K^+ channel pore is formed by four subunits, symmetrically arranged around a central axis that coincides with the ion-translocation pathway. Presently, structures are available for a number of K^+ channels, including the eukaryotic voltage gated K^+ channel, $K_v 1.2$ and all these structures

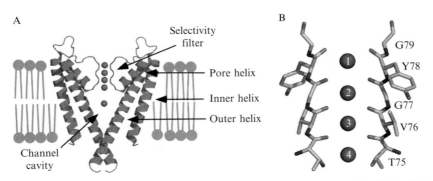

Figure 7.1 The K^+ channel pore. (A) The KcsA channel is shown (pdb:1k4c). Only two opposite subunits of the tetrameric protein are shown. The selectivity filter, residues 75 to 79, is colored red. K^+ ions in the channel are depicted as magenta spheres. (B) Close-up view of the selectivity filter in stick representation with the front and the back subunits removed. The four K^+-binding sites in the selectivity filter are shown and the ion-binding sites are numbered 1 through 4. (See Color Insert.)

indicate a similar fold for the pore domain of K^+ channels (Jiang et al., 2002, 2003; Kuo et al., 2003; Long et al., 2005).

The narrowest portion of the ion-conduction path through the pore domain is referred to as the selectivity filter. In the selectivity filter, a row of four K^+ binding sites is created using four sequential, main chain carbonyl oxygen atoms (from the sequence T-V-G-Y-G) and the side-chain hydroxyl oxygen atom of the threonine residue (Fig. 7.1B) (Doyle et al., 1998; Zhou et al., 2001). The stretch of amino acids that forms the selectivity filter is highly conserved in K^+ channels and is referred to as the signature sequence (Doyle et al., 1998). The high degree of sequence conservation predicts a similar structure for the selectivity filter of all K^+ channels (Shealy et al., 2003). This is indeed observed in the structures of the K^+ channels presently available (Jiang et al., 2002, 2003; Kuo et al., 2003; Long et al., 2005; Zhou et al., 2001).

Investigating the molecular basis of selectivity filter function requires the ability to modify the ion-binding sites in the selectivity filter (Muralidharan and Muir, 2006). Protein modification is generally accomplished by traditional site-directed mutagenesis (SDM). However, SDM is limited to the set of natural amino acids, which limits the kind of changes that can be introduced. Further, the changes that can be introduced are confined to the amino acid side chain. Therefore, SDM is of limited utility in investigating the ion-binding sites in the K^+ selectivity filter as they are formed mainly by the backbone atoms. This limitation of SDM was the motivation behind the development of a chemical synthesis for the K^+ channel, KcsA. Chemical synthesis, in contrast to SDM, gives us an almost limitless variety of modifications that can be introduced and, importantly, allows us to manipulate the protein backbone (Kent, 1988; Dawson and Kent, 2000).

Incorporation of unnatural amino acids and certain peptide backbone modifications into proteins can also be achieved using the nonsense suppression mutagenesis approach (Beene et al., 2003; Mendel et al., 1995). Recent advances in this approach now enable the in vivo incorporation of the unnatural amino acid (Wang and Schultz, 2004). Although the nonsense suppression mutagenesis approach is very powerful, a distinct advantage of using chemical synthesis for protein modification is that chemical synthesis provides greater freedom in the kinds of amino acids and peptide backbone modifications that we can incorporate. For example, chemical synthesis can be used to incorporate D-amino acids or to substitute amide bonds with olefins (Fu et al., 2005; Valiyaveetil et al., 2004).

The size of the protein has been the major factor that has limited the use of chemical synthesis to investigate proteins. Protein size is a limitation, as chemical synthesis is carried out using solid phase peptide synthesis (SPPS), which is limited to the synthesis of peptides around 50 to 60 amino acids in length. The synthesis of peptides longer than approximately 60 amino acids is not efficient and in most cases results in very low yields (Kent, 1988). One of the important advances in the field of protein synthesis was the development of the native chemical ligation reaction (NCL) (Dawson and Kent, 2000; Dawson *et al.*, 1994; Muir, 2003). NCL involves a reaction between a peptide with a C-terminal thioester and a peptide with an N-terminal Cys (Fig. 7.2). The NCL reaction links the peptides together with a native peptide bond at the ligation site. Therefore, we can synthesize a protein from a number of suitably sized peptides by using the NCL reaction to link the peptides together to obtain the desired protein. In the field of membrane proteins, chemical synthesis has so far been reported for the HCV protease cofactor protein NS4A, the influenza M2 channel, and the bacterial mechanosensitive channel MscL by taking advantage of the NCL reaction (Bianchi *et al.*, 1999; Clayton *et al.*, 2004; Kochendoerfer *et al.*, 1999).

The NCL reaction can also be used for protein semisynthesis in which the protein is assembled from a synthetic peptide and protein segment or segments obtained by recombinant means (Muir *et al.*, 1998). This application of the NCL reaction, referred to as expressed protein ligation (EPL), has greatly extended the size limits of proteins that can be modified using chemical synthesis (Muir, 2003). We have used the EPL approach for the semisynthesis of the KcsA K^+ channel.

In this chapter, we describe the protocols that are used in our laboratory for the semisynthesis of the KcsA K^+ channel. As an application of the

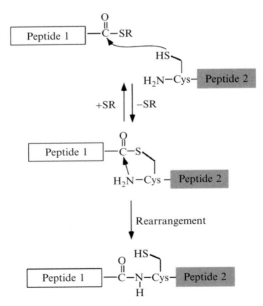

Figure 7.2 The NCL reaction. The mechanism of the NCL reaction involves a transthioesterification followed by intramolecular rearrangement, resulting in linking the peptides via a normal peptide bond.

protocols, we describe the semisynthesis of a KcsA channel with a substitution of Phe103 with *para*-fluoro-phenylalanine (pF–Phe). The Phe103 residue lines the cavity region of the KcsA pore, which contains a binding site for ions and a number of blockers that target K⁺ channels (Fig. 7.1A) (Yohannan *et al.*, 2007; Zhou and MacKinnon, 2004). Substitution of Phe103 with pF–Phe will provide a spectroscopic probe in the cavity region to investigate the environment in the cavity region and to monitor the binding of blockers to the channel.

2. EXPERIMENTAL PROTOCOLS

2.1. Synthetic design

Each subunit of the tetrameric KcsA channel is 160 amino acids long. However, structural studies on the KcsA channel have been carried out using a truncated form of the channel lacking the C-terminal 35 amino acids (Doyle *et al.*, 1998). We have selected this truncated form of the KcsA channel for semisynthesis because its smaller size (125 amino acids compared to 160 amino acids) makes it an easier synthetic target. The synthesis consists of two steps: the assembly of the KcsA polypeptide and the *in vitro* folding of the KcsA polypeptide to the native state (Fig. 7.3). The 125 amino acid long KcsA polypeptide can be synthesized from two component peptides by a single ligation reaction. We have previously identified residue 70 as a position at which a Cys substitution is well tolerated and was therefore selected as the ligation site (Valiyaveetil *et al.*, 2006). The synthesis of the KcsA polypeptide therefore requires a peptide thioester corresponding to residues 1 to 69 (which we refer to as the N-peptide) and a peptide with an N-terminal Cys corresponding to residues 70 to 125 (which we refer to as the C-peptide). The regions of interest in the KcsA channel, such as the ion-binding sites in the selectivity filter (residues 75 to 79), or the cavity region (residues 100 to 108) are contained within the C-peptide. Therefore, to use chemical synthesis for the modification of these regions of interest, we use solid-phase peptide synthesis (SPPS) for the C-peptide. The N-peptide thioester is obtained by recombinant means.

The truncated form of the KcsA channel is, however, not functional in single channel measurements. A mutational analysis of the KcsA channel identified an A98G substitution that increased the open probability of the KcsA channel and also allowed functional measurements to be carried out on the truncated form of the channel (Valiyaveetil *et al.*, 2004). The A98G substitution is therefore incorporated into the semisynthetic channels to enable electrophysiological measurements. We also routinely incorporate the amino acid substitutions (Q58A, T61S, R64D) that render the KcsA channel sensitive to the peptide toxin, agitoxin₂ (AgTx₂) (MacKinnon *et al.*,

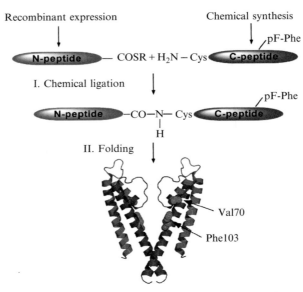

Figure 7.3 Semisynthesis of KcsA. The KcsA polypeptide residues (1 to 123) is obtained by the expressed protein ligation of a recombinantly expressed N-peptide thioester (1 to 69) and a chemically synthesized peptide (70 to 123) with an N-terminal cysteine. The KcsA polypeptide obtained by the ligation reaction is folded to the native tetrameric state. Two subunits of the KcsA tetramer are shown. Val70, the ligation site, and Phe103, which is substituted with pF-Phe is depicted in space fill. (See Color Insert.)

1998). This is useful as we can then use block by $AgTx_2$ to confirm that the electrical activity observed is due to ion movement through the KcsA channel.

The second step in the synthesis is the *in vitro* folding of the protein to the native state. *In vitro* folding is necessary because the synthesis steps only provide the unfolded polypeptide, which has to be folded to the native state for further characterization. The KcsA channel can be folded *in vitro* from the unfolded state to the native state using lipid vesicles (Valiyaveetil *et al.*, 2002). We use this lipid based folding protocol to convert the semisynthetic KcsA polypeptide to the native state for functional measurements and structural studies.

2.2. Synthesis of the KcsA 70–123 (pF-Phe103) C-peptide

The 54-amino-acid C-peptide (CE**TATTV**GYGDLYP**VT**LWGRL-**VAVVVM**V**AGIT**SFGL**VT**AALA**T**WFVGREQERRG) is assembled by manual solid-phase peptide synthesis using a slightly modified version of *in situ* neutralization/2-(1H-benzo-triazol-1-yl-1,1,3,3-tetramethyuronium hexafluorophosphate (HBTU) activation (Schnolzer *et al.*, 1992).

The C-peptide sequence consists of 17 β-branched amino acids (bold), which show slow and incomplete couplings (Kent, 1988). As a precaution, the residues are double coupled using HBTU in DMSO. Non-β-branched amino acid residues that are present in sections of the sequence containing stretches of β-branched amino acids (underlined) are also double coupled using HBTU in DMSO. The Phe103→p-fluorophenylalanine (pF-Phe) substitution is introduced during the peptide synthesis by coupling pF-Phe instead of Phe at position 103. Following chain assembly, the cleavage of the peptide from the resin and side-chain deprotection is carried out using anhydrous HF. The crude peptide obtained is dissolved in 50% trifluoroethanol (TFE): buffer A (H_2O + 0.1% trifluoroacetic acid, TFA) and purified by RP-HPLC on a linear 60 to 100% buffer B (90% acetonitrile, 10% H_2O, 0.1% TFA) gradient on a C4 column (Fig. 7.4). Fractions containing the purified peptide are identified by ES-MS, pooled, and lyophilized.

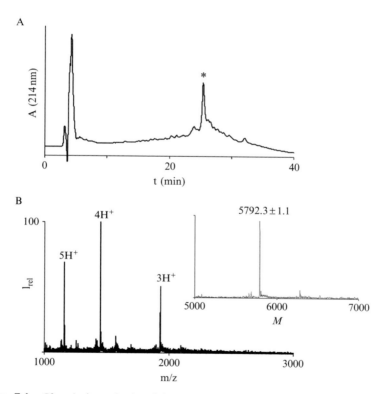

Figure 7.4 Chemical synthesis of the C-peptide. (A) Analytical RP-HPLC of the crude obtained after the HF cleavage reaction (gradient 50 to 100% Buffer B on a C4 column). The C-peptide is marked with an asterisk. (B) ES-MS of the C-peptide (inset, reconstructed spectrum, expected mass = 5793.6).

2.3. Generation of the N-peptide thioester

We have developed a sandwich fusion approach for recombinant expression of the KcsA 1-69 N-peptide thioester (Valiyaveetil *et al.*, 2002). In the fusion construct, residues 1 to 69 of the KcsA channel are sandwiched between glutathione-S-transferase (GST, from the plasmid pGEX4T-2, available from <u>GE Healthcare</u>) at the N-terminus and gyrA intein (from the plasmid pTXB3, available from New England Biolabs) at the C-terminus (Fig. 7.5). In the fusion construct, the gyrA intein is used to generate a thioester at the C-terminus of the N-peptide (Blaschke *et al.*, 2000). The expression of a membrane spanning peptide fused to an intein is toxic to the cells and subsequently results in very low protein expression (Valiyaveetil *et al.*, 2002). To circumvent this problem, we direct expression of the fusion protein to inclusion bodies. This is accomplished by appending GST at the N-terminus of the fusion protein. A thrombin site followed by a His_6 linker and a factor X_a site is introduced between the GST tag and the N-terminus of the N-peptide.

Expression of the fusion construct is carried out in *E. coli* BL21 cells. Growth of cells for protein production is carried out in LB medium. Cells are induced at a cell density of 1.0 O.D. unit (600 nm) with 1 m*M* IPTG. Following induction, cell growth is carried out for 2 h at 37 °C with vigorous shaking. Expression of the fusion protein is confirmed by SDS-PAGE of a small aliquot of the cells before and after induction (Fig. 7.6A).

The expression of the fusion protein as inclusion bodies considerably simplifies its purification. For isolation of the inclusion bodies, the cells are

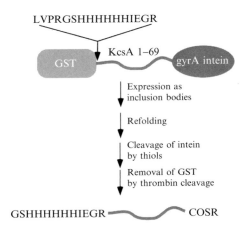

Figure 7.5 Expression of membrane spanning thioester peptides. The sandwich fusion strategy used for the overexpression of the N-peptide thioester is shown.

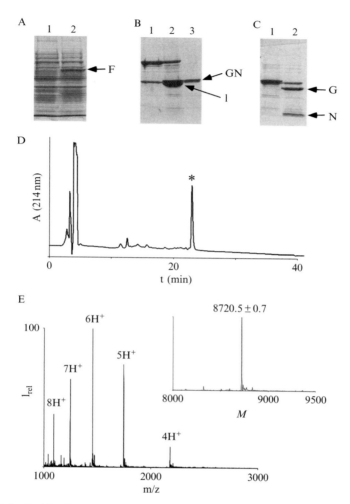

Figure 7.6 Purification of the N-peptide thioester. (A) SDS-PAGE gel showing the overexpression of the N-peptide sandwich fusion protein (F). Lane 1: uninduced cells; lane 2: induced cells. (B) SDS-PAGE gel showing the cleavage of the fusion protein using MESNA. Lane 1: the fusion protein (F) after partial purification; lane 2: cleavage of the fusion protein with MESNA releases the gyr A intein (I) and generates a MESNA thioester at the C-terminus of the GST-N-peptide fusion (GN); lane 3: the flow-through from the chitin column. The chitin column binds the intein (I), so the flow through contains the GST-N-peptide thioester. (C) SDS- PAGE showing the proteolysis of the GST-N-peptide thioester fusion protein by thrombin to release the GST tag (G) from the N-peptide thioester (N) with lane 1: before thrombin addition; lane 2: after proteolysis with thrombin. (D) Analytical RP-HPLC of the thrombin cleavage reaction after the TCA precipitation and acetone washes (gradient 50 to 100% buffer B on a C4 column). The N-peptide thioester is marked with an asterix. (E) ES-MS of the N-peptide thioester (inset, reconstructed spectrum, expected mass = 8721.1).

suspended in 50 mL of 50 mM Tris, pH 7.5, 200 mM NaCl per liter of bacterial culture and lysed by sonication. Triton X-100 is added to 1% (v/v), and the lysate is stirred at RT for 30 min. The cell lysate is then centrifuged (12,000g, 10 min, RT) to obtain the Triton X-100 insoluble fraction, which consists mainly of the inclusion bodies. The inclusion bodies are washed with 2 M urea to remove contaminants and then solubilized using 1% N-lauryl sarcosine and 6 M urea. Any insoluble particles are removed by centrifugation (12,000g, 10 min, RT). The solubilized fusion protein is reduced by DTT (2 mM) and then folded by removal of the denaturants, urea, and N-lauryl sarcosine by dialysis. The dialysis is carried out at 4 °C against 20 vol of dialysis buffer (50 mM Tris, pH 7.5, 0.2 M NaCl) containing 1% Triton X-100. A minimum of 4 h to O/N dialysis is carried out. Following dialysis, the folded fusion protein is treated with 2-mercaptoethanesulfonic acid (MESNA), which results in cleavage of the gyrA intein from the fusion protein and the generation of a MESNA thioester at the C-terminus of the GST-N-peptide fusion (Fig. 7.6B). For intein cleavage, MESNA (solid) is added to a concentration of 0.1 M and the reaction is carried out at RT with gentle stirring. Almost complete cleavage of the intein is observed after O/N incubation (Fig. 7.6B). The cleaved gyrA intein and any uncleaved fusion protein are removed by passage through a chitin resin column (New England Biolabs) (Fig. 7.6B). This step takes advantage of the chitin binding domain (CBD) present at the C-terminus of the gyrA intein construct used. The chitin resin binds approximately 2 mg of protein per milliliter of resin. To determine the volume of the chitin resin to be used, we estimate the amount of the gyrA intein and uncleaved fusion present from a Coomassie-stained gel of the cleavage reaction. The GST-N-peptide fusion protein thioester obtained after passage through the chitin column is proteolysed using thrombin to remove the GST tag (Fig. 7.6C). Thrombin (Roche Diagnostics) is added at 0.5 units per milligram of GST-KcsA. The N-peptide thioester peptide released by proteolysis is purified using RP-HPLC. The detergent Triton X-100, used in prior steps, strongly interferes with the HPLC purification and so is removed before HPLC. For Triton X-100 removal, we first use a tricholoroacetic acid purification step (10% at 4 °C, 30 min.) in which both the proteins and the detergent are precipitated. The precipitate is then washed with ice-cold acetone, which solubilizes only the Triton X-100, thereby efficiently removing the Triton X-100. The peptide precipitate after the acetone washes is dissolved in 50% TFE: buffer A and the N-peptide thioester is purified by preparative RP-HPLC on a 50 to 100% buffer B gradient on a C4 preparative column for 60 min (Fig. 7.6D). The presence of the desired peptide is confirmed by ES-MS of the HPLC fractions (Fig. 7.6E). The fractions containing the pure peptide are pooled and lyophilized. This protocol generally yields 2 to 3 mg of N-peptide thioester per liter of bacterial culture.

2.4. The ligation reaction

The ligation reaction to generate the KcsA polypeptide is carried out in detergents, which are required to keep the peptides soluble under the ligation conditions. We routinely use sodium dodecylsulfate (SDS), as it is very effective in solubilizing the peptides. The use of SDS, however, interferes with a subsequent HPLC analysis/purification step. Other detergents such as Fos–12, decyl maltoside (DM), and octyl glucoside can also be used in the ligation reactions. These detergents show reduced interference with HPLC analysis and are preferred if an HPLC purification step is required post-ligation. The KcsA ligation reaction is carried out in 0.1 M sodium phosphate buffer at pH 7.4 using 2% thiophenol as a catalyst. For ease of dissolution, the peptides are co-lyophilized with SDS, and the resulting peptide-detergent powder easily solubilizes on the addition of the ligation buffer. The ligation reaction is carried out at 37 °C and the extent of ligation is monitored by SDS-PAGE. We observe significant reaction within 2 to 4 h, though we generally allow the ligation reaction to continue O/N (Fig. 7.7A). After the O/N incubation, the reaction mixture is reduced by the addition of DTT to 0.1 M. A purification step for the ligation product is not carried out, and the reaction mixture is directly used for *in vitro* folding.

2.5. Folding of the semisynthetic KcsA channel

In vitro folding of the KcsA channel takes place when the unfolded KcsA polypeptide is mixed with lipid vesicles (Valiyaveetil *et al.*, 2002). Although different types of lipid vesicles support channel folding, we generally use asolectin vesicles, mainly for cost considerations. For formation of lipid vesicles, the asolectin pellets (Avanti Polar Lipids) are dissolved in cyclohexane and then lyophilized. The lyophilized lipid powder is resuspended in the lipid buffer (50 mM 2-(N-Morpholino)ethanesulfonic acid, pH 6.4, 0.3 M KCl, and 10 mM DTT) at 20 mg/mL and sonicated for the formation of lipid vesicles. *In vitro* folding is carried out by dilution of the ligation reaction mixture into the lipid vesicles and assayed by SDS gel electrophoresis. The folded KcsA molecules migrate as a tetramer on an SDS gel while the unfolded KcsA migrates as a monomer. Therefore, SDS gel electrophoresis provides an easy assay for detecting folding. We carry out a number of test dilutions to determine the optimal dilution for effective folding and a folding reaction for the ligation mixture is carried out using the optimal conditions identified (Fig. 7.7B).

For purification of the folded semisynthetic KcsA, the folding reaction is dialyzed (2x) against 100 vol of 50 mM Tris, pH 7.5, 0.3 M KCl, for removal of DTT. The folding mixture is then solubilized by the addition of 40 mM DM (Anatrace), and the proteins are separated from the lipids on a cobalt affinity column (Clontech) (Fig. 7.7C). The folded tetrameric semisynthetic KcsA channel is then separated from the unfolded monomeric

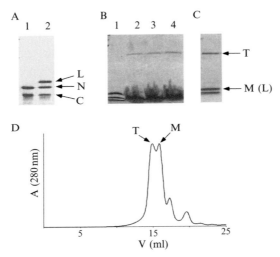

Figure 7.7 Semisynthesis of pF-Phe103-KcsA. (A) SDS-PAGE gel showing the ligation reaction between the N-peptide thioester (N) and the C-peptide (C) to form the KcsA polypeptide (L). Lane 1: 0 min; lane 2: after 12 h. (B) SDS-PAGE gel showing the folding of the KcsA polypeptide using lipid vesicles. Lane 1: no lipids; lane 2: 1:8 dilution, lane 3: 1:9 dilution; lane 4: 1:10 dilution of the ligation reaction into lipid vesicles. (C) SDS-PAGE gel showing the semisynthetic KcsA after partial purification using a Co^{2+} affinity column. In panels (B) and (C), the unfolded monomeric (M, which corresponds to the ligation product) and the folded tetrameric KcsA (T) are indicated. (D) Size-exclusion chromatography on a Superdex S200 column (buffer: 50 mM Tris, 150 mM KCl, 5 mM decylmaltoside) is used to separate the tetrameric folded semisynthetic KcsA channels (T) from the unfolded protein (M).

protein and the unreacted peptides by size–exclusion chromatography on a Superdex S200 column (GE Healthcare) using a column buffer consisting of 50 mM Tris pH 7.5, 150 mM KCl, and 5 mM DM (Fig. 7.7D).

2.6. Functional characterization of the semisynthetic KcsA

The semisynthetic KcsA channels are reconstituted into lipid vesicles for functional characterization. For reconstitution into lipid vesicles, the purified protein is mixed with a lipid solution containing 1-palmitoyl-2-oleoyl-phosphatidylethanolamine (POPE, 7.5 mg/mL, Avanti Polar Lipids) and 1-palmitoyl-2-oleoyl-phosphatidylglycerol (POPG, 2.5 mg/mL) in 10 mM HEPES, pH 7.5, 450 mM KCl solubilized by 34 mM CHAPS (Anatrace) (Heginbotham *et al.*, 1998). Lipid vesicles were formed by removal of the detergent by dialysis against 1000 vol of 10 mM HEPES, pH 7.5, 450 mM KCl. The lipid vesicles obtained are flash frozen in liquid nitrogen and stored in small aliquots at $-80\ ^{\circ}$C.

Functional characterization of the semisynthetic KcsA channels is carried out using planar lipid bilayers (Morais-Cabral *et al.*, 2001). Lipid vesicles

pF-Phe103

Phe103

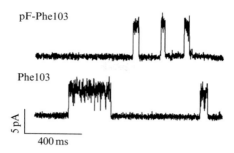

5 pA

400 ms

Figure 7.8 Electrophysiological analysis of semisynthetic KcsA. Single-channel traces for the pF-Phe103 KcsA channels obtained by semisynthesis and the corresponding wild-type KcsA channels recorded at +150 mV in 150 mM KCl are shown.

containing the semisynthetic KcsA channel are fused with lipid bilayers of POPE (15 mg/mL) and POPG (5 mg/mL) in decane painted over a 250-μm hole in a polystyrene partition separating chambers containing the internal and external solutions. To induce fusion, a salt gradient is maintained across the bilayer with the external solution containing 150 mM KCl and 10 mM HEPES, pH 7.0, and the internal solution containing 20 mM KCl and 10 mM succinate, pH 4.0. Following channel fusion, the KCl concentration of the internal solution is adjusted to 150 mM by addition of a requisite amount of a 3 M KCl solution. The membrane voltage is controlled and the current recorded by an Axopatch 200B amplifier with a Digidata 1322A analog-to-digital converter and Axoclamp software (Axon Instruments, Union City, CA).

The pF-Phe103 KcsA channels are functional and a single channel recording for the mutant channel and the corresponding wild-type control is shown (Fig. 7.8). The mutant channel shows a chord conductance of 43 pS at + 150 mV, which corresponds to the cord conductance for the wild-type channel under these recording conditions. The functional similarity is expected, as the Phe to pF-Phe substitution is not expected to cause any structural distortion. In the pF-Phe substitution, we have introduced a spectroscopic probe into the channel cavity, which will enable spectroscopic measurements to investigate the environment in the channel cavity and to monitor the binding of ions and blockers to the channel cavity.

3. APPLICATION OF SEMISYNTHESIS IN INVESTIGATING THE SELECTIVITY FILTER OF THE K⁺ CHANNELS

3.1. D-Ala substitution in the selectivity filter (Valiyaveetil *et al.*, 2004, 2006)

The K⁺ channel-selectivity filter consists of two absolutely conserved glycine residues (see Fig. 7.1B). Semisynthesis was used to demonstrate that the first conserved glycine residue in the selectivity filter (Gly77) can be replaced by D-Ala with retention of structure and function. The crystal structure of the Gly77→D-Ala mutant indicated that the structure of the selectivity filter with the D-Ala substitution was essentially identical to the wild-type structure. In contrast, substitution of Gly 77 with L-Ala (or any other L-amino acid) results in a nonfunctional channel. The results indicate that the Gly residue in the selectivity filter plays the role of a surrogate D-amino acid. The absolute conservation of the Gly residue in the K^+ selectivity filter results from Gly being the only natural amino acid that can play the role of a D-amino acid surrogate.

The selectivity filter of a K^+ channel undergoes a conformational change at low concentrations of K^+. To understand the role of this conformation change, the semisynthesis procedure was used to generate a channel that did not undergo this change. This was accomplished by replacing the first conserved Gly in the selectivity filter with D-Ala. Using crystallography, it was demonstrated that the presence of D-Ala locks the selectivity filter into a single conformation. Functional characterization of the mutant showed that it was capable of conducting Na^+ in the absence of K^+, in contrast to the native channel. This indicates that the K^+-specific conformational change of the selectivity filter is fundamental to the selectivity mechanism of K^+ channels.

3.2. Ester substitution in the selectivity filter (Valiyaveetil *et al.*, 2006)

The selectivity filter of K^+ channels consists of four contiguous ion-binding sites that are equivalent in terms of binding K^+ (see Fig. 7.1B). Semisynthesis was used to selectively perturb one ion-binding site through the use of an amide-to-ester substitution. X-ray crystallography was used to demonstrate that, in the ester mutant, the structure of the selectivity filter is largely unperturbed, but the distribution of K^+ and other permeant ions within the filter are altered. A functional analysis indicated that the ester substitution substantially altered ion-conduction properties. The results suggest that the energetic optimization of the selectivity filter ion-binding sites is necessary for obtaining maximal flux through the channel.

4. SUMMARY

We have described the semisynthesis of the KcsA channel. The semisynthesis involves an expressed protein ligation of a recombinantly expressed peptide thioester and a chemically synthesized peptide with an N-terminal cysteine to assemble the KcsA polypeptide, which is then folded *in vitro* to the native state. These protocols allow us to use chemical synthesis to manipulate the selectivity filter and other functionally important regions of the channel, like the channel cavity. While the protocols have been described for the KcsA K$^+$ channel, we anticipate that similar protocols will be applicable for the semisynthesis of other integral membrane proteins.

ACKNOWLEDGMENT

This research was supported by a Scientist Development Grant (0835166N) to F.I.V. from the American Heart Association, a Pew Scholar Award to F.I.V. and a Seed Grant from the Medical Research Foundation of Oregon.

REFERENCES

Beene, D. L., Dougherty, D. A., and Lester, H. A. (2003). Unnatural amino acid mutagenesis in mapping ion channel function. *Curr. Opin. Neurobiol.* **13**, 264–270.

Bianchi, E., Ingenito, R., Simon, R. J., and Pessi, A. (1999). Engineering and chemical synthesis of a transmembrane protein: The HCV protease cofactor protein NS4A. *J. Am. Chem. Soc.* **121**, 7698–7699.

Blaschke, U. K., Silberstein, J., and Muir, T. W. (2000). Protein engineering by expressed protein ligation. *Methods. Enzymol.* **328**, 478–496.

Clayton, D., Shapovalov, G., Maurer, J., Dougherty, D., Lester, H., and Kochendoerfer, G. (2004). Total chemical synthesis and electrophysiological characterization of mechanosensitive channels from *Escherichia coli* and Mycobacterium tuberculosis. *Proc. Natl. Acad. Sci. USA* **101**, 4764–4769.

Dawson, P. E., and Kent, S. B. H. (2000). Synthesis of native proteins by chemical ligation. *Annu. Rev. Biochem.* **69**, 923–960.

Dawson, P. E., Muir, T. W., Clarklewis, I., and Kent, S. B. H. (1994). Synthesis of proteins by native chemical ligation. *Science* **266**, 776–779.

Doyle, D. A., Cabral, J. M., Pfuetzner, R. A., Kuo, A. L., Gulbis, J. M., Cohen, S. L., Chait, B. T., and MacKinnon, R. (1998). The structure of the potassium channel: Molecular basis of K$^+$ conduction and selectivity. *Science* **280**, 69–77.

Fu, Y., Bieschke, J., and Kelly, J. W. (2005). E-olefin dipeptide isostere incorporation into a polypeptide backbone enables hydrogen bond perturbation: Probing the requirements for Alzheimer's amyloidogenesis. *J. Am. Chem. Soc.* **127**, 15366–15367.

Heginbotham, L., Kolmakova-Partensky, L., and Miller, C. (1998). Functional reconstitution of a prokaryotic K$^+$ channel. *J. Gen. Physiol.* **111**, 741–749.

Hille, B. (2001). Ion Channels of Excitable Membranes. Third ed.; Sinauer Associates, Inc: Sunderland, MA.

Jiang, Y., Lee, A., Chen, J., Ruta, V., Cadene, M., Chait, B., and MacKinnon, R. (2003). X-ray structure of a voltage-dependent K$^+$ channel. *Nature* **423**, 33–41.

Jiang, Y. X., Lee, A., Chen, J. Y., Cadene, M., Chait, B. T., and MacKinnon, R. (2002). Crystal structure and mechanism of a calcium-gated potassium channel. *Nature* **417**, 515–522.

Kent, S. (1988). Chemical synthesis of peptides and proteins. *Annu. Rev. Biochem.* **57**, 957–989.

Kochendoerfer, G. G., Salom, D., Lear, J. D., Wilk-Orescan, R., Kent, S. B. H., and DeGrado, W. F. (1999). Total chemical synthesis of the integral membrane protein influenza A virus M2: Role of its C-terminal domain in tetramer assembly. *Biochemistry* **38**, 11905–11913.

Kuo, A., Gulbis, J. M., Antcliff, J. F., Rahman, T., Lowe, E. D., Zimmer, J., Cuthbertson, J., Ashcroft, F. M., Ezaki, T., and Doyle, D. A. (2003). Crystal structure of the potassium channel KirBac1.1 in the closed state. *Science* **300**, 1922–1926.

Long, S., Campbell, E., and Mackinnon, R. (2005). Crystal structure of a mammalian voltage-dependent Shaker family K$^+$ channel. *Science* **309**, 897–903.

MacKinnon, R., Cohen, S. L., Kuo, A. L., Lee, A., and Chait, B. T. (1998). Structural conservation in prokaryotic and eukaryotic potassium channels. *Science* **280**, 106–109.

Mendel, D., Cornish, V. W., and Schultz, P. G. (1995). Site-directed mutagenesis with an expanded genetic code. *Annu. Rev. Biophys. Biomol. Struct.* **24**, 435–462.

Morais-Cabral, J. H., Zhou, Y. F., and MacKinnon, R. (2001). Energetic optimization of ion conduction rate by the K$^+$ selectivity filter. *Nature* **414**, 37–42.

Muir, T. (2003). Semisynthesis of proteins by expressed protein ligation. *Annu. Rev. Biochem.* **72**, 249–289.

Muir, T. W., Sondhi, D., and Cole, P. A. (1998). Expressed protein ligation: A general method for protein engineering. *Proc. Natl. Acad. Sci. USA* **95**, 6705–6710.

Muralidharan, V., and Muir, T. W. (2006). Protein ligation: An enabling technology for the biophysical analysis of proteins. *Nat. Methods* **3**, 429–438.

Schnolzer, M., Alewood, P., Jones, A., Alewood, D., and Kent, S. B. (1992). *In situ* neutralization in Boc-chemistry solid phase peptide synthesis. Rapid, high yield assembly of difficult sequences. *Int. J. Pept. Protein. Res.* **40**, 180–193.

Shealy, R. T., Murphy, A. D., Ramarathnam, R., Jakobsson, E., and Subramaniam, S. (2003). Sequence-function analysis of the K$^+$-Selective family of ion channels using a comprehensive alignment and the KcsA channel structure. *Biophys. J.* **84**, 2929–2942.

Valiyaveetil, F., MacKinnon, R., and Muir, T. (2002). Semisynthesis and folding of the potassium channel KcsA. *J. Am. Chem. Soc.* **124**, 9113–9120.

Valiyaveetil, F., Sekedat, M., MacKinnon, R., and Muir, T. (2006). Structural and functional consequences of an amide-to-ester substitution in the selectivity filter of a potassium channel. *J. Am. Chem. Soc.* **128**, 11591–11599.

Valiyaveetil, F., Zhou, Y., and MacKinnon, R. (2002). Lipids in the structure, folding, and function of the KcsA K$^+$ channel. *Biochemistry* **41**, 10771–10777.

Valiyaveetil, F. I., Leonetti, M., Muir, T. W., and MacKinnon, R. (2006). Ion selectivity in a semisynthetic K$^+$ channel locked in the conductive conformation. *Science* **314**, 1004–1007.

Valiyaveetil, F. I., Sekedat, M., Mackinnon, R., and Muir, T. W. (2004). Glycine as a D-amino acid surrogate in the K($^+$)-selectivity filter. *Proc. Natl. Acad. Sci. USA* **101**, 17045–17049.

Valiyaveetil, F. I., Sekedat, M., Muir, T. W., and MacKinnon, R. (2004). Semisynthesis of a functional K$^+$ channel. *Angew. Chem. Int. Ed. Engl.* **43**, 2504–2507.

Wang, L., and Schultz, P. G. (2004). Expanding the genetic code. *Angew. Chem. Int. Ed. Engl.* **44**, 34–66.

Yohannan, S., Hu, Y., and Zhou, Y. (2007). Crystallographic study of the tetrabutylammonium block to the KcsA K$^+$ channel. *J. Mol. Biol.* **366**, 806–814.

Zhou, Y., and MacKinnon, R. (2004). Ion binding affinity in the cavity of the KcsA potassium channel. *Biochemistry* **43**, 4978–4982.

Zhou, Y., Morais-Cabral, J., Kaufman, A., and MacKinnon, R. (2001). Chemistry of ion coordination and hydration revealed by a K$^+$ channel-Fab complex at 2.0 A resolution. *Nature* **414**, 43–48.

CHAPTER EIGHT

Segmental Isotopic Labeling of Proteins for Nuclear Magnetic Resonance

Dongsheng Liu,* Rong Xu,* *and* David Cowburn*

Contents

Abstract

Nuclear magnetic resonance (NMR) spectroscopy has emerged as one of the principle techniques of structural biology. It is not only a powerful method for elucidating the three-dimensional structures under near physiological

* New York Structural Biology Center, New York, USA

Methods in Enzymology, Volume 462
ISSN 0076-6879, DOI: 10.1016/S0076-6879(09)62008-5

conditions but also a convenient method for studying protein-ligand interactions and protein dynamics. A major drawback of macromolecular NMR is its size limitation, caused by slower tumbling rates and greater complexity of the spectra as size increases. Segmental isotopic labeling allows for specific segment(s) within a protein to be selectively examined by NMR, thus significantly reducing the spectral complexity for large proteins and allowing for the application of a variety of solution-based NMR strategies. Two related approaches are generally used in the segmental isotopic labeling of proteins: expressed protein ligation and protein *trans*-splicing. Here, we describe the methodology and recent application of expressed protein ligation and protein *trans*-splicing for NMR structural studies of proteins and protein complexes. We also describe the protocol used in our lab for the segmental isotopic labeling of a 50-kDa protein Csk (C-terminal Src kinase) using expressed protein ligation methods.

1. INTRODUCTION

Nuclear magnetic resonance (NMR) spectroscopy has emerged as one of the principle techniques of structural biology. It is not only a powerful method for elucidating the three-dimensional structures under near physiological conditions but also a convenient method for studying protein-ligand interactions and protein dynamics. As molecular weight increases, the size limitations of NMR become apparent—a slower tumbling rate reduces resolution and adds greater complexity of the spectra. Slower tumbling of the macromolecule means a faster transverse relaxation rate, thereby resulting in poor signal-to-noise (s/n) ratio of the spectrum. A larger size implies the appearance of more signals in the spectrum, which complicates the assignment of individual signals. In recent years, higher-magnetic-field NMR machines have been constructed to increase resolution and signal to noise. Cryogenic probes are widely used to increase the signal to noise significantly. The development of multinuclear multidimensional experiments based on sophisticated pulse schemes has greatly enhanced the utility of NMR for the study of macromolecular structure and dynamics (Mittermaier and Kay, 2006; Wuthrich, 2003).

In the sample preparation aspect, the structural investigation of proteins by NMR has heavily depended on the incorporation of stable isotopes by increased (mainly ^{15}N, ^{13}C) or decreased (mainly ^{1}H) occurrence of an NMR-active isotope. The combination of ^{15}N, ^{13}C triple-resonance spectroscopy with deuteration, however, holds the most promise for addressing the molecular weight limitations currently imposed on structure determination by solution NMR (Mittermaier and Kay, 2006). Although using deuterium labeling and TROSY type experiments (Wider, 2005; Wuthrich, 2003) addresses the issue of line width, NMR methods still have been applied most routinely to proteins under approximately

20 kDa, to avoid the issues of spectral overlap and broad line widths (Card and Gardner, 2005). The segmental labeling of proteins, in which segments of the sequence with arbitrary isotopic composition are recombined, represents a significant step to reduce the spectra complexity.

The principal applications of segmental labeling (Cowburn and Muir, 2001) may be identified as (1) assignment and structure determination of large proteins and their complexes by reducing their spectra overlap, (2) formation of labeled proteins from readily producible fragments when the complete product is impractical, or cytotoxic, for expressions and/or labeling, (3) segmental labeling of a segment known to be a major epitope for interaction with a ligand in a fuller sequence context, (4) observation of domain orientation and ligand perturbation in large systems using residual dipolar couplings and/or relaxation methods, (5) introduction of nonnatural or modified amino acids (e.g., phosphotyrosine), which cannot be included in normal expression methods, and (6) introduction of NMR unobservable tags (e.g., solubility-enhancement tags or lanthanide-binding tags). Although there are other methods for making segmental isotopic-labeled protein or RNA samples (Kim *et al.*, 2002; Varadan *et al.*, 2004), the main focus here is expressed protein ligation and protein *trans*-splicing methods.

Expressed protein ligation and protein *trans*-splicing have been applied to a large number of different protein-engineering problems. The successful implementations in segmental isotopic labeling for NMR are still somewhat limited. This may be a result of several factors. If the overall aim is reduction of spectral complexity, several issues must be addressed. First, a project involving segmental labeling will require more time and reagents than a conventional approach. If the target protein is composed of a single domain, normally a refolding procedure is required, no matter whether *trans*-splicing or the expressed protein ligation method is used. If the target protein contains several independent folding domains, the ligation can be conducted at native conditions; a refolding procedure is potentially not required. Compared with protein-peptide ligation, protein-protein ligation is generally slow. Because of this, purification methods are required remove the unligated precursors (Harris, 2006; Shi and Muir, 2005; Vitali *et al.*, 2006; Zuger and Iwai, 2005).

2. SEGMENTAL ISOTOPIC LABELING USING EXPRESSED PROTEIN LIGATION

2.1. Overview

Expressed protein ligation is a protein-engineering approach that allows for recombinant and synthetic polypeptides to be chemoselectively and regioselectively joined together (Fig. 8.1A) (Muir *et al.*, 1998). It is a modified version of native chemical ligation (NCL) (Dawson *et al.*, 1994) in which at

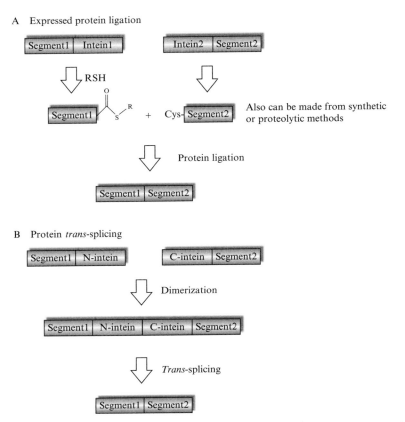

Figure 8.1 (A) Segmental isotopic labeling using expressed protein ligation. Intein-based fusion approaches allow the straightforward introduction of α-thioester and N-terminal Cys. N-terminal Cys can also be made by cleavage of the leader sequence and chemical synthesis methods. (B) Segmental labeling using protein *trans*-splicing. The N- and C-terminal pieces (referred to in the figure as N-intein and C-intein) individually have no activity but, when combined, associate noncovalently to give a functional protein-splicing element. Conserved residues around the splicing junction are a Cys/Ser at the N-terminus of intein, an His-Asn sequence at the C-terminus of intein, and a Ser/Thr/Cys at the N-terminus of C-terminal extein (segment 2).

least one of the building blocks are made by recombinant DNA methods. The first step of this process is the chemoselective transthioesterification of an unprotected protein C-terminal α-thioester with an N-terminal Cys of a second segment. The so-formed thioester spontaneously undergoes an S→N-acyl transfer to form a native peptide bond, and the resulting peptide product is obtained. The reaction can proceed in aqueous conditions at neutral pH, and internal Cys residues within both peptide segments are permitted. Although this procedure is relatively simple, segmental labeling using expressed protein ligation had not been applied until the introduction of the intein–mediated purification with an affinity chitin-binding tag (IMPACT) system (New England Biolabs) (Chong et al., 1997; Southworth et al., 1999). The IMPACT system contains a series of expression vectors (e.g., pTXBx, pTYBx, pTWINx) that allow for the fusion of the cleavable intein tag to either the C-terminus (C-terminal fusion) or N-terminus (N-terminal fusion) of a target protein (Xu and Evans, 2001). It provides a practical way to generate the reactive C-terminal α-thioester and N-terminal Cys segments. The combination of IMPACT system with NCL provides a powerful method for the segmental labeling of proteins for the NMR studies (Table 8.1).

2.2. Ligation site

For segmental labeling of a protein requiring refolding, a ligation site can be selected within the domain (Ottesen et al., 2008). For a ligation without refolding process, the ligation point is always chosen to be located within the linker region that connects the two domains (Vitali et al., 2006). In both cases, the residues flanking the ligation site are of special significance. One of these is the N-terminal Cys residue of the C-terminal segment, the other is the C-terminal residue forming the α-thioester for the N-terminal portion (Fig. 8.1A). Because the Cys residue at the junction is generally regarded as obligatory for EPL, the primary consideration is to choose a ligation site before a Cys residue. In the absence of Cys at the desired ligation site, mutation usually is introduced at sites with similar residues such as Ala or Ser. In comparing a homologous family of similar proteins, it may be desirable to mutate highly conserved residues.

Attention should also be paid to the residue at the C-terminus of the α-thioester segment (X position). Efficiency of the protein-intein thiolysis step varies, and different inteins have their own set of preferences at this position (Chong et al., 1998). For example, if the target protein is fused with Mxe GyrA intein from pTWIN1 vector, C-terminal Asp, Pro, Gly, and Ala should be avoided. If the target protein is fused with Sce VMA intein from pTYB1, pTYB3 vector, C-terminal Asp, and Arg should be avoided because of the in vivo cleavage; Asn, Cys, and Pro should also be avoided because of the possible block of cleavage. If the Mth RIR1 intein from pTWIN2 vector was fused to the C-terminus of the target protein, it is recommended that a Gly or Ala be present at the C-terminus of the target protein, whereas Asp and Pro are not recommended (Xu and Evans, 2001).

An additional consideration is that the ligation reaction rate highly relies on the molar concentration of both precursors. For the segmental labeling of protein, the ligation speed is then critical because the molar concentration of both segments cannot be very high. Although all 20 amino acids are potentially suitable for ligation, Val, Ile, and Pro represent bad choices for the α-thioester

Table 8.1 Selected examples of segmental isotopic labeling using expressed protein ligation methods

Name/Total size	Linker	Method for α-thioester	Method for N-Cys	Comments
Abl SH3–SH2 domain/ 18 kDa	G–C	*Sce* VMA from pTYB2	Factor Xa cleavage	First segmental labeled sample using expressed protein ligation methods. Thiophenol and benzyl mercaptan were used as catalysts. ^{1}H–^{15}N HSQC acquired (Xu et al., 1999)
σ factor σ^A and $\Delta 1.1$-σ^A/47 and 32 kDa	G–CG	*Mxe* GyrA from pTXB1	Factor Xa cleavage	Cleavage and ligation on the chitin beads. Ligation reaction was rapid and efficiency was high (Camarero et al., 2002)
Mxe GyrA inteine/22 kDa	F–C	Peptide synthesis	Factor Xa cleavage	Ligation buffer: (6 M Gdm·Cl/0.1 M NaPi at pH 8) containing 3% MESNA and 2% ethanethiol. Room temperature 5 h. ^{1}H–^{15}N HSQC, 2D H(N)CO used (Romanelli et al., 2004)
G protein α subunit/ 40 kDa	K–C	*Mxe* GyrA from pTXB3	Peptide synthesis	On column cleavage and ligation with 1% MESNA. 4 °C, 24 h (Grant et al., 2006)
PTB RRM3–RRM4/ 24 kDa	G–C	*Mxe* GyrA from pTWIN1	*Ssp* DnaB from pTWIN1	Ligation on column at 37 °C for 16 h. 130 interdomain NOEs were unambiguously identified (Vitali et al., 2006)
Npl3p RM1–RRM2/ 21 kDa	Y–C Or G–C	*Mxe* GyrA from pTWIN1	*Ssp* DnaB from pTWIN1	Ligation conducted at native or denatured conditions. ^{1}H–^{15}N HSQC suggested no domain interaction (Skrisovska and Allain, 2008)

hnRNP L RRM3-RRM4/ 25 kDa	F-C	*Mxe* GyrA from pTWIN1	*Ssp* DnaB from pTWIN1	Refolding required before ligation. On column ligation. High ligation efficiency. Interdomain NOE obtained (Skrisovska and Allain, 2008)
DHFR/19 kDa	L-C	Peptide synthesis	Met aminopeptidase	A single ^{13}C residue in an active site loop was generated. Ligation conducted at denatured state (Ottesen *et al.*, 2008)
c-Crk-II/34 kDa	G-C G-C	*Mxe* GyrA from pTXB1 *Sce* VMA from pTYB2	Factor Xa cleavage	Sequential ligation of three recombinant polypeptides. The incorporation of isotopes in the central domain of the protein. (Blaschke *et al.*, 2000; Ottesen *et al.*, 2008)

because of slow ligation rates. Rapid ligation rates are observed when the α-thioester residue X is His, Cys, or Gly (Hackeng *et al.*, 1999).

2.3. Synthesis of a segment with C-terminal α-thioester

Although chemical synthetic methods are feasible (e.g., Romanelli *et al.*, 2004), the recombinant intein fusion expression method is the most widely used in the generation of C-terminal α-thioester for segmental labeling, because isotopic labeling is more easily obtained and expression is more feasible for longer segments. Currently, three modified inteins from the IMPACT system can be used to generate segment C-terminal α-thioester: *Mxe* GyrA(198 aa, from pTWIN1, pTXB1,3) (Evans *et al.*, 1998), *Mth* RIR1(134 aa, from pTWIN2) (Evans *et al.*, 1999), and *Sce* VMA(454 aa, from pTYB1,2,3,4) (Chong and Xu, 1997). Cloning a target gene into these vectors results in the fusion of the C-terminus of the target protein to the N-terminus of an intein, which is mutated to undergo thiol-induced cleavage at its N terminus (Xu and Evans, 2001). As noted earlier, the C-terminal residue of the target protein is critical because the *in vivo* and thiol-induced cleavage at the intein can be dramatically affected by it.

2.4. Synthesis of a segment with N-terminal cysteine

Three methods generally used in the preparation of segments with N-terminal Cys are chemical synthesis, cleavage of a precursor sequence expressed with the Cys residue adjacent to a cleavage site, and intein-mediated recombinant methods (Muralidharan and Muir, 2006). Several cleavage methods can be used: (1) methionyl aminopeptidase (Camarero *et al.*, 2001; Gentle *et al.*, 2004; Iwai and Pluckthun, 1999), (2) factor Xa protease (Camarero *et al.*, 2002; Romanelli *et al.*, 2004; Xu *et al.*, 1999), (3) TEV protease (Tolbert and Wong, 2004; Tolbert *et al.*, 2005), (4) cyanogen bromide (Macmillan and Arham, 2004), (5) SUMO protease (Weeks *et al.*, 2007), (6) leader peptidase (Hauser and Ryan, 2007), (7) enterokinase (Hosfield and Lu, 1999), and (8) thrombin. We have found that the leader sequence LVPRC (modified from the optimal thrombin cleavage sequence LVPRGS) can also be cleaved by thrombin with lower efficiency. Thus, thrombin cleavage can be used in the preparation of protein segments with C-terminal Cys (Liu *et al.*, 2008). Each cleavage method has its limitations. For example, nonspecific proteolysis can be observed when there are sequences in the protein similar to those of the cleavage site; cyanogen bromide cleavage requires denaturation and no other Met residues in the target protein; endogenous methionyl aminopeptidase is often inefficient and results in low yields of the desired material.

N-terminal fusion with an intein can also be exploited to prepare N-terminal Cys segments. The modified inteins *Ssp* DnaB (Mathys *et al.*, 1999), *Mth* RIR1 (Evans *et al.*, 1999), and *Mxe* GyrA (Southworth *et al.*, 1999) have all been engineered to generate N-terminal Cys proteins. The pTWIN1 and pTWIN2 vector from IMPACT system can be used for this purpose with the *Ssp* DnaB N-terminal fusion. A target protein sequence was fused in-frame to the C-terminus of the intein. Following expression in *E. coli*, the fusion protein was bound to a chitin resin and the cleavage of intein fusion is induced in a pH- and temperature-dependent fashion. Sometimes the protein undergoes significant *in vivo* cleavage, thus causing the loss of the affinity tag (Mathys *et al.*, 1999).

2.5. Ligation protocol

The isolated α-thioester and N-terminal Cys segments can be ligated in the presence of many additives, such as denaturents, chaotropes, or detergents. The optimum pH of the ligation reaction is 7 to 8, and the presence of a moderate concentration of the thiol reagent (e.g., 2-mercaptoethanesulfonic acid (MESNA)) is always required. The thiol-induced cleavage and ligation can also be carried out simultaneously by adding both thiol and N-terminal Cys segment (Camarero *et al.*, 2002; Vitali *et al.*, 2006) (Table 8.1). A reactive C-terminal thioester can be generated on the target protein when an appropriate thiol compound is used to induce cleavage of the intein. MESNA is an attractive reagent, as it provides higher ligation efficiency and is odorless and extremely soluble in aqueous solutions (Ayers *et al.*, 1999; Evans *et al.*, 1998). Johnson and Kent (2006) have compared the use of a number of thiol compounds and found a highly effective and practical catalyst MPAA (4-carboxylmethyl)thiophenol). Ligation in the presence of MPAA showed an order of magnitude (i.e. ≈10-fold) rate increase over the standard catalyst mix used for ligation. The authors also suggested that MPAA is a much more effective alternative to MESNA for use in the semisynthesis of protein by expressed protein ligation. A recent study has suggested that reaction of DTT-thioester protein and Tris(tris(hydroxymethyl) aminomethane) buffer can take place, and a Tris adduct formed, which is fused to the C-terminus of the protein by a stable amide bond (Peroza and Freisinger, 2008). Although the reactions of other thioesters/thiol systems with Tris are not reported, caution is appropriate in the use of Tris buffer.

3. SEGMENTAL LABELING USING PROTEIN *TRANS*-SPLICING

The ligation of proteins segments can also be accomplished by *trans*-splicing, which relies on the high affinity and catalytic activity of the two halves of an intein to ligate the two extein sequence (Fig. 8.1B). The N- and C-terminal pieces individually have no splicing activity but, when combined, they associate noncovalently to give a functional protein-splicing element.

The *trans*-splicing reaction is initiated by binding of the intein fragments together, whereas expressed protein ligation reaction lacks this favorable binding. Several split inteins have been studied, including native split *Ssp* DnaE (Evans *et al.*, 2000; Martin *et al.*, 2001), artificially split PI-*Pfu*I (Otomo *et al.*, 1999b; Yagi *et al.*, 2004; Yamazaki *et al.*, 1998), *Mtu* RecA (Lew *et al.*, 1998; Mills and Perler, 2005; Shingledecker *et al.*, 1998), *Psp* Pol-1 (Southworth *et al.*, 1998), *Ssp* DnaB (Brenzel *et al.*, 2006), and *Sce* VMA (Mootz *et al.*, 2003; Schwartz *et al.*, 2007). Among these split inteins, *Ssp* DnaB, *Sce* VMA, and *Ssp* DnaE are reported to be active in protein *trans*-splicing under native conditions, while others require a denaturation-refolding treatment of the mixed intein halves (Otomo *et al.*, 1999a,b; Yagi *et al.*, 2004; Yamazaki *et al.*, 1998). The naturally occurring *Ssp* DnaE intein has been shown to mediate protein *trans*-splicing *in vivo* under native conditions and to provide a segmental isotope method in the living cells (Zuger and Iwai, 2005). It is recommended to include several residues flanking the target protein sequence to enhance the ligation efficiency (Xu and Evans, 2001). Table 8.2 summarizes the typical use of protein *trans*-splicing methods for the segmental labeling for NMR sample preparations.

4. MULTIPLE SEGMENT ASSEMBLY

For extremely large systems, segmental labeling involving more than two domains may be necessary. Both expressed protein ligation (Blaschke *et al.*, 2000; Cotton *et al.*, 1999; Ottesen *et al.*, 2008) and protein *trans*-splicing (Otomo *et al.*, 1999a; Shi and Muir, 2005) methods can be used in this task. Using expressed protein ligation, a protecting group on N-terminal Cys of a central segment should be introduced to avoid self-ligation and/or cyclization when the central segment(s) is ligated with the C-terminal segment. The protection group can be a protease-cleavable peptide. After the central segment is ligated with the C-terminal segment, the protecting group is removed and the resulted protein ligated with N-terminal segment α-thioester. For the *trans*-splicing protocol, two separate inteins, which should be orthogonal to each other, may be used to ligate N- and C-terminal segments to both the ends of the middle segment.

5. SEGMENTAL LABELING OF C-TERMINAL SRC KINASE(CSK)

5.1. Overview

Csk has been shown to be important in the regulation of neural development, T cell development and regulation, and cytoskeletal organization (Cole *et al.*, 2003; Roskoski, 2005). Topographically, Csk is similar to the

Table 8.2 Selected examples of segmental isotopic labeling using protein *trans*-splicing methods

Name/Total size	Linker	Intein	Comments
αC/10 kDa	GGG–TG	PI-*Pfu*I	First protein to be segmental isotope labeled. Denatured precursors refolding. *Trans*-splicing at 70 °C (Yamazaki et al., 1998). Ligation yield was improved by some modification of the protocol (Otomo et al., 1999b). ^{1}H–^{15}N HSQC, ^{15}N edited NOESY used
MBP/42 kDa	GGG–TG	PI-*Pfu*I	Denatured precursors refolding. *Trans*-splicing at 70 °C. Ligation protocol optimized. 2D ^{1}H–^{15}N HSQC used (Otomo et al., 1999b)
MBP/42 kDa	GGG–TG TNP–CGE	PI-*Pfu*I PI-*Pfu*II	Central-segment isotope labeling through protein *trans*-splicing using two separate inteins (Otomo et al., 1999a)
H$^+$-ATPase β subunit/ 52 kDa	GGG–TG	PI-*Pfu*I	Denatured precursors refolding. *Trans*-splicing at 70 °C, 1 h. Triple resonance 3D, RDC experiments used (Yagi et al., 2004)
GB1-CBD/ 15 kDa	GS-CFNKGT	*Ssp* DnaE	Ligation *in vivo*. Based on a dual expression system that allows for sequential expression of two precursor fragments in different media. ^{1}H–^{15}N HSQC, HNCACB, 2D-[^{1}H–^{15}N]-HNCO used (Zuger and Iwai, 2005)

Src family kinase, having a SH3 and SH2 domain pair followed by a catalytic domain. A proline-rich sequence from the protein tyrosine phosphatases (murine PEP, (Cloutier and Veillette, 1996), human *PTPN22*/Lyp (Bottini *et al.*, 2004)) can bind to the Csk-SH3 domain. Csk is a 50-kDa protein that is not good target for traditional NMR studies because of its size and self-association. Segmental labeling is a promising approach to overcoming the size issue. From preliminary studies, we knew that the refolding of Csk was difficult, so we decided use expressed protein ligation for segmental labeling. We chose the pTWIN1 expression vector (New England Biolabs) for the expression of both segments with a view to obtain segmentally labeled, full-length Csk. At the same time, this approach provides SH32 and kinase domain separately and conveniently. We also needed a fusion system to overcome the poor expression and folding of the kinase domain; pTWIN1 vectors are designed for protein purification or for the isolation of proteins with an N-terminal cysteine and/or a C-terminal thioester. It has a modified *Ssp* DnaB intein as N-terminal fusion (intein 1) and an *Mxe* GyrA intein as C-terminal fusion (intein 2). The presence of the N- and C-terminal chitin binding domain (CBD) facilitates purification.

Csk can be divided into three domains: SH3, SH2, and kinase domain (Fig. 8.2). The ligation point is chosen to be located within the SH2-kinase linker region that connects the two domains, residues V172 to E194 (Holtrich *et al.*, 1991; Partanen *et al.*, 1991). Because there is no Cys in the linker region, we have to find a similar residue that can be mutated to Cys with minimal change. Ser and Ala are good choices for mutation to Cys. A178, A179, and S186 can be considered potential mutation sites. A178 and S186 were not used because the proceeding residues, V177 and

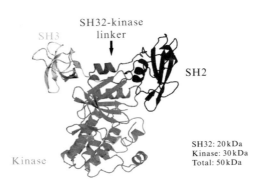

SH32-kinase linker
SH3-**SH2**-VMEGTVA-AQDEFYRSGWALNMKE-kinase

Figure 8.2 Structure of full-length Csk and the sequence of SH32-kinase linker (modeled from 1K9A (Ogawa *et al.*, 2002)). The SH3, SH2, and kinase domain are indicated. A178 was chosen to form an α-thioester and A179 to mutate into Cys. In the structure of Csk, A178 and A179 are located at the N-terminal of a small helix αBC.

R185, would likely result in low efficiency for the ligation (Hackeng *et al.*, 1999). Thus, we chose A179 to mutate to Cys and A178 to form an α-thioester. In the crystal structure of Csk, A178 and A179 are located at the N-terminal of a small helix αBC (Ogawa *et al.*, 2002), and we anticipated that the mutations would lead to minimal structural perturbations. In the following steps, the SH32 and kinase domain sequence were cloned into the expression vector, which allowed the generation of SH32-intein2-CBD and CBD-intein1-kinase fusion proteins, respectively.

5.2. Cloning Csk SH32 and kinase gene to expression vector

The Csk SH32 domain gene, containing residues M1 to A178 was amplified from human full-length Csk gene by polymerase chain reaction (PCR) using primers 5'-GGT GGT CAT ATG TCA GCA ATA CAG GCC (SH3_intein_fw) and 5'-GGT GGT TGC TCT TCC GCA CGC CAC TGT GCC CTC CAT (SH2_intein_rev) with Nde I and Sap I cleavage sites. A stop codon should not be included in the reverse primer for the sequential expression of *Mxe* GyrA intein (intein2). The Sap I site would be lost after cloning, so no additional residue was introduced between SH32 and intein2. All of the PCR reactions were in 50-μl final volume with pfu polymerase (Novagen), 0.5 mM of all four dNTPs (Sigma), 0.2 μM of each oligonucleotide, and 50 ng of plasmid DNA. PCR was carried out in a Mastercycler (Eppendorf) machine, and DNA was amplified by 30 cycles under standard conditions with annealing at 60 °C and elongation at 72 °C. The PCR product was purified, digested with Nde I and Sap I, and ligated to the similarly digested pTWIN1 vector. After transformation of the plasmid into *E. coli* DH5a, the clone was validated by DNA sequencing using T7 universal primer (5'-TAA TAC GAC TCA CTA TAG GG). The vector was named pTWIN1-SH32. The corresponding protein product was named SH32-intein2-CBD (Fig. 8.3).

The Csk kinase domain gene containing residues A179 to L450 with an A179C mutation, with or without C-terminal His-tag, was amplified from full-length Csk gene by PCR using primers containing Sap I and BamH I sites. For the cloning of kinase without His-tag, primers: 5'-GGT GGT TGC TCT TCC AAC TGC AGG GAT GAG TTC TAC CGC (kinase_intein_fw) and 5'-GGT GGT GGA TCC TTA CAG GTG CAG CTC GTG GGT (kinase_intein_rev) were used. For the cloning of kinase with C-terminal His-tag, primers: kinase_intein_fw and 5'- GGT GGT GGA TCC TTA ATG GTG ATG GTG ATG GTG CAG GTG CAG CTC GTG GGT (kinase_intein_his_rev) were used. A stop codon should be included in the reverse primer. The Sap I site is lost after cloning, so no additional residue was introduced between *Ssp* DnaB intein (intein1) and the kinase domain. The amplified gene was cloned into the expression vector pTWIN1, and the sequences were confirmed by DNA sequencing

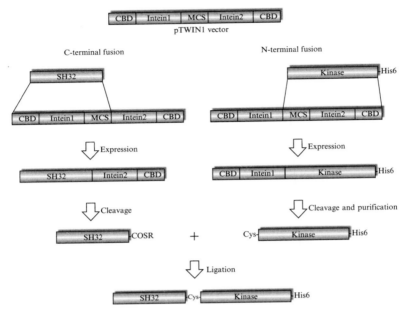

Figure 8.3 Segmental labeling of human Csk (50 kDa) protein. Genes of SH32 or kinase domain are cloned into pTWIN1 vector as C-terminal fusion and N-terminal fusion, respectively. Intein2-CBD was fused to the C-terminal of SH32, which allows the isolation of SH32 using the thio-inducible cleavage activity of intein2. A reactive C-terminal thioester can be generated on SH32 when MESNA is used as the thiol reagent to induce cleavage of the intein. Kinase domain was fused with CBD-intein1 to permit the preparation of kinase with an N-terminal cysteine. His-tag was added to facilitate the purification because of the *in vivo* cleavage. The final ligation of SH32 with α-thioester and kinase with N-terminal cysteine form the segmental labeled full-length Csk.

using *Ssp* DnaB forward primer (5′-ACT GGG ACT CCA TCG TTT CT) and T7 terminator reverse primer (5′-TAT GCT AGT TAT TGC TCA G). The designed fusion proteins without and with a C-terminal His-tag are named pTWIN1-kinase and pTWIN1-kinase-His, respectively (Fig. 8.3). The corresponding protein products of pTWIN1-kinase and pTWIN1-kinase-his are CBD-intein1-kinase and CBD-intein1-kinase-his, respectively. Both of the fusion proteins contain A179C mutations.

5.3. Expression and purification Csk SH32 with C-terminal *Mxe* GyrA intein (intein2)

For the expression of SH32-intein2-CBD protein, the plasmid pTWIN1-SH32 was transformed into BL21-CondonPlus (DE3)-RIL Component Cell (Stratagene, 230245). A fresh cell harboring pTWIN-SH32 was inoculated in 20 ml of LB medium containing 100 μg/ml ampicillin and cultured at 37 °C until the optical density (A600) reached 0.8. The whole culture was then transferred to 1 L of the same medium and cultured at 37 °C until the optical density (A600) reached 0.6. The cells were induced with 0.5 mM IPTG at 30 °C. For the isotopic labeling, M9 media with 0.1% $^{15}NH_4Cl$ and/or 0.2% [U-^{13}C] glucose was used. After induction at 30 °C overnight, the cell pellet was collected by centrifugation at 4000g for 20 min at 4 °C, and then resuspended in 35 ml of buffer A (50 mM Tris-HCl, pH 7.5 200 mM NaCl) and stored at −80 °C for future use. No *in vivo* cleavage of the SH32-intein2-CBD was found during the foregoing processing procedure. As a practical note, in systems that are susceptible to *in vivo* degradation, we recommend that the cell growth and protein purification steps be performed on the same day to improve yields.

Cells containing the SH32-intein2-CBD protein were passed through a French press cell twice to break the cells. The cell lysate was centrifuged at 12,000g for 20 min at 4 °C. The clarified cell extract was loaded to 6 ml of chitin beads (New England Biolabs S6651L), which preequilibrated with buffer A. The column was washed with 50 ml of buffer A to remove the unbound protein. The column bound with protein SH32-intein2-CBD was stored at 4 °C for the ligation with kinase domain. The fusion protein was stable for at least 4 days at 4 °C with no thiol reducing agent added. The long-term storage should be at −80 °C. For the purification of the SH32 domain, the cleavage of the intein-tag was induced by equilibrating the chitin beads with 50 mM DTT, 50 mM KPi, pH 7.2. After 24 h of on-column cleavage at room temperature, the target protein was eluted from the column by 40 ml of buffer B (20 mM Tris-HCl, pH 8.0) and loaded onto a Mono Q column equilibrated with buffer B at a flow rate of 1 ml/min (Fig. 8.4A). The elution was conducted by linear gradient to 60% buffer C (20 mM Tris-HCl, pH 8.0, 1.0 M NaCl) in 60 min. The major peak near 34% of buffer C was collected and identified as SH32.

5.4. Expression and purification of Csk kinase with N-terminal *Ssp* DnaB intein (intein1)

For the expression of CBD-intein1-kinase or CBD-intein1-kinase-his, the plasmid pTWIN1-kinase or pTWIN1-kinase-his was transformed into BL21-CondonPlus (DE3)-RIL Component Cell (Stratagene, 230245). A fresh cell harboring pTWIN-SH32 was inoculated in 20 ml of LB medium containing 100 μg/ml ampicillin and cultured at 37 °C until the optical density (A600) reached 0.8. The whole culture was then transferred to 1 L of the same medium and cultured at 37 °C until the optical density (A600) reached 0.6. The cells were induced with 0.2 mM IPTG at 15 °C for 16 h (for LB or nondeuterated M9 media) or 40 h

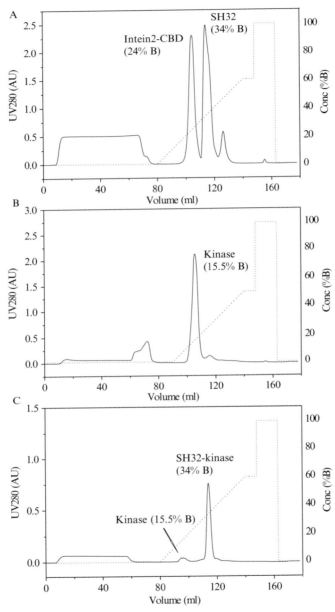

Figure 8.4 Mono Q elution profile of selected product. Loading buffer: 20 m*M* Tris-HCl, pH 8.0 (buffer B); elution buffer: 20 m*M* Tris-HCl, pH 8.0, 1.0 *M* NaCl (buffer C), flow rate of 1 ml/min. The elution was conducted by linear gradient to 50 to 60% buffer C. (A) SH32 domain. (B) Kinase domain. (C) Final purification step of SH32-kinase ligation product.

(when deuteration applied). After centrifugation at 4000g for 20 min, the cells were resuspended in 35 ml of buffer A and stored at $-80\ ^\circ$C for future use.

Partial degradation was found for the two fusion proteins after induction. Two bands, 60 kDa and 30 kDa, appeared on the SDS-PAGE gel and corresponded to the fusion and degradation products, respectively. The calculated molecular weight for the CBD-intein1-kinase, CBD-intein1, and kinase were 60, 30, and 30 kDa, respectively. Because the cell growth in D_2O is much slower than the H_2O media, the degradation was more severe. This makes it impossible to use chitin beads to purify the kinase protein. We used CBD-intein1-kinase-his protein in the ligation experiments. Although the degradation still remains, both nondegraded fusion kinase and degraded kinase domain can bind the metal affinity beads.

For the purification of the C-terminal kinase protein, cells containing the CBD-intein1-kinase-his and its degradation products were passed through a French pressure cell twice, at 4 $^\circ$C, to break the cells. The lysate was centrifuged at 12,000g for 20 min at 4 $^\circ$C to remove the cell debris. The clarified cell extract was loaded to 10 ml of TALON beads charged with Co^{2+} and preequilibrated with buffer A. The column was washed with 50 ml of buffer A to remove the unbound protein, whereas the degraded and intein1 fusion kinase-his can bind to the column. The bound protein was eluted by buffer A with 150 mM imidazole and the total eluate volume was about 10 ml. The purified protein can be stored at 4 $^\circ$C for several weeks. For long-term storage, glycerol was added to final 25% (v/v) and stored at $-80\ ^\circ$C. In the case of kinase used in ligation, no additional purification procedures were required. If the kinase was directly used in the NMR experiments without ligation, additional purification procedures are needed. For the purification of the kinase domain, the pH of elusion protein (mixture of degraded kinase-his and fusion CBD–chitin-kinase-his protein) was adjusted to 7.0 with 1 M Tris-HCl (pH 6.5) buffer and stored at 15 $^\circ$C for 2 days to let the fusion protein degrade. The degraded protein mixture was passed through a chitin column equilibrated with buffer A. CBD-intein1 bind to the column and the flow through was collected, exchanged to buffer B, and loaded onto a Mono Q column equilibrated with buffer B at a flow rate of 1 ml/min. The elution was conducted by linear gradient to 50% buffer C in 50 min. The peak near 15.5% of buffer C was collected and identified as kinase-his with an N-terminal cysteine (Fig. 8.4B).

5.5. Ligation of SH32 domain with testing peptide

An easy way to test whether the α-thioester can be formed after the MESNA thiolysis is to use a fluorescent peptide. A model peptide NH_2-CGRGRGRK [fluorescein]-$CONH_2$ was chemically synthesized by New England Peptide. The observed molecular weight was 1246 Da. This peptide was used to monitor reactivity of SH32 (Ayers *et al.*, 1999; Xu *et al.*, 1999). The peptide was designed to be highly soluble, and the fluorescein probe was incorporated to provide a convenient way to monitor the progress of the ligation reaction by using SDS-PAGE and other methods. In preliminary ligation studies, we investigated whether the peptide could be ligated with Csk SH32. Nearly quantitative ligation of the synthetic peptide to the corresponding protein domain with C-terminal thioester was observed when a large access of peptide was added (molar ratio 1:10) as indicated by SDS-PAGE. Typically, 20 μl of chitin beads with loaded SH32-intein2-CBD were used. To the beads was added a solution of synthetic peptide (1 mg/ml) in buffer A, along with 3% MESNA. The suspension was gently agitated at room temperature overnight, the supernatant removed, and analysis performed by SDS-PAGE. The ligation of the synthetic peptide to the recombinant protein was evidenced by the production of a highly fluorescent 21-kDa protein band on SDS-PAGE. These studies thus established that the C-terminal thioester was successfully formed during the MESNA thiolysis. After 48 h of storage at 15 °C, the SH32 α-thioester failed to ligate with the synthetic peptide.

5.6. Ligation of SH32 with kinase domain

The kinase mixture (CBD-intein1-kinase-his, kinase-his) from the TALON beads was concentrated to about 1 mM using Amicon Ultra centrifugal filter (Millipore, UFC901024) at 4 °C. The kinase protein mixture was flushed to the chitin column bound with SH32-intein2-CBD in the presence of buffer A. The molar ratio of SH32 α-thioester and N-terminal cysteine kinase was estimated to be 1:1 from the SDS-PAGE of the solution. Then, 0.5 mM EDTA and 200 mM of MESNA were added to the system and the column stored at 15 °C for 72 h. During this time, three reactions occurred: thiolysis of SH32-intein2-CBD to form SH32 α-thioester; degradation of the fusion CBD-intein1-kinase-his to form kinase-his, and reaction of SH32 α-thioester with kinase-his with an N-terminal cysteine. SDS-PAGE was used to monitor the ligation process. The speed of the MESNA-induced cleavage depends on the concentration of MESNA and the period of incubation. We increased the MESNA to 200 mM to increase the cleavage speed, as the Ala at the C-terminal of the SH32 is generally not recommended in the pTWIN1 C-terminal fusion. On-column ligation has been reported to increase the ligation efficiency (Vitali *et al.*, 2006). During 3 days of ligation, about 20 to 25% of the SH32 and kinase protein form the ligated segmental labeled protein.

5.7. Purification of ligation product

Because of the low ligation efficiency, removing the unligated protein from the ligated sample is critical. Figure 8.5 shows the combination of the Co^{2+} column and the Q column for the purification procedure. After 72 h of ligation, the ligation mixture was eluted from the chitin column using

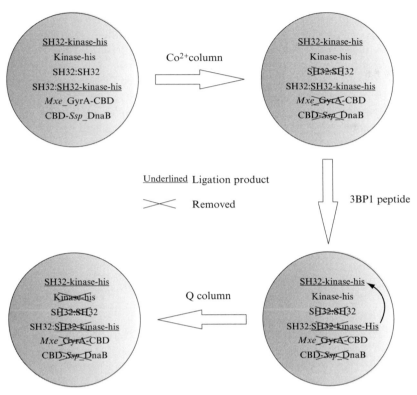

Figure 8.5 Purification procedure used in the isolation of the ligated product. The molecule in the first circle is the total product after the ligation experiment. After a Co^{2+} iron column, the protein with a C-terminal His-tag is bound to the column. The proteins without His-tag are removed. One problem is the dimerization of the SH32; although SH32 cannot bind to the Co^{2+} column, it can bind to the ligated sample that is bound to the column. PEP-3BP1 peptide was used to disrupt the dimerization via SH3 domain. In the final purification step, as the SH2 domain is highly negative charged, it can bind to the Q column at high salt concentrations (see Fig. 8.4).

buffer A. The amount of protein was monitored using the Coomassie Plus-200 protein assay reagent (Pierce, 23238). Elution was continued until no more protein was detected, typically about 30 ml. The eluate was dialysed against 1.0 L of buffer A at 4 °C for 4 h, twice, to completely remove the MESNA and EDTA. The solution was then loaded to 10 ml of TALON beads charged with Co^{2+} and preequilibrated with buffer A. The column was washed with 50 ml of buffer A to remove the unbound protein (mainly inteins and SH32). The peptide PEP-3BP1 is used to reduce the self-association of the SH3 domain (Borchert et al., 1994; Ghose et al., 2001). The protein tyrosine phosphatase PEP proline-rich peptide (PEP-3BP1, residues 605 to 629, SRRTDDEIPPPLPERTPESFIVVEE) was expressed and purified as described (Ghose *et al.*, 2001). The column was washed with another 10 ml of buffer A with 1 m*M* PEP-3BP1 peptide in it to remove the SH32 domain bind with the ligation product, as a result of the dimerization of SH3 domain. The bound protein (mainly SH32-kinase-his, kinase-his) was eluted by buffer A with 150 m*M* imidazole, and the total eluate volume was about 10 ml. Then 40 ml of buffer B was added to the eluate to reduce the salt concentration, and the total 50 ml eluate was loaded onto a Mono Q column equilibrated with buffer B at a flow rate of 1 ml/min (Fig. 8.4C). The elution was conducted by linear gradient to 60% buffer C (20 m*M* Tris-HCl, pH 8.0, 1.0 M NaCl) in 60 min. The major peak near 34% of buffer C was collected and identified as SH32-kinase-his ligation product.

5.8. NMR spectroscopy

^{1}H-^{15}N HSQC spectra were acquired on the segmental labeled Csk sample at 25 °C, 800 MHz. Figure 8.6A is the ^{1}H-^{15}N HSQC of [^{15}N-SH32]-Kinase, and about 160 amide peaks can be identified from this spectrum. Figure 8.6B is ^{1}H-^{15}N HSQC SH32-[^{15}ND-Kinase], and about 240 amide

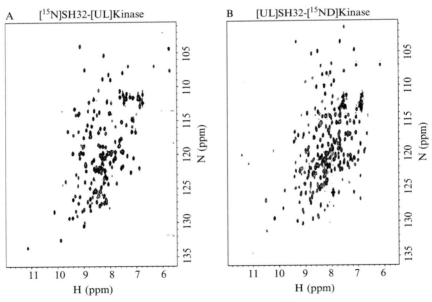

Figure 8.6 ^{1}H-^{15}N HSQC spectra at 800 MHz of (A) [^{15}N-SH32]-Kinase; (B) SH32-[^{15}N, ^{2}H-Kinase]. All of the samples are concentrated to 100 μ*M*. The buffer contains 200 μ*M* nonlabeled PEP-3BP1 peptide, 50 m*M* Tris-HCl (pH 7.5), 10 m*M* DTT, 1.0 m*M* EDTA, 0.01% (w/v) NaN$_3$, 5% D$_2$O, 0.1 m*M* DSS. NMR spectra were taken at 25 °C.

peaks can be identified from this spectrum. All NMR experiments were performed on Bruker 800 MHz, US[2] spectrometers, which were equipped with triple-resonance cryoprobes. Protein solutions were prepared in the following buffer conditions: 50 mM Tris-HCl (pH 7.5), 10 mM DTT, 1.0 mM EDTA, 0.01% (w/v) NaN3, 5% D2O, 0.1 mM DSS (4,4-dimethyl-4-silapentane-1-sulfonate). All the NMR samples were sealed in the NMR tube under a nitrogen atmosphere. Pulsed-field gradient techniques with a WATERGATE pulse sequence were used for all H_2O experiments and resulted in a good suppression of the solvent signal. The ^1H chemical shifts were referenced to internal DSS. In the 2D ^1H-^{15}N HSQC experiments, 512 complex points were collected in the ^1H dimension and 128 complex points in the 15N dimension. The corresponding sweep widths were 14 and 36 ppm in the ^1H and ^{15}N dimensions, respectively. The ^{15}N dimension was multiplied by a cosine-bell window function and zero filled to 512 points before Fourier transformation using Topspin 1.4. The ^{15}N chemical shift was referenced indirectly using the $^1H/^{15}N$ frequency ratios of the zero point 0.101329118 (^{15}N) (Live et al., 1984; Wishart et al., 1995).

ACKNOWLEDGMENTS

We are grateful to Dr. Michael Goger, Dr. Kaushik Dutta, and Dr. Shibani Bhattacharya for their assistance in the NMR instrumentation and to Professor Tom Muir for advice and discussion. Supported by NIH grants EB001991 and GM47021.

REFERENCES

Ayers, B., Blaschke, U. K., Camarero, J. A., Cotton, G. J., Holford, M., and Muir, T. W. (1999). Introduction of unnatural amino acids into proteins using expressed protein ligation. Biopolymers 51, 343–354.

Blaschke, U. K., Cotton, G. J., and Muir, T. W. (2000). Synthesis of multi-domain proteins using expressed protein ligation: Strategies for segmental isotopic labeling of internal regions. Tetrahedron 56, 9461–9470.

Borchert, T. V., Mathieu, M., Zeelen, J. P., Courtneidge, S. A., and Wierenga, R. K. (1994). The Crystal-structure of human csksh3: Structural diversity near the Rt-Src and N-Src loop. Febs Lett. 341, 79–85.

Bottini, N., Musumeci, L., Alonso, A., Rahmouni, S., Nika, K., Rostamkhani, M., MacMurray, J., Meloni, G. F., Lucarelli, P., Pellecchia, M., Eisenbarth, G. S., Comings, D., et al. (2004). A functional variant of lymphoid tyrosine phosphatase is associated with type I diabetes. Nat. Genet. 36, 337–338.

Brenzel, S., Kurpiers, T., and Mootz, H. D. (2006). Engineering artificially split inteins for applications in protein chemistry: Biochemical characterization of the split Ssp DnaB intein and comparison to the split Sce VMA intein. Biochemistry-Us 45, 1571–1578.

Camarero, J. A., Fushman, D., Sato, S., Giriat, I., Cowburn, D., Raleigh, D. P., and Muir, T. W. (2001). Rescuing a destabilized protein fold through backbone cyclization. *J. Mol. Biol.* **308,** 1045–1062.

Camarero, J. A., Shekhtman, A., Campbell, E. A., Chlenov, M., Gruber, T. M., Bryant, D. A., Darst, S. A., Cowburn, D., and Muir, T. W. (2002). Autoregulation of a bacterial sigma factor explored by using segmental isotopic labeling and NMR. *Proc. Natl. Acad. Sci. USA* **99,** 8536–8541.

Card, P. B., and Gardner, K. H. (2005). Identification and optimization of protein domains for NMR studies. *Method Enzymol.* **394,** 3–16.

Chong, S. R., Mersha, F. B., Comb, D. G., Scott, M. E., Landry, D., Vence, L. M., Perler, F. B., Benner, J., Kucera, R. B., Hirvonen, C. A., Pelletier, J. J., Paulus, H., *et al.* (1997). Single-column purification of free recombinant proteins using a self-cleavable affinity tag derived from a protein splicing element. *Gene* **192,** 271–281.

Chong, S. R., Williams, K. S., Wotkowicz, C., and Xu, M. Q. (1998). Modulation of protein splicing of the *Saccharomyces cerevisiae* vacuolar membrane ATPase intein. *J. Biol. Chem.* **273,** 10567–10577.

Chong, S. R., and Xu, M. Q. (1997). Protein splicing of the *Saccharomyces cerevisiae* VMA intein without the endonuclease motifs. *J. Biol. Chem.* **272,** 15587–15590.

Cloutier, J. F., and Veillette, A. (1996). Association of inhibitory tyrosine protein kinase p50 (csk) with protein tyrosine phosphatase PEP in T cells and other hemopoietic cells. *EMBO J.* **15,** 4909–4918.

Cole, P. A., Shen, K., Qiao, Y. F., and Wang, D. X. (2003). Protein tyrosine kinases Src and Csk: A tail's tale. *Curr. Opin. Chem. Biol.* **7,** 580–585.

Cotton, G. J., Ayers, B., Xu, R., and Muir, T. W. (1999). Insertion of a synthetic peptide into a recombinant protein framework: A protein biosensor. *J. Am. Chem. Soc.* **121,** 1100–1101.

Cowburn, D., and Muir, T. W. (2001). Segmental isotopic labeling using expressed protein ligation. *Method Enzymol.* **339,** 41–54.

Dawson, P. E., Muir, T. W., Clarklewis, I., and Kent, S. B. H. (1994). Synthesis of proteins by native chemical ligation. *Science* **266,** 776–779.

Evans, T. C., Benner, J., and Xu, M. Q. (1998). Semisynthesis of cytotoxic proteins using a modified protein splicing element. *Protein Sci.* **7,** 2256–2264.

Evans, T. C., Benner, J., and Xu, M. Q. (1999). The cyclization and polymerization of bacterially expressed proteins using modified self-splicing inteins. *J. Biol. Chem.* **274,** 18359–18363.

Evans, T. C., Martin, D., Kolly, R., Panne, D., Sun, L., Ghosh, I., Chen, L. X., Benner, J., Liu, X. Q., and Xu, M. Q. (2000). Protein trans-splicing and cyclization by a naturally split intein from the dnaE gene of Synechocystis species PCC6803. *J. Biol. Chem.* **275,** 9091–9094.

Gentle, I. E., De Souza, D. P., and Baca, M. (2004). Direct production of proteins with N-terminal cysteine for site-specific conjugation. *Bioconjugate Chem.* **15,** 658–663.

Ghose, R., Shekhtman, A., Goger, M. J., Ji, H., and Cowburn, D. (2001). A novel, specific interaction involving the Csk SH3 domain and its natural ligand. *Nat. Struct. Biol.* **8,** 998–1004.

Grant, J. E., Guo, L. W., Vestling, M. M., Martemyanov, K. A., Arshavsky, V. Y., and Ruoho, A. E. (2006). The N-terminus of GTP gamma S-activated transducin alpha-subunit interacts with the C terminus of the cGMP phosphodiesterase gamma-subunit. *J. Biol. Chem.* **281,** 6194–6202.

Hackeng, T. M., Griffin, J. H., and Dawson, P. E. (1999). Protein synthesis by native chemical ligation: Expanded scope by using straightforward methodology. *Proc. Natl. Acad. Sci. USA* **96,** 10068–10073.

Harris, T. K. (2006). Discovering new drug-targeting sites on flexible multidomain protein kinases: Combining segmental isotopic and site-directed spin labeling for nuclear magnetic resonance detection of interfacial clefts. *Methods Mol. Biol.* **316,** 199–225.

Hauser, P. S., and Ryan, R. O. (2007). Expressed protein ligation using an N-terminal cysteine containing fragment generated *in vivo* from a pe1B fusion protein. *Protein Expres. Purif.* **54,** 227–233.

Holtrich, U., Brauninger, A., Strebhardt, K., and Rubsamenwaigmann, H. (1991). 2 Additional protein-tyrosine kinases expressed in human lung — 4th member of the fibroblast growth-factor receptor family and an intracellular protein-tyrosine kinase. *Proc. Natl. Acad. Sci. USA* **88,** 10411–10415.

Hosfield, T., and Lu, Q. (1999). Influence of the amino acid residue downstream of (Asp)(4) Lys on enterokinase cleavage of a fusion protein. *Anal. Biochem.* **269,** 10–16.

Iwai, H., and Pluckthun, A. (1999). Circular beta-lactamase: Stability enhancement by cyclizing the backbone. *Febs Lett.* **459,** 166–172.

Johnson, E. C. B., and Kent, S. B. H. (2006). Insights into the mechanism and catalysis of the native chemical ligation reaction. *J. Am. Chem. Soc.* **128,** 6640–6646.

Kim, I., Lukavsky, P. J., and Puglisi, J. D. (2002). NMR study of 100 kDa HCV IRES RNA using segmental isotope labeling. *J. Am. Chem. Soc.* **124,** 9338–9339.

Lew, B. M., Mills, K. V., and Paulus, H. (1998). Protein splicing *in vitro* with a semisynthetic two-component minimal intein. *J. Biol. Chem.* **273,** 15887–15890.

Liu, D. S., Xu, R., Dutta, K., and Cowburn, D. (2008). N-terminal cysteinyl proteins can be prepared using thrombin cleavage. *Febs Lett.* **582,** 1163–1167.

Live, D. H., Davis, D. G., Agosta, W. C., and Cowburn, D. (1984). Long-range hydrogen-bond mediated effects in peptides: N-15 NMR-study of gramicidin-S in water and organic-solvents. *J. Am. Chem. Soc.* **106,** 1939–1941.

Macmillan, D., and Arham, L. (2004). Cyanogen bromide cleavage generates fragments suitable for expressed protein and glycoprotein ligation. *J. Am. Chem. Soc.* **126,** 9530–9531.

Martin, D. D., Xu, M. Q., and Evans, T. C. (2001). Characterization of a naturally occurring trans-splicing intein from Synechocystis sp PCC6803. *Biochemistry-Us* **40,** 1393–1402.

Mathys, S., Evans, T. C., Chute, I. C., Wu, H., Chong, S. R., Benner, J., Liu, X. Q., and Xu, M. Q. (1999). Characterization of a self-splicing mini-intein and its conversion into autocatalytic N- and C-terminal cleavage elements: Facile production of protein building blocks for protein ligation. *Gene* **231,** 1–13.

Mills, K. V., and Perler, F. B. (2005). The mechanism of intein-mediated protein splicing: Variations on a theme. *Protein Peptide Lett.* **12,** 751–755.

Mittermaier, A., and Kay, L. E. (2006). Review - New tools provide new insights in NMR studies of protein dynamics. *Science* **312,** 224–228.

Mootz, H. D., Blum, E. S., Tyszkiewicz, A. B., and Muir, T. W. (2003). Conditional protein splicing: A new tool to control protein structure and function *in vitro* and *in vivo*. *J. Am. Chem. Soc.* **125,** 10561–10569.

Muir, T. W., Sondhi, D., and Cole, P. A. (1998). Expressed protein ligation: A general method for protein engineering. *Proc. Natl. Acad. Sci. USA* **95,** 6705–6710.

Muralidharan, V., and Muir, T. W. (2006). Protein ligation: An enabling technology for the biophysical analysis of proteins. *Nat. Methods* **3,** 429–438.

Ogawa, A., Takayama, Y., Sakai, H., Chong, K. T., Takeuchi, S., Nakagawa, A., Nada, S., Okada, M., and Tsukihara, T. (2002). Structure of the carboxyl-terminal Src kinase, Csk. *J. Biol. Chem.* **277,** 14351–14354.

Otomo, T., Ito, N., Kyogoku, Y., and Yamazaki, T. (1999a). NMR observation of selected segments in a larger protein: Central-segment isotope labeling through intein-mediated ligation. *Biochemistry-Us* **38,** 16040–16044.

Otomo, T., Teruya, K., Uegaki, K., Yamazaki, T., and Kyogoku, Y. (1999b). Improved segmental isotope labeling of proteins and application to a larger protein. *J. Biomol. Nmr.* **14,** 105–114.

Ottesen, J. J., Bar-Dagan, M., Giovani, B., and Muir, T. W. (2008). An amalgamation of solid phase peptide synthesis and ribosomal peptide synthesis. *Biopolymers* **90,** 406–414.

Partanen, J., Armstrong, E., Bergman, M., Makela, T. P., Hirvonen, H., Huebner, K., and Alitalo, K. (1991). Cyl encodes a putative cytoplasmic tyrosine kinase lacking the conserved tyrosine autophosphorylation site (Y416src). *Oncogene* **6,** 2013–2018.

Peroza, E. A., and Freisinger, E. (2008). Tris is a non-innocent buffer during intein-mediated protein cleavage. *Protein Expres. Purif.* **57,** 217–225.

Romanelli, A., Shekhtman, A., Cowburn, D., and Muir, T. W. (2004). Semisynthesis of a segmental isotopically labeled protein splicing precursor: NMR evidence for an unusual peptide bond at the N-extein-intein junction. *Proc. Natl. Acad. Sci. USA* **101,** 6397–6402.

Roskoski, R. (2005). Src kinase regulation by phosphorylation and dephosphorylation. *Biochem. Bioph. Res. Co.* **331,** 1–14.

Schwartz, E. C., Saez, L., Young, M. W., and Muir, T. W. (2007). Post-translational enzyme activation in an animal via optimized conditional protein splicing. *Nat. Chem. Biol.* **3,** 50–54.

Shi, J. X., and Muir, T. W. (2005). Development of a tandem protein trans-splicing system based on native and engineered split inteins. *J. Am. Chem. Soc.* **127,** 6198–6206.

Shingledecker, K., Jiang, S. Q., and Paulus, H. (1998). Molecular dissection of the *Mycobacterium tuberculosis* RecA intein: Design of a minimal intein and of a trans-splicing system involving two intein fragments. *Gene* **207,** 187–195.

Skrisovska, L., and Allain, F. H. (2008). Improved segmental isotope labeling methods for the NMR study of multidomain or large proteins: Application to the RRMs of Npl3p and hnRNP L. *J. Mol. Biol.* **375,** 151–164.

Southworth, M. W., Adam, E., Panne, D., Byer, R., Kautz, R., and Perler, F. B. (1998). Control of protein splicing by intein fragment reassembly. *EMBO J.* **17,** 918–926.

Southworth, M. W., Amaya, K., Evans, T. C., Xu, M. Q., and Perler, F. B. (1999). Purification of proteins fused to either the amino or carboxy-terminus of the *Mycobacterium xenopi* gyrase A intein. *Biotechniques* **27,** 110–120.

Tolbert, T. J., Franke, D., and Wong, C. H. (2005). A new strategy for glycoprotein synthesis: Ligation of synthetic glycopeptides with truncated proteins expressed in *E-coli* as TEV protease cleavable fusion protein. *Bioorgan. Med. Chem.* **13,** 909–915.

Tolbert, T. J., and Wong, C. H. (2004). Conjugation of glycopeptide thioesters to expressed protein fragments: Semisynthesis of glycosylated interleukin-2. *Methods Mol. Biol.* **283,** 255–266.

Varadan, R., Assfalg, N., Haririnia, A., Raasi, S., Pickart, C., and Fushman, D. (2004). Solution conformation of Lys(63)-linked di-ubiquitin chain provides clues to functional diversity of polyubiquitin signaling. *J. Biol. Chem.* **279,** 7055–7063.

Vitali, F., Henning, A., Oberstrass, F. C., Hargous, Y., Auweter, S. D., Erat, M., and Allain, F. H. T. (2006). Structure of the two most C-terminal RNA recognition motifs of PTB using segmental isotope labeling. *EMBO J.* **25,** 150–162.

Weeks, S. D., Drinker, M., and Loll, P. J. (2007). Ligation independent cloning vectors for expression of SUMO fusions. *Protein Expres. Purif.* **53,** 40–50.

Wider, G. (2005). Nmr techniques used with very large biological macromolecules in solution. *Method Enzymol.* **394,** 382–398.

Wishart, D. S., Bigam, C. G., Yao, J., Abildgaard, F., Dyson, H. J., Oldfield, E., Markley, J. L., and Sykes, B. D. (1995). H-1, C-13 and N-15 chemical-shift referencing in biomolecular Nmr. *J. Biomol. Nmr.* **6,** 135–140.

Wuthrich, K. (2003). NMR studies of structure and function of biological macromolecules (Nobel Lecture). *J. Biomol. Nmr.* **27**, 13–39.

Xu, M. Q., and Evans, T. C. (2001). Intein-mediated ligation and cyclization of expressed proteins. *Methods* **24**, 257–277.

Xu, R., Ayers, B., Cowburn, D., and Muir, T. W. (1999). Chemical ligation of folded recombinant proteins: Segmental isotopic labeling of domains for NMR studies. *Proc. Natl. Acad. Sci. USA* **96**, 388–393.

Yagi, H., Tsujimoto, T., Yamazaki, T., Yoshida, M., and Akutsu, H. (2004). Conformational change of H+-ATPase beta monomer revealed on segmental isotope labeling NMR spectroscopy. *J. Am. Chem. Soc.* **126**, 16632–16638.

Yamazaki, T., Otomo, T., Oda, N., Kyogoku, Y., Uegaki, K., Ito, N., Ishino, Y., and Nakamura, H. (1998). Segmental isotope labeling for protein NMR using peptide splicing. *J. Am. Chem. Soc.* **120**, 5591–5592.

Zuger, S., and Iwai, H. (2005). Intein-based biosynthetic incorporation of unlabeled protein tags into isotopically labeled proteins for NMR studies. *Nat. Biotechnol.* **23**, 736–740.

CHAPTER NINE

SEMISYNTHESIS OF MEMBRANE-ATTACHED PRION PROTEINS

Nam Ky Chu* *and* Christian F. W. Becker*

Contents

Abstract

Conversion of cellular prion protein (PrP^{C}) into the pathological conformer (PrP^{Sc}) is the hallmark of prion diseases and has been studied extensively by using recombinantly expressed PrP (rPrP). Because of the inherent difficulties of expressing and purifying posttranslationally modified rPrP variants only a limited amount of data is available for membrane-associated PrP and its

* Technische Universität München, Department of Chemistry, Protein Chemistry Group, Garching, Germany

Methods in Enzymology, Volume 462
ISSN 0076-6879, DOI: 10.1016/S0076-6879(09)62009-7

behavior *in vitro* and *in vivo*. Protein semisynthesis provides two alternative routes to access multimilligram amounts of membrane-attached rPrP, which are described in detail here. In both cases, rPrP fused to a C-terminal extension comprising either the *Mycobacterium xenopi* GyrA mini-intein or the *Synechocystis sp.* DnaE N-terminal split intein is expressed in *E. coli*. Protein purification was followed by reaction with chemically synthesized palmitoylated membrane anchor peptides to yield rPrP^{Palm} or with a chemically synthesized glycosylphosphatidylinositol (GPI) anchor to give rPrP^{GPI}. Solubility problems encountered with synthetic membrane anchors were overcome by either incorporating a polyethylene glycol-based C-terminal tag (removable by specific proteolysis) or by direct incorporation into liposomes.

The new rPrP^{Palm} variants studied by a variety of *in vitro* methods exhibited a high affinity to liposomes and an increased lag phase during aggregation when compared to rPrP. Similar results were obtained for rPrP^{GPI}, in which only one alkyl chain is sufficient for quantitative membrane attachment. *In vivo* studies demonstrated that double lipidated rPrP^{Palm} is efficiently taken up into the membranes of mouse neuronal and human epithelial kidney cells.

1. INTRODUCTION

The conversion of cellular prion protein (PrP^C) into its pathological isoform PrP^{Sc} causes transmissible spongiform encephalopathies (TSEs) (Aguzzi and Polymenidou, 2004; Chesebro, 2003; Collinge, 2001; Prusiner *et al.*, 1998; Weissmann *et al.*, 1996). Humans and other mammalian species are affected by these infectious neurodegenerative disorders, in which both PrP isoforms exhibit significant differences in their biophysical properties, though their primary structure is identical (Hope *et al.*, 1986; Meyer *et al.*, 1986; Pan *et al.*, 1993; Safar *et al.*, 1993). Characteristic features are a disulfide bond between α–helices 2 and 3 (Turk *et al.*, 1988), the occurrence of a mixture of non-, mono-, and diglycosylated PrP variants and a C–terminal glycosylphosphatidylinositol (GPI) anchor (Stahl *et al.*, 1987, 1992).

Numerous studies have indicated that membrane association via the GPI anchor strongly influences conversion of cellular PrP into its pathogenic isoform PrP^{Sc} (Critchley *et al.*, 2004; Gorodinsky and Harris, 1995; Morillas *et al.*, 1999; Taylor and Hooper, 2006). However, the speculation that GPI anchoring might contribute to the pathogenicity of PrP is controversial (Chesebro *et al.*, 2005; Elfrink *et al.*, 2008; Lewis *et al.*, 2006). *In vitro* experiments with raft preparations from PrP infected and noninfected cells indicate that colocalization of PrP^C and PrP^{Sc} is a prerequisite for conversion of PrP^C into PrP^{Sc} (Baron and Caughey, 2003; Baron *et al.*, 2002; Eberl *et al.*, 2004; Hicks *et al.*, 2006).

Even though this seems to be direct evidence for the influence of membrane attachment of PrP on the process of conversion, the majority of *in vitro* studies on function, structure, folding, and stability of PrP use either recombinant protein lacking the GPI anchor or heterogeneous protein preparations from mammalian cell lines. This is because isolation of homogenous GPI-anchored PrP is currently impossible, and thus the exact role of the GPI anchor cannot be assessed directly with the materials currently available. Here we describe the use of two general strategies based on expressed protein ligation and protein trans–splicing for the construction of homogenous membrane-anchored proteins carrying either a peptidic membrane anchor or a synthetic GPI anchor, with a particular focus on the prion protein.

2. CHEMICAL SYNTHESIS OF MEMBRANE ANCHORS

To target rPrP to membranes, we have tested a variety of different lipidated peptide sequences, of which the ones with two palmitoyl modifications lead to a high affinity for liposomes and cell membranes (Fig. 9.1A) (Olschewski *et al.*, 2007; Zacharias *et al.*, 2002). Furthermore, peptidic membrane anchors provided sufficient flexibility for incorporation of a fluorescence label and a protease cleavage site (tobacco etch virus [TEV] protease, amino acids ENLYFQ) for controlled release of rPrP from its membrane anchor. Peptide synthesis was carried out by standard Fmoc-based chemistry on Wang resin (Atherton and Sheppard, 1989). Incorporation of palmitoyl groups and fluorescent dyes was achieved by applying combinations of orthogonal protecting groups for lysine side chains such as ivDde and Mmt. Initial tests revealed that such double-lipidated peptides can be synthesized in sufficient quality; however, purification and subsequent use in ligation reactions was hampered by severely decreased solubility in a variety of aqueous and nonaqueous buffer systems. These problems were overcome by introducing a short oligoethylene glycol–like solubilization tag at the C-terminus of the peptides, which significantly improved the handling properties of the peptides (Becker *et al.*, 2004; Rose and Vizzavona, 1999). As a result of these improved handling properties, purified membrane anchor peptides were obtained with yields of up to 40%, based on the amount of crude peptide and ligation reactions that could be carried out in the absence of detergent and organic solvents, also leading to dramatically improved yields (up to 4-fold). Even though such peptide-based membrane anchors lead to tight membrane association (artificial membranes and cell membranes), they are not comparable in structure to native GPI anchors with their complex carbohydrates and attached lipids (Fig. 9.1D) (Paulick *et al.*, 2007). To study the biological relevance of GPI anchor attachment for rPrP and its conversion into PrP[Sc], we have used a cysteine-modified synthetic GPI anchor prepared by P. Seeberger's group.

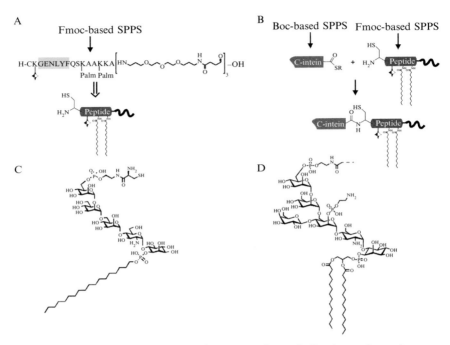

Figure 9.1 Membrane anchors used for expressed protein ligation and protein trans-splicing reactions to generate membrane-anchored rPrP. (A) Amino acid sequence of one peptidic membrane anchor equipped with two palmitoyl groups, a fluorescent dye and a C-terminal polymer tag for solubilization. (B) Scheme of native chemical ligation of a peptide based membrane anchor with the chemically synthesized DnaE^C-intein segment for use in protein trans-splicing. (C) Chemically synthesized GPI anchor carrying a C18 alkyl chain and a cysteine residue for NCL linked to the ethanolamine moiety of the GPI. (D) For direct comparison, a natively occurring GPI anchor is shown here.

The synthetic GPI was designed on the basis of information available about GPI anchors found in native PrP, and it consists of the core pseudopenta-saccharide typically found in mammalian GPIs (Fig. 9.1C) (Stahl *et al.*, 1992). Details about this synthesis can be found in Becker *et al.* (2008).

3. SEMISYNTHESIS OF rPrP^{PALM} BY EXPRESSED PROTEIN LIGATION

This synthesis strategy relies on the expression of recombinant murine PrP (rPrP, amino acids 90 to 232) in fusion with the *Mxe* GyrA mini-intein and two affinity tags (His-tag and chitin-binding domain), which allow for straightforward purification of the fusion construct from cell lysates (Fig. 9.2A).

3.1. Cloning procedure

Mouse rPrP (aa 90–232) DNA was PCR amplified and cloned into a modified pTXB3 vector (New England Biolabs) containing the *Mxe* GyrA mini-intein and a chitin-binding domain (CBD) with an additional 6xHis tag in between them using *Nco*I and *Sap*I restriction enzymes. Thus, the intein was C-terminally fused to rPrP(90–232). This construct was transformed into XL1 Blue cells (Invitrogen) and tested by PCR

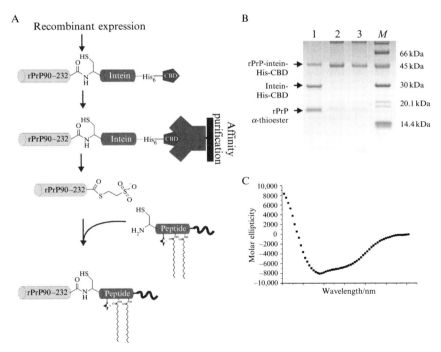

Figure 9.2 Expressed protein ligation of rPrP α-thioester with membrane anchor peptides. (A) Schematic representation of rPrP-intein-His-CBD fusion protein purification and cleavage with MESNA. Addition of a membrane anchor peptide with an N-terminal cysteine leads to rPrP^{Palm}. (B) SDS-PAGE of the rPrP-intein-His-CBD fusion protein after expression (lane 2, 8 μl and lane 3, 4 μl) and after inducing the cleavage reaction in 4 *M* urea buffer (pH 8.0) by adding 250 m*M* MESNA for 20 h (lane 1). The band with an apparent molecular weight of 16 kDa corresponds to rPrP α-thioester, whereas the band at 27 kDa represents the GyrA intein-His-CBD. (C) CD spectrum of folded rPrP^{Palm} in 20 m*M* NaOAc at pH 5.0.

amplification. The DNA from positive clones was checked by DNA sequencing and retransformed into BL21(DE3) RIL cells (Invitrogen).

3.2. Bacterial expression and protein purification

For expression, BL21(DE3) RIL cells were inoculated into 10 ml of LB medium containing 0.1 mg/ml of ampicillin (overnight at 37 °C). The entire culture was transfer into 1 L of warm LB/ampicillin medium and grown at 37 °C until the OD_{600nm} reached 0.6. Expression was induced by addition of isopropyl-β-D-thiogalactopyranoside (Sigma) to a final concentration of 1 mM, and the cultures were grown for an additional 4 h. Cells were harvested by centrifugation, resuspended in lysis buffer (1xPBS, 0.1 mM EDTA, pH 7.4), and disrupted by a M110S Microfluidizer (Microfluids Corporation, Newton). Analysis of pellet and supernatant revealed that the overexpressed fusion protein was deposited in inclusion bodies that required solubilization in chaotrope-containing buffers. The pellet was washed with PBS buffer containing 0.1% Triton X-100 to remove membrane debris and subsequently solubilized in a buffer containing 8 M Gdn-HCl and 50 mM Tris-HCl at pH 8.0 over 4 h at 4 °C. The 6xHis tag introduced between the intein and CBD in this construct allowed for purification of denatured rPrP-intein fusion protein by loading the solubilized inclusion bodies onto a Ni-NTA column (QIAGEN) equilibrated with 6 M Gdn-HCl, 50 mM Tris-HCl at pH 8.0 (Fig. 9.2A). The column was washed with 5 vol of the previously described buffer containing 30 mM imidazole. The rPrP-intein fusion protein was eluted with 0.25 M imidazole in the buffer. Fractions containing the fusion protein were pooled and checked by SDS-PAGE and by electrospray MS to unequivocally identify the recombinantly expressed rPrP with its C-terminal intein, His tag, and CBD fusion.

3.3. Intein cleavage and purification

Generating rPrP with a C-terminal α-thioester moiety for native chemical ligation reactions with synthetic membrane anchors required cleavage of the scissile Ser-Cys amide bond between rPrP and the GyrA intein that was induced with an excess of alkyl thiols (here mercaptoethansulfonate sodium salt, MESNA) (Dawson et al., 1994) (Fig. 9.2A). To achieve partial folding of the denatured fusion protein at 100 to 200 mM concentrations, a dialysis step was carried out against a buffer containing 4 M urea, 50 mM Tris-HCl, and 50 mM NaCl at pH 8.0. Addition of 0.25 M MESNA to this solution induced cleavage of the rPrP with C-terminal α-thioester from the fusion protein. However, because of the presence of 4 M urea, required to solubilize the fusion protein, the intein was only partially refolded and the thiolysis reaction proceeded slowly requiring incubation times of more than

12 h to give 50% cleavage yields. SDS–PAGE analysis of the cleavage reaction revealed the presence of a 17-kDa band representing rPrP α-thioester, a band at 27 kDa for the cleaved off GyrA intein–His–CBD part of the fusion protein and the remaining rPrP-intein-His-CBD fusion protein at 45 kDa (Fig. 9.2B). All components were identified by mass spectrometry. No cleavage was observed in Gdn–HCl-containing buffers and lowering the urea concentration below 4 M led to precipitation of the fusion protein and diminished cleavage yields. Overnight incubation at room temperature achieved a cleavage efficiency of 50%. Substantial amounts of carbamylation were observed after prolonged incubation times for the cleavage reaction, which did not increase cleavage yields beyond 60%. The resulting rPrP α-thioester was purified via a C4-RP-HPLC column (Vydac) by applying a linear gradient of 35% acetonitrile + 0.08% TFA in water + 0.1% TFA to 70% acetonitrile + 0.08% TFA in water + 0.1% TFA over 60 min. Fractions containing the protein were identified by ESI-MS, pooled, lyophilized and either stored at −20 °C or directly used in native chemical ligation reactions with membrane anchors.

3.4. Native chemical ligation reactions

The rPrP α-thioester was dissolved in 8 M urea, 300 mM NaP_i, 17 mg/ml dodecyl-phosphocholine (DPC) and 1% (v/v) thiophenol at pH 7.8 for ligation reactions with peptidic membrane anchors at concentrations of 1 mM rPrP α-thioester and 1.5 to 2 equivalents of membrane anchor peptides (Clayton et al., 2004; Olschewski et al., 2007). Ligation reactions were quenched after 24 h at room temperature by addition of 3 vol equivalents of ligation buffer and 20% (v/v) of β-mercaptoethanol for 20 min. The ligation mixtures were purified by RP-HPLC on a C4 column (Vydac) using linear gradients of 50% acetonitrile + 0.08% TFA in water + 0.1% TFA to 90% acetonitrile + 0.08% TFA in water + 0.1% TFA over 60 min. Collected fractions were analyzed by ESI-MS, pooled accordingly, and lyophilized to give 30% of rPrP^{Palm}.

Ligation of rPrP α-thioester with synthetic, cysteine-modified GPI anchor (Fig. 9.3A) proceeded more efficiently in 6 M Gdn–HCl buffer with 200 mM NaP_i at pH 7.8 in the presence of 1% (v/v) thiophenol (Dawson et al., 1994). The reaction yielded 50% rPrP^{GPI}, which was purified on a C4-RP-HPLC column (Vydac) using a similar procedure as that described for rPrP^{Palm}. Purification and analysis were aggravated by the physicochemical properties of rPrP^{GPI}. The synthetic GPI anchor conferred lipophilic properties to the protein, while the large hydrophilic glycan ensures solubility in water, thus leading to an amphiphilic compound. However, RP-HPLC in combination with ion exchange leads to pure rPrP^{GPI} that was unequivocally identified by MALDI-MS (Fig. 9.3B). Excess GPI anchor lacking the β-mercaptoethanol protecting group was

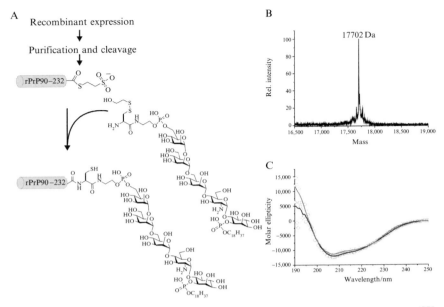

Figure 9.3 Expressed protein ligation with a synthetic GPI anchor leading to rPrP^{GPI}. (A) Schematic representation of the reaction starting with rPrP α-thioester and cysteine-modified GPI. (B) MALDI-MS of purified rPrP^{GPI} (calculated molecular mass: 17704 Da). (C) CD spectrum of folded rPrP^{GPI} in 20 mM NaOAc at pH 5.5 (black circles) in comparison to rPrP (grey triangles) at 0.2 mg/ml concentration.

recovered and recycled to improve the overall synthesis efficiency. No addition of detergents or lipids was required during this ligation reaction as that which was described for peptidic membrane anchors in native chemical ligation reactions (Durek *et al.*, 2004; Grogan *et al.*, 2005).

3.5. Folding of rPrP^{Palm} and rPrP^{GPI}

The proteins were incubated in 6 M Gdn-HCl, 50 mM Tris-HCl at pH 7.5 containing 5 mM glutathione (GSH) and 0.5 mM oxidized glutathione (GSSG) for 1 h at concentrations of 0.1 μM. Gel filtration was used to exchange the Gdn-HCl containing buffer against a buffer with 20 mM NaOAc at pH 5.5 (folding buffer) containing 20 mM OG in case of rPrP^{Palm} and no detergent in case of rPrP^{GPI} (Riek *et al.*, 1996). Folding yields of 80 to 90% were achieved, and the resulting protein solutions were concentrated to 10 μM. CD spectroscopy indicated an α-helical secondary structure indicative for PrP^C. The observed molar ellipticities for folded rPrP^{Palm} and rPrP^{GPI} are comparable to PrP^C obtained from expression in bacterial and eukaryotic systems and indistinguishable from folded rPrP without a GPI anchor (Fig. 9.3C).

3.6. Comments

Handling of rPrP samples for prolonged times in urea–containing buffers can lead to carbamylation of the protein documented by a mass increase of 43 Da. This modification can be avoided by using freshly prepared urea buffers and the highest purity of urea available.

4. SEMISYNTHESIS OF rPrP^{PALM} BY PROTEIN TRANS-SPLICING

The second strategy avoids generation and separation of a rPrP α–thioester species and is based on protein trans-splicing by expressing rPrP in fusion with the N-terminal segment of the *Synechocystis sp.* DnaE split intein (DnaE^N) and chemical synthesis of its C-terminal segment (DnaE^C) linked to membrane anchor peptides (Giriat and Muir, 2003; Martin *et al.*, 2001). Both DnaE segments spontaneously associate when correctly folded and form a functional intein, which undergoes a series of transesterification steps to give the desired modified rPrP^{Lip} (Fig. 9.4A) (Evans *et al.*, 2000). The advantages of this strategy lie within the reduced handling time in urea-containing buffers and the increased yields of the ligation reaction as described subsequently.

4.1. Cloning procedure

Mouse PrP 90–232 DNA was cloned into a modified pTXB1 vector containing the DnaE N-terminal split intein sequence (DnaE^N) from *Synechocystis sp.* and an additional C-terminal 6xHis tag using *Nde*I and *Sap*I restriction enzymes (rPrP-DnaE^N-His). For efficient expression, this vector was transferred into *E. coli* BL21(DE3) RIL.

4.2. Bacterial expression and protein purification

As described previously, protein expression was induced with IPTG and gave ≈20 mg rPrP-DnaE^N-His fusion protein per 1 L of *E. coli* culture. Cells were harvested by centrifugation and lysed in the microfluidizer. Then rPrP-DnaE^N-His was deposited in inclusion bodies and had to be solubilized in 8 *M* Gdn-HCl buffer for 4 to 6 h prior to Ni-NTA (QIAGEN) purification. The obtained fractions were dialyzed using Slide-A-Lyzer cassettes with a molecular weight cutoff of 3.5 kDa (Pierce) against 6 *M* Gdn-HCl, 50 m*M* Tris-HCl at pH 8.0 to remove the high amount of imidazole (≈300 m*M*) for further refolding of the protein.

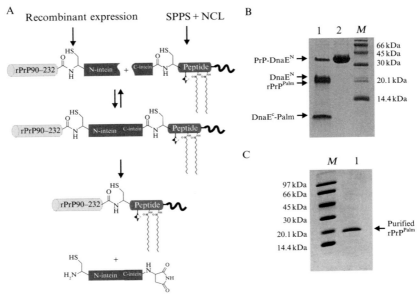

Figure 9.4 Access to rPrP^{Palm} by protein trans-splicing. (A) Scheme of protein trans-splicing with recombinantly produced rPrP-DnaE^N-His and synthetic DnaE^C-membrane anchor. On association of the intein segments, a series of transthioesterification steps leads to release of the desired product rPrP^{Palm} and the by-product containing both DnaE segments with a C-terminal aspartimide. (B) SDS-PAGE gel of purified rPrP-DnaE^N-His (lane 2). Lane 1 represents a trans-splicing reaction between rPrP-DnaE^N-His and DnaE^C-membrane anchor after 24 h. The rPrP^{Palm} band is similar in size to the DnaE^N-His band. (C) SDS-PAGE of purified rPrP^{Palm} indicating the high homogeneity of this sample.

4.3. Refolding of rPrP-DnaE^N-His

The rPrP-DnaE^N-His in 6 M Gdn-HCl buffer was diluted 1:10 to a final concentration of 0.1 mg/ml into a buffer containing 0.6 M L-arginine, 50 mM Tris-HCl (pH 8.6), and 5 mM GSH/0.5 mM GSSG. The resulting solution was incubated for 12 h at 4 °C. L-arginine and the GSH/GSSG redox agents were removed by dialysis against 50 mM Tris-HCl buffer (pH 7.5). The secondary structure of refolded PrP-DnaE^N-His was checked by CD spectroscopy.

4.4. Synthesis of DnaE^C-membrane anchor peptides

To use the membrane anchor peptide for the protein trans-splicing strategy, the C-terminal domain of the DnaE split intein (DnaE^C) had to be connected with the N-terminus of the peptides. This linkage was established by ligating the DnaE^C-peptide thioester (synthesized by Boc-SPPS (Schnölzer *et al.*, 1992) with a C-terminal α-thioester (Hackeng *et al.*, 1999)) to the membrane anchor peptide (Fig. 9.1B). Boc-based SPPS yielded DnaE^C α-thioester with purified yields of 25% (36 amino acids: MVKVIGRRSL GVQRIFDIGL PQDHNFLLAN GAIAAN). The ligation reaction was tested under a variety of conditions. A buffer system containing 6 M Gdn-HCl and 300 mM NaP$_i$ at pH 7.5 with 1% (v/v) thiophenol as ligation mediator using concentrations of 5 to 6 mM of C- and N-terminal peptides gave the best results with yields of ≈60% (Dawson *et al.*, 1994). Reactions were quenched by addition of 3 vol equivalents of ligation buffer and 20% (v/v) of β-mercaptoethanol. The addition of detergent did not increase reaction yields when peptides containing a solubilization tag were used. The purified fusion peptides consisting of the DnaE^C domain and the membrane anchor peptide were obtained in reasonable purity and with yields between 18 and 30%. Purification was achieved by RP-HPLC as described previously for the membrane anchor peptides. Product containing fractions were identified by ESI-MS and lyophilized for use in trans-splicing reactions or storage at −20 °C.

4.5. Trans-splicing with rPrP-DnaE^N-His

Protein trans-splicing of rPrP-DnaE^N and DnaE^C-membrane anchors was carried out at equimolar concentrations of 5 μM in 50 mM Tris-HCl at pH 8 (trans-splicing buffer) in the presence of 50 mM MESNA and optional addition of OG or DOPC liposomes. Reactions were stopped by addition of SDS sample buffer for SDS-PAGE analysis or by transfer into 6 M Gdn-HCl buffer. Purification was achieved by inverse affinity chromatography using Ni-NTA fast flow sepharose (Qiagen) in 6 M GdnHCl buffer. Unreacted rPrP-DnaE^N-His and DnaE^N-His were retained on the Ni-NTA material. Subsequent size-exclusion chromatography (Superdex 75, Pharmacia) allowed for separation of rPrP^{Palm} from unreacted DnaE^C-membrane anchor and DnaE^C not in complex with DnaE^N-His.

4.6. Folding of rPrP^{Palm}

Folding of rPrP^{Palm} obtained by protein trans-splicing was achieved as described earlier for rPrP^{Palm} prepared by expressed protein ligation. Combined yields of the trans-splicing reaction between rPrP-DnaE^N and membrane anchor peptide linked to DnaE^C as well as the folding of the resulting splicing product rPrP^{Palm} were typically about 20%, leading to 2 mg of rPrP^{Palm} from 1 L of *E. coli* culture.

4.7. Comments

To achieve high trans–splicing yields, it was necessary to add low concen-trations (5 to 10 mM) of reducing agents such as MESNA, DTT or TCEP to the reaction mixtures. In the absence of these reducing agents, trans-splicing yields are reduced by 20 to 50%, most likely because of oxidative modification of side-chain thiol groups required for successful trans-splicing. Such protein trans-splicing reactions can also be carried out with both components (rPrP-DnaE^N-His and DnaE^C-membrane anchor pep-tide) in buffers with 4 M urea. However, the reaction proceeds much slower under such conditions (4 days compared to 1 day), and carbamylation can be observed. Trans-splicing reactions were also carried out with the DnaE^C-membrane anchor peptide construct attached to DOPC vesicles and subsequent addition of rPrP-DnaE^N, leading to similar yields as described for detergent-containing buffers.

We also studied the influence of short amino acid sequences of the native extein sequences from the DnaE^C-split intein, which have been reported to be required for efficient trans-splicing yields (Martin *et al.*, 2001). A direct comparison of the 36-amino-acid DnaE^C intein with a DnaE^C intein contain-ing three original extein residues (amino acids CFN) provided no evidence for more efficient trans-splicing when the extra residues were included.

5. LIPOSOME ATTACHMENT AND AGGREGATION OF rPrP^{Palm} AND rPrP^{GPI}

Both strategies described here lead to recombinant α-helical rPrP with C-terminal double palmitoylated membrane anchors (rPrP^{Palm}) or a syn-thetic GPI anchor (rPrP^{GPI}) in multimilligram amounts with total yields of $\approx$10 to 20% based on the expression of rPrP fusion proteins in *E. coli*. The resulting rPrP variants constitute a first example of applying EPL and protein trans-splicing toward proteins that, because of their disposition to aggregate, are inherently difficult to handle and play an important role in conformational protein diseases. A series of denaturation and renaturation steps was required to obtain functional rPrP-intein fusion proteins as well as folded rPrP^{Palm} and rPrP^{GPI}, respectively. The folded rPrP^{GPI} and rPrP^{Palm} preparations were stable at pH 5.0 or 5.5, respectively for up to 48 h at 4 °C and could also be stored at -80 °C for prolonged periods of time without any indication of aggregation. The folding procedure used here is based on results obtained for nonlipidated PrP samples (Riek *et al.*, 1996). No difference between secondary structure of membrane-bound rPrP^{Palm}, rPrP^{GPI}, and rPrP was found in our experiments, which is consistent with data reported previously for membrane associated variants of PrP (Baron and Caughey, 2003; Eberl *et al.*, 2004; Hicks *et al.*, 2006).

5.1. Vesicle attachment of rPrP^{Palm} and rPrP^{GPI}

Reconstitution of rPrP^{Palm} and rPrP^{GPI} into small unilamellar vesicles (SUVs) was achieved by dilution of a DDM (n-dodecyl-β-D-maltoside) containing solution of rPrP^{Palm} into PBS buffer containing DOPC SUVs and subsequent removal of DDM by treatment with biobeads (Bio–Rad). Formation of small unilamellar vesicles (SUVs) was achieved by hydration of DOPC films, sonication, and subsequent centrifugation at 100,000g for 30 min at 22 °C (Beckman Coulter Optima).

Membrane insertion of rPrP^{Palm} with two palmitoyl modifications and rPrP^{GPI} with only one C18 lipid chain is almost quantitative. Unambiguous proof that rPrP^{Palm} is attached to DOPC vesicles via its two C-terminal palmitoyl chains was provided by sucrose density gradient centrifugation. The rPrP^{Palm}-containing fractions were obtained at densities of about 10%. Treating the vesicles for 6 h with TEV protease removed rPrP from its lipid anchor, and it is therefore found in the top layer of the sucrose gradient. This data clearly indicates that proteolytic removal of the palmitoyl anchor from rPrP^{Palm} leads to detachment of rPrP from the vesicles and proves that the rPrP^{Palm}-vesicle interaction is mediated by the alkyl chains of the palmitoyl groups. Therefore, unspecific hydrophobic interactions of rPrP with vesicle membranes can be ruled out.

5.2. Aggregation assays

Aggregation into a proteinase K (PK) resistant rPrP variant (rPrP^{res}) can be induced by transferring rPrP, rPrP^{Palm}, and rPrP^{GPI} at 1 μM concentrations into chaotrope-containing mixtures such as 3 M urea/1 M Gdn–HCl in buffer with 150 mM NaCl and 20 mM NaP$_i$ at pH 6.8 and at 37 °C (aggregation buffer) (Bocharova et al., 2005). Formation of rPrP^{res} is observed in Gdn–HCl/urea containing buffers after 15 h for rPrP in the presence and absence of liposomes and after 15 h for rPrP^{Palm}, and after 12 h for rPrP^{GPI} in the absence of SUVs. Formation of PK–resistant material is significantly retarded for SUV-attached rPrP^{Palm} and rPrP^{GPI} (same concentrations at 37 °C). Resistant material is not observed after 12 to 15 h, only after 36 h. The results are confirmed by using a thioflavin T fluorescence assays (Kaneko et al., 1997), which indicate a much shorter lag phase for rPrP in buffers with and without SUVs (aggregation starts after 30 to 60 min at 37 °C) when compared to rPrP^{Palm} or rPrP^{GPI} in buffer with SUVs (aggregation starts after 12 to 15 h at 37 °C). For this assay, aliquots were withdrawn during incubation of rPrP, rPrP^{Palm}, and rPrP^{GPI} in aggregation buffer and diluted into 5 mM sodium acetate buffer (at pH 5.5) to a final concentration of 0.3 μM of rPrP, rPrP^{Palm}, and rPrP^{GPI}, respectively. ThT (Sigma) was added to a final concentration of 10 μM. The emission spectra (from 460 nm to 520 nm) were recorded with excitation at 445 nm on a LS 55 Spectrometer (PerkinElmer). The fluorescence intensities at emission maximum (488 nm) were determined and plotted against time. The obtained results indicate that membrane anchoring of rPrP^{Palm} and rPrP^{GPI} reduces the speed of large aggregate formation.

6. CONCLUSION

The modification of rPrP via expressed protein ligation and protein trans-splicing with membrane anchors based on either flexible palmitoylated peptides or nativelike cysteine-carrying synthetic GPIs paves the way for detailed analyses of the influence of membrane association on conversion of PrP^C into PrP^{res} or PrP^{Sc}. A complex posttranslational modification, such as a GPI anchor attachment, has been studied only in model systems so far (Paulick and Bertozzi, 2008; Paulick et al., 2007; Tanaka et al., 2003), and the described synthesis route for homogeneous preparations of rPrP^{GPI} now allows for detailed studies of the influence of this modification on protein structure and function. In vitro and in vivo experiments using GPI-anchored PrP will help elucidate the influence of GPI-mediated membrane association on the conversion into pathogenic PrP^{Sc}.

Both synthesis strategies could be transferred to the many members of the class of membrane-attached proteins, especially the 1% of all eukaryotic proteins that carry GPI anchors (Walsh et al., 2005).

REFERENCES

Aguzzi, A., and Polymenidou, M. (2004). Mammalian prion biology: One century of evolving concepts. Cell 116, 313–327.
Atherton, B., and Sheppard, R. C. (1989). Solid phase peptide synthesis: A practical approach. IRL Press, Oxford University Press, Oxford.
Baron, G. S., and Caughey, B. (2003). Effect of glycosylphosphatidylinositol anchor-dependent and -independent prion protein association with model raft membranes on conversion to the protease-resistant isoform. J. Biol. Chem. 278, 14883–14892.
Baron, G. S., Wehrly, K., Dorward, D. W., Chesebro, B., and Caughey, B. (2002). Conversion of raft associated prion protein to the protease-resistant state requires insertion of PrP-res (PrPSc) into contiguous membranes. EMBO J. 21, 1031–1040.
Becker, C. F., Liu, X., Olschewski, D., Castelli, R., Seidel, R., and Seeberger, P. H. (2008). Semisynthesis of a glycosylphosphatidylinositol-anchored prion protein. Angew. Chem. Int. Ed. Engl. 47, 8215–8219.
Becker, C. F., Oblatt-Montal, M., Kochendoerfer, G. G., and Montal, M. (2004). Chemical synthesis and single channel properties of tetrameric and pentameric TASPs (template-assembled synthetic proteins) derived from the transmembrane domain of HIV virus protein u (Vpu). J. Biol. Chem. 279, 17483–17489.

Bocharova, O. V., Breydo, L., Parfenov, A. S., Salnikov, V. V., and Baskakov, I. V. (2005). *In vitro* conversion of full-length mammalian prion protein produces amyloid form with physical properties of PrP(Sc). *J. Mol. Biol.* **346,** 645–659.

Chesebro, B. (2003). Introduction to the transmissible spongiform encephalopathies or prion diseases. *Br. Med. Bull.* **66,** 1–20.

Chesebro, B., Trifilo, M., Race, R., Meade-White, K., Teng, C., Lacasse, R., Raymond, L., Favara, C., Baron, G., Priola, S., Caughey, B., Masliah, E., *et al.* (2005). Anchorless prion protein results in infectious amyloid disease without clinical scrapie. *Science* **308,** 1435–1439.

Clayton, D., Shapovalov, G., Maurer, J. A., Dougherty, D. A., Lester, H. A., and Kochendoerfer, G. G. (2004). Total chemical synthesis and electrophysiological characterization of mechanosensitive channels from Escherichia coli and Mycobacterium tuberculosis. *Proc. Natl. Acad. Sci. USA* **101,** 4764–4769.

Collinge, J. (2001). Prion diseases of humans and animals: Their causes and molecular basis. *Annu. Rev. Neurosci.* **24,** 519–550.

Critchley, P., Kazlauskaite, J., Eason, R., and Pinheiro, T. J. T. (2004). Binding of prion proteins to lipid membranes. *Biochem. Biophys. Res. Commun.* **313,** 559–567.

Dawson, P. E., Muir, T. W., Clark-Lewis, I., and Kent, S. B. H. (1994). Synthesis of proteins by native chemical ligation. *Science* **266,** 776–779.

Durek, T., Alexandrov, K., Goody, R. S., Hildebrand, A., Heinemann, I., and Waldmann, H. (2004). Synthesis of fluorescently labeled mono- and diprenylated Rab7 GTPase. *J. Am. Chem. Soc.* **126,** 16368–16378.

Eberl, H., Tittmann, P., and Glockshuber, R. (2004). Characterization of recombinant, membrane-attached full-length prion protein. *J. Biol. Chem.* **279,** 25058–25065.

Elfrink, K., Ollesch, J., Stohr, J., Willbold, D., Riesner, D., and Gerwert, K. (2008). Structural changes of membrane-anchored native PrP(C). *Proc. Natl. Acad. Sci USA* **105,** 10815–10819.

Evans, T. C., Martin, D., Kolly, R., Panne, D., Sun, L., Ghosh, I., Chen, L. X., Benner, J., Liu, X. Q., and Xu, M. Q. (2000). Protein trans-splicing and cyclization by a naturally split intein from the dnaE gene of Synechocystis species PCC6803. *J. Biol. Chem.* **275,** 9091–9094.

Giriat, I., and Muir, T. W. (2003). Protein semi-synthesis in living cells. *J. Am. Chem. Soc.* **125,** 7180–7181.

Gorodinsky, A., and Harris, D. A. (1995). Glycolipid-anchored proteins in neuroblastoma-cells form detergent-resistant complexes without caveolin. *J. Cell Biol.* **129,** 619–627.

Grogan, M. J., Kaizuka, Y., Conrad, R. M., Groves, J. T., and Bertozzi, C. R. (2005). Synthesis of lipidated green fluorescent protein and its incorporation in supported lipid bilayers. *J. Am. Chem. Soc.* **127,** 14383–14387.

Hackeng, T. M., Griffin, J. H., and Dawson, P. E. (1999). Protein synthesis by native chemical ligation: Expanded scope by using straightforward methodology. *Proc. Natl. Acad. Sci. USA* **96,** 10068–10073.

Hicks, M. R., Gill, A. C., Bath, I. K., Rullay, A. K., Sylvester, I. D., Crout, D. H., and Pinheiro, T. J. T. (2006). Synthesis and structural characterization of a mimetic membrane-anchored prion protein. *FEBS J.* **273,** 1285–1299.

Hope, J., Morton, L. J. D., Farquhar, C. F., Multhaup, G., Beyreuther, K., and Kimberlin, R. H. (1986). The major polypeptide of scrapie-associated fibrils (Saf) has the same size, charge-distribution and N-terminal protein-sequence as predicted for the normal brain protein (Prp). *EMBO J.* **5,** 2591–2597.

Kaneko, K., Wille, H., Mehlhorn, I., Zhang, H., Ball, H., Cohen, F. E., Baldwin, M. A., and Prusiner, S. B. (1997). Molecular properties of complexes formed between the prion protein and synthetic peptides. *J. Mol. Biol.* **270,** 574–586.

Lewis, P. A., Properzi, F., Prodromidou, K., Clarke, A. R., Collinge, J., and Jackson, G. S. (2006). Removal of the glycosylphosphatidylinositol anchor from PrP(Sc) by cathepsin D does not reduce prion infectivity. *Biochem. J.* **395,** 443–448.

Martin, D. D., Xu, M. Q., and Evans, T. C., Jr. (2001). Characterization of a naturally occurring trans-splicing intein from Synechocystis sp. PCC6803. *Biochemistry* **40,** 1393–1402.

Meyer, R. K., Mckinley, M. P., Bowman, K. A., Braunfeld, M. B., Barry, R. A., and Prusiner, S. B. (1986). Separation and properties of cellular and scrapie prion proteins. *Proc. Natl. Acad. Sci. USA* **83,** 2310–2314.

Morillas, M., Swietnicki, W., Gambetti, P., and Surewicz, W. K. (1999). Membrane environment alters the conformational structure of the recombinant human prion protein. *J. Biol. Chem.* **274,** 36859–36865.

Olschewski, D., Seidel, R., Miesbauer, M., Rambold, A. S., Oesterhelt, D., Winklhofer, K. F., Tatzelt, J., Engelhard, M., and Becker, C. F. W. (2007). Semisynthetic murine prion protein equipped with a GPI anchor mimic incorporates into cellular membranes. *Chem. Biol.* **14,** 994–1006.

Pan, K. M., Baldwin, M., Nguyen, J., Gasset, M., Serban, A., Groth, D., Mehlhorn, I., Huang, Z. W., Fletterick, R. J., Cohen, F. E., and Prusiner, S. B. (1993). Conversion of alpha-helices into beta-sheets features in the formation of the scrapie prion proteins. *Proc. Natl. Acad. Sci. USA* **90,** 10962–10966.

Paulick, M. G., and Bertozzi, C. R. (2008). The glycosylphosphatidylinositol anchor: A complex membrane-anchoring structure for proteins. *Biochemistry* **47,** 6991–7000.

Paulick, M. G., Wise, A. R., Forstner, M. B., Groves, J. T., and Bertozzi, C. R. (2007). Synthetic analogues of glycosylphosphatidylinositol-anchored proteins and their behavior in supported lipid bilayers. *J. Am. Chem. Soc.* **129,** 11543–11550.

Prusiner, S. B., Scott, M. R., DeArmond, S. J., and Cohen, F. E. (1998). Prion protein biology. *Cell* **93,** 337–348.

Riek, R., Hornemann, S., Wider, G., Billeter, M., Glockshuber, R., and Wüthrich, K. (1996). NMR structure of the mouse prion protein domain PrP(121- 231). *Nature* **382,** 180–182.

Rose, K., and Vizzavona, J. (1999). Stepwise solid-phase synthesis of polyamides as linkers. *J. Am. Chem. Soc.* **121,** 7034–7038.

Safar, J., Roller, P. P., Gajdusek, D. C., and Gibbs, C. J. (1993). Thermal-stability and conformational transitions of scrapie amyloid (prion) protein correlate with infectivity. *Protein Sci.* **2,** 2206–2216.

Schnölzer, M., Alewood, P., Jones, A., Alewood, D., and Kent, S. B. H. (1992). *In situ* neutralization in Boc-chemistry solid phase peptide synthesis. Rapid, high yield assembly of difficult sequences. *Int. J. Pept. Protein Res.* **40,** 180–193.

Stahl, N., Baldwin, M. A., Hecker, R., Pan, K. M., Burlingame, A. L., and Prusiner, S. B. (1992). Glycosylinositol phospholipid anchors of the scrapie and cellular prion proteins contain sialic-acid. *Biochemistry* **31,** 5043–5053.

Stahl, N., Borchelt, D. R., Hsiao, K., and Prusiner, S. B. (1987). Scrapie prion protein contains a phosphatidylinositol glycolipid. *Cell* **51,** 229–240.

Tanaka, Y., Nakahara, Y., Hojo, H., and Nakahara, Y. (2003). Studies directed toward the synthesis of protein-bound GPI anchor. *Tetrahedron* **59,** 4059–4067.

Taylor, D. R., and Hooper, N. M. (2006). The prion protein and lipid rafts (Review). *Mol. Membr. Biol.* **23,** 89–99.

Turk, E., Teplow, D. B., Hood, L. E., and Prusiner, S. B. (1988). Purification and properties of the cellular and scrapie hamster prion proteins. *Eur. J. Biochem.* **176,** 21–30.

Walsh, C. T., Garneau-Tsodikova, S., and Gatto, G. J., Jr. (2005). Protein posttranslational modifications: The chemistry of proteome diversifications. *Angew. Chem. Int. Ed Engl.* **44,** 7342–7372.

Weissmann, C., Fischer, M., Raeber, A., Bueler, H., Sailer, A., Shmerling, D., Rulicke, T., Brandner, S., and Aguzzi, A. (1996). The role of PrP in pathogenesis of experimental scrapie. *Cold Spring Harb. Symp. Quant. Biol.* **61,** 511–522.
Zacharias, D. A., Violin, J. D., Newton, A. C., and Tsien, R. Y. (2002). Partitioning of lipid–modified monomeric GFPs into membrane microdomains of live cells. *Science* **296,** 913–916.

CHAPTER TEN

USE OF INTEIN-MEDIATED PROTEIN LIGATION STRATEGIES FOR THE FABRICATION OF FUNCTIONAL PROTEIN ARRAYS

Souvik Chattopadhaya,* Farhana B. Abu Bakar,* *and* Shao Q. Yao*,†

Contents

Abstract

This section introduces a simple, rapid, high-throughput methodology for the site-specific biotinylation of proteins for the purpose of fabricating functional protein arrays. Step-by-step protocols are provided to generate biotinylated proteins using *in vitro*, *in vivo*, or cell-free systems, together with useful hints for troubleshooting. *In vitro* and *in vivo* biotinylation rely on the chemoselective

* Department of Biological Sciences, NUS MedChem Program of the Office of Life Sciences, National University of Singapore, Singapore
† Department of Chemistry, NUS MedChem Program of the Office of Life Sciences, National University of Singapore, Singapore

Methods in Enzymology, Volume 462
ISSN 0076-6879, DOI: 10.1016/S0076-6879(09)62010-3

native chemical ligation (NCL) reaction between the reactive α-thioester group at the C-terminus of target proteins, generated via intein-mediated cleavage, and the added cysteine biotin. The cell-free system uses a low concentration of biotin-conjugated puromycin. The biotinylated proteins can be either purified or directly captured from crude cellular lysates onto an avidin-functionalized slide to afford the corresponding protein array. The methods were designed to preserve the activity of the immobilized protein such that the arrays provide a highly miniaturized platform to simultaneously interrogate the functional activities of thousands of proteins. This is of paramount significance, as new applications of microarray technologies continue to emerge, fueling their growth as an essential tool for high-throughput proteomic studies.

1. INTRODUCTION

One of the key steps in the fabrication of functional protein arrays (Hall *et al.*, 2007; Tao *et al.*, 2007) is the immobilization of full-length proteins/ protein domains in a manner that preserves the stability and functional integrity of the immobilized proteins. Functional protein arrays have been used to study protein–protein, protein–DNA, protein–small molecule interactions, and biochemical assays of whole proteomes (Chattopadhaya and Yao, 2006; Hall *et al.*, 2004; Zhu *et al.*, 2001). Recent advances have made a plethora of methods available for protein immobilization—the choice of the method depends on downstream applications (Uttamchandani *et al.*, 2006). Random immobilization can be achieved via the use of amine-, aldehyde- (MacBeath and Schreiber, 2000), or epoxide-derivatized surfaces (Zhu *et al.*, 2000) or even via passive adsorption onto gel pads (Guschin *et al.*, 1997), poly-L-lysine-, or nitrocellulose-coated surfaces. Although straightforward, such methods may give rise to proteins that exhibit weaker activities or binding affinities, as the linkage chemistry may affect protein folding and surface accessibility by interacting proteins and/or ligands (LaBear and Ramachandran, 2005; Vijayendran and Leckband, 2001).

To obviate this drawback, an alternative approach is to uniformly orient the proteins on the chip surface via site-specific immobilization strategies. The bulk of such methodologies, thus far, have focused on the use of peptide tags and protein fusions to the N- or C-terminus of the target proteins, which are then tethered onto cognate affinity surfaces. For example, a nickel nitrilotriacetic acid (Ni-NTA)-coated surface was used to fabricate the yeast proteome array, containing 5800 chimeric yeast proteins expressed with an amino-terminal glutathione-S-transferase-polyhistidine tag (i.e. GST-(His)$_6$ tag (Zhu *et al.*, 2001)). Tirrell and colleagues have made use of heterodimeric leucine zipper domains (ZR and ZE) to immobilize target proteins having ZR as the affinity tag onto slides coated with the ZE capture domain (Zhang *et al.*, 2005). In other approaches, fusions of target protein to cutinase (Hodneland *et al.*, 2002) and human DNA repair protein O^6-alkylguanine transferase (AGT; Watzke *et al.*, 2006) have been site-specifically immobilized onto phosphonate-containing self-assembled monolayer (SAM) and modified glass surfaces, respectively. Walsh and colleagues have made use of the enzyme Sfp phosphopantetheinyl transferase to site-specifically biotinylate fusion proteins having a peptide carrier protein (PCP) tag followed by immobilization on avidin-functionalized slides (Yin *et al.*, 2004). These methods, however, are not without inherent limitations; either the binding is relatively weak and sensitive to pH and some common buffer components (for His_6-Ni-NTA; Paborsky *et al.*, 2004) or the size of the fusion protein (e.g., hAGT is 207 amino acids, PCP is 80-120 amino acids) is too large and may potentially interfere with biochemical studies, such as binding interactions with other proteins, ligands, and/or small molecules, thereby giving rise to false positives. Apart from the use of tags and fusions, site-specific protein immobilization has also been achieved using selective chemical reactions such as Diels–Alder ligation (de Araújo *et al.*, 2006), Staudinger reaction (Soellner *et al.*, 2003; Watzke *et al.*, 2006), and 1,3-dipolar cycloaddition reactions (Lin *et al.*, 2006). Although elegant, each of these methods requires elaborate modifications of the slide surface and extensive manipulations of the expressed protein before its immobilization, thereby limiting the generality of the applications. For these reasons, general and efficient methods for site-specific immobilization that overcome these shortcomings are needed.

2. Intein-Mediated Protein Ligation Strategies

Intein-based methodologies were initially developed as a tag-free, affinity-based protein purification system (Xu *et al.*, 2000). The method has since been widely adopted in the field of protein engineering to generate proteins with unnatural functionalities via the native chemical ligation reaction (Dawson *et al.*, 1994; Muir, 2003; Muralidharan and Muir, 2006). More recently, it has been extended for the site-specific modification of proteins with polyamide nucleic acids (Lovrinovic *et al.*, 2003), carbohydrates, fluorescent probes (Muralidharan *et al.*, 2004), affinity tags (Tolbert and Wong, 2000), total synthesis of proteins (Bang and Kent, 2004; Sohma *et al.*, 2007), and the generation of photo-caged proteins (Hahn and Muir, 2004; Pellois and Muir, 2005; Schwartz *et al.*, 2007). The use of inteins is advantageous as self-splicing is autocatalytic and traceless (i.e., no fragment of the intein is present in the mature protein).

We have further expanded the application of intein-mediated protein ligation strategies into the realm of microarrays (Fig. 10.1 and 10.2).

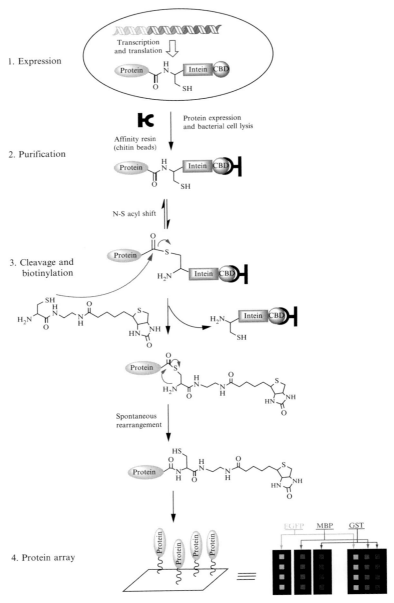

1. Expression

2. Purification

3. Cleavage and biotinylation

4. Protein array

Figure 10.1 Schematic representation of the different steps of *in vitro* purification, on–column biotinylation, and immobilization on avidin slides to generate functional protein arrays. (See Color Insert.)

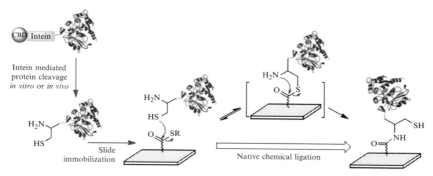

Figure 10.2 Site-specific immobilization of N-terminal cysteine containing proteins, generated by intein-mediated protein splicing on thioester slides via the chemoselective native chemical ligation reaction.

Our approach entails the generation of proteins that are site-specifically biotinylated at the C-terminus via the expressed protein ligation (EPL) reaction between the thioester moiety at the C-terminus of target proteins and cysteine-biotin. The proteins are then site-specifically immobilized on avidin-coated slides. The benefits of this strategy are that (1) proteins are immobilized in their native (Fig. 10.1) state characterized by the absence of bulky fusion tags; (2) it is a one-pot reaction in which thiolysis and ligation are done essentially in the same step; (3) the method is scalable and high-throughput as it does not involve tedious downstream protein purification steps; (4) versatile because the strategies can be used in conjugation with *in vitro*, *in vivo*, and cell-free systems, thereby providing a wider choice of methods to obtain biotinylated proteins; (5) biotin and avidin have extremely high affinity ($K_d \approx 10^{-15}$ M), thereby ensuring stable immobilization; (6) uniform protein density at each spot can be achieved as each molecule of avidin binds to four biotin molecules. Besides our methods, EPL strategies for site-specific covalent protein tethering have been reported elsewhere, whereby proteins bearing the C-terminal α-thioester moiety were covalently immobilized on slides modified with cysteine-containing polyethylene glycol linker (Camarero *et al.*, 2004). In an alternative approach, the same group used protein mediated by the split (Fig. 10.3) intein DnaE to immobilize proteins via their C-termini (Kwon *et al.*, 2006).

2.1. N-terminal intein fusions for protein immobilization

Apart from C-terminal modifications, intein-mediated protein splicing can also be used to generate N-terminal cysteine-containing proteins. In this method, a cysteine residue is introduced in the target proteins just after the cleavage site of the N-terminal intein tag (Fig. 10.2). Following *in vitro* or

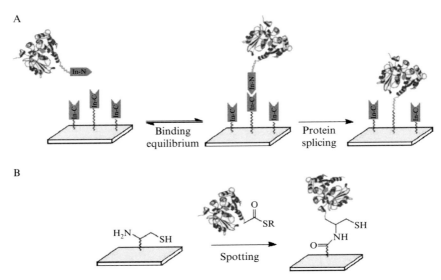

Figure 10.3 (A) Site-specific immobilization of proteins through split-intein mediated protein trans-splicing (In-N and In-C denotes N- and C-intein respectively). (B) Selective immobilization of proteins α-thioester recombinantly generated via intein-mediated protein expression onto cysteine-containing modified glass surfaces.

in vivo cleavage, proteins are immobilized via the free N-terminal cysteine residue onto thioester-functionalized slides in a site-specific manner (Girish *et al.*, 2005). Besides intein, alternative approaches to generate N-terminal cysteine containing proteins have made use of the highly selective cysteine protease: tobacco etch virus NIa (TEV) protease (Tolbert and Wong, 2002). The canonical TEV recognition and cleavage sequence (ENLYFQ↓G/S) was appended to the N-terminus of the target protein, except that the P_1' position was altered from a glycine/serine to cysteine. This change did not alter the TEV activity and specificity. Consequently, following enzymatic cleavage, N-terminal cysteine-containing proteins could be obtained easily. Furthermore, because TEV protease can be expressed in an active form in bacteria (Kapust and Waugh, 2000), the method can be extended to generate N-terminal cysteine-containing proteins *in vivo*.

In the following sections, a step-by-step protocol for C-terminal protein biotinylation using *in vitro*, *in vivo*, and the cell-free systems is provided (Table 10.1). Some insights have also been provided for the generation of N-terminal cysteine-containing proteins. Finally, a section on the preparation of avidin and thioester slides together with necessary troubleshooting guidelines has been included (Fig. 10.4). The chemical synthesis of cysteine-biotin has been reported elsewhere (Chattopadhaya *et al.*, 2006) and has not been included herein.

Table 10.1 Comparison between *in vitro*, *in vivo* and cell-free biotinylation strategies

	In vitro protein biotinylation	*In vivo* protein biotinylation	Cell-free protein biotinylation
DNA template	Plasmid DNA	Plasmid DNA	Plasmid DNA or PCR products can be used Compatible with Gateway cloning strategy
Cleavage of intein fusion protein	Requires minimal *in vivo* cleavage	Maximum *in vivo* cleavage is a necessity	NA
	On-column cleavage can be regulated by altering buffer pH, temperature, incubation time, concentration of MESNA and/or cysteine-biotin	Extent of *in vivo* cleavage cannot be regulated	NA
Protein purification	Requires additional steps of protein purification	No purification steps are involved	No purification is required
Yield and purity of biotinylated proteins	Highest yield Relatively pure	Lesser yield than *in vitro* Biotinylated proteins are present within a mixture of host proteins	Lesser yield than *in vitro* Biotinylated proteins present in cellular lysate

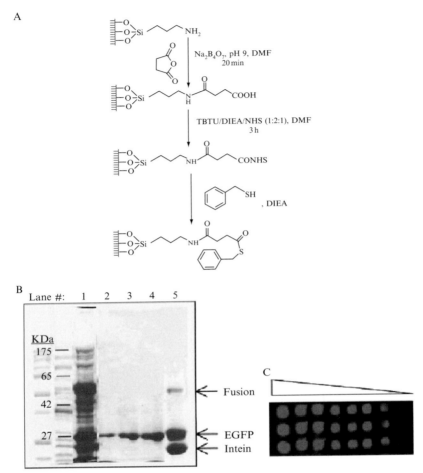

Figure 10.4   (A) Scheme showing the preparation of thioester slides. (B) Expression and *in vitro* purification of N-terminal cysteine-containing EGFP by intein-mediated cleavage on a chitin column. Lane 1: Total cell lysate containing expressed CBD-intein-EGFP fusion. Lane 2 to 4: First three elute fractions. Lane 5: Proteins retained on the column after elution. (C) Detection of N-terminal cysteine containing EGFP on thioester slides. (L to R) 1, 0.5, 0.2, 0.1, 0.05, 0.02, 0.01 mg/ml of EGFP was spotted (in triplicates) and detection was done using Cy5-labeled anti-EGFP antibody. (See Color Insert.)

2.2. C-terminal intein fusions for protein biotinylation and immobilization

For C-terminal biotinylation, we made use of specific interaction between the C-terminal thioester moiety of target protein(s) and a small molecule, cysteine-biotin 150. The thioester group is generated by cleavage of the intein tag and proteins can be biotinylated by using with *in vitro*, *in vivo* or even cell-free methods. Following biotinylation at the C-terminal, proteins are then site-specifically tethered onto avidin-coated slides. The benefits of this strategy are: (1) proteins are immobilized in their native state characterized by the absence of "bulky" fusion tags; (2) it is a "one-pot" reaction in which thiolysis and ligation are done essentially in the same step; (3) the method is scalable and high-throughput as it does not involve tedious downstream protein purification steps; (4) versatile since the strategies can be used in conjugation with *in vitro*, *in vivo* and cell-free systems thereby providing a wider choice of methods to obtain biotinylated proteins; (5) biotin and avidin have extremely high affinity ($K \sim 10^{-15}$ M) thereby ensuring stable immobilization; (6) uniform protein density at each spot can be achieved as each molecule of avidin binds to four biotin molecules. Besides our methods, EPL strategies for site-specific covalent protein tethering have been reported elsewhere whereby proteins bearing the C-terminal α-thioester moiety were covalently immobilized on slides modified with cysteine-containing polyethylene glycol linker (Camarero *et al.*, 2004) (Fig. 10.3A). In an alternative approach, the same group used protein splicing mediated by the split intein DnaE, to immobilize proteins via their C-termini (Kwon *et al.*, 2006) (Fig. 10.3B).

3. *IN VITRO*, *IN VIVO* AND CELL FREE STRATEGIES FOR PROTEIN BIOTINYLATION AT THE C-TERMINAL

In vitro protein biotinylation (Fig. 10.5) involves the expression of target protein having a C-terminal intein and chitin binding domain (CBD) fusion. Following expression, affinity purification and biotinylation of target proteins are carried out in the same step such that C-terminally biotinylated proteins can be obtained directly from crude cellular lysates without the need of any intervening purification steps (Lesaicherre *et al.*, 2002). Simple addition of cysteine-biotin together with a thiol, 2-mercaptoethanesulfonic acid (MESNA), during purification on chitin column results in on–column self cleavage of the fusion protein and, at the same time, installs a reactive α-thioester group at the C-terminus of the target protein. Subsequently, the thioester reacts with cysteine-biotin to form a peptide bond, thus generating the desired protein, which is exclusively biotinylated at the C-terminus. On elution, the biotinylated proteins can be directly spotted on avidin-functionalized slides. The strategy is particularly amenable for protein expression and purification in 96-well formats, thereby allowing for potential high-throughput generation of biotinylated proteins (Lue *et al.*, 2004). Perhaps the only rate-limiting step of this approach is that the intein-

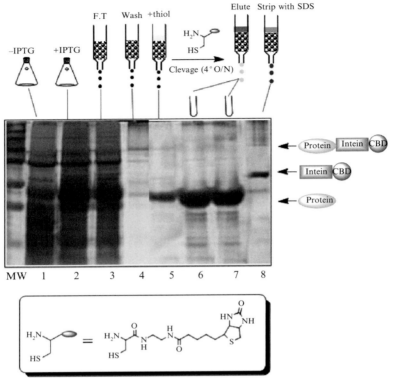

Figure 10.5 SDS-PAGE analysis of samples collected at different stages of *in vitro* purification and on-column biotinylation of maltose-binding protein-intein fusion using a chitin column. Lane 1: Total cellular lysate of un-induced samples; Lane 2: Cellular lysate of samples after overnight induction; Lane 3: Flow-through of clarified lysate; Lane 4: Chitin column wash; Lane 5: Flow-through after quick flush of column with MESNA and cysteine-biotin; Lanes 6 and 7: First two elute fractions after overnight cleavage at 4 °C; Lane 8: Proteins that remain bound to chitin beads after stripping with SDS. (Inset: structure of cysteine-biotin).

mediated chemical ligation is a relatively slow process, requiring overnight incubations and a high concentration of cysteine-biotin.

For *in vivo* protein biotinylation, previous attempts had relied on the use of proteins fused with a short 15-amino acid peptide Avitag, which is recognized and biotinylated by biotin ligase: a 35.5-kDa monomeric enzyme encoded by the *birA* gene in *E. coli* (Cronan and Reed, 2000). However, only a limited amount of biotin ligase is expressed within the cells, making this method inefficient and toxic because of a concomitant decrease in biotinylation of an essential endogenous lipid synthetic protein. Our approach entails the simple addition of the cell-permeable cysteine-biotin to the bacterial culture overexpressing the intein fusion protein.

Following a brief incubation, cells are lysed and biotinylated proteins are captured directly on avidin slides without further purification (Lue *et al.*, 2004). Direct immobilization of fusion proteins from crude cell lysates makes it feasible to fabricate protein array in high-throughput fashion by eliminating time-consuming and costly purification steps. It is important to note that this approach is not lethal to cells because no endogenous biotin ligase is involved. In addition, under optimal conditions, routinely 20 to 40% of target proteins expressed by the host cell can be biotinylated. More recent observations have indicated that, for proteins that are biotinylated poorly, a simple switch in the intein used may result in a 2- to 10-fold improvement for both *in vivo* and *in vitro* biotinylation (Tan *et al.*, 2004) Fig. 10.6.

A significant milestone in the fabrication of protein chips was the use of cell-free systems for protein expression. In this approach termed nucleic acid programmable protein array (NAPPA), *de novo* protein synthesis using an *in vitro* transcription and translation system and affinity capture was done in a single step (Ramachandran *et al.* 2004). For this purpose, a mixture of avidin and anti-GST antibodies was co-immobilized with biotinylated plasmids at predefined positions on the array. The plasmids encoded for the target protein together with a C-terminal GST tag. Following expression, proteins were captured *in situ* via the GST tag. The use of cell-free

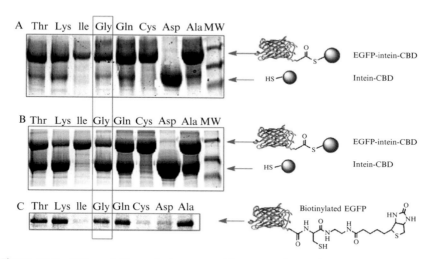

Figure 10.6 Effect of C-terminal amino acid on on-column cleavage and biotinylation. Samples from different steps of *in vitro* purification and biotinylation and analyzed by SDS-PAGE. (A) Proteins bound to chitin beads before on-column cleavage. This shows the extent of *in vivo* cleavage. (B) Proteins bound to chitin beads after cysteine-biotin and MESNA addition, which indicates the extent of on-column cleavage. (C) Biotinylated EGFP after elution from column. Glycine is the preferred residue, as it gave minimal *in vivo* cleavage and maximal biotinylation efficiency.

systems is advantageous as it can circumvent the problems associated with *in vivo* protein expression: mainly formation of inclusion bodies, mRNA degradation, proteolysis due to protein misfolding, and difficulties in the expression of toxic proteins. Such systems have also been known to allow for specific and efficient C-terminal protein modifications (Miyamoto-Sato *et al.*, 2000; Tao and Zhu, 2006). Based on these observations, our cell-free protein biotinylation approach makes use of the rational intervention by the antibiotic puromycin during the terminal steps of protein biosynthesis (Fig. 10.7). Puromycin resembles the $3'$ end of aminoacyl tRNA and competitively inhibits protein synthesis by blocking the action of peptidyl transferase. At low concentrations, puromycin serves as a low-molecular-weight acceptor substrate in the peptidyl transfer reaction and gets incorporated when a stop codon is presented at the ribosomal A-site, followed by the release of the resulting protein-puromycin adduct. Using a biotin-conjugated puromycin analogue, $5'$-biotin-dc-Pmn and commercial cell-free systems, we were able to obtain biotinylated proteins using both plasmids and PCR products as DNA templates within a matter of hours (Tan *et al.*, 2004). Furthermore, we have shown that the strategy is compatible with high-throughput cloning/expression procedures such as the Gateway (Invitrogen) cloning strategy. This is significant, as it paves the way for parallel

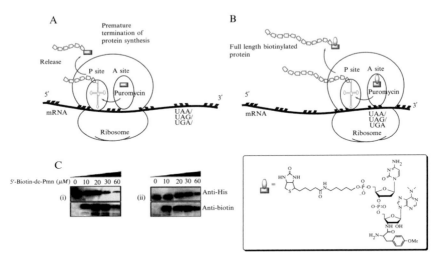

Figure 10.7 Schematic for cell-free biotinylation of protein using $5'$-biotin-dc-Pmn. (A) At high concentrations, puromycin inhibits protein synthesis. (B) At low concentrations, it is incorporated into newly synthesized proteins in response to a stop codon to give full length proteins at the specifically biotinylated at the C-terminal. (C) Expression and biotinylation of GFP using plasmid DNA (i) or PCR product (ii) as template in cell-free systems; anti-His (top row) immunoblots show the amount of protein synthesized while antibiotin blots (bottom row) show the extent of biotinylation. (Inset: structure of $5'$-biotin-dc-Pmn).

assay of whole proteomes as expression-ready ORF collections for high-throughput protein production are now available for human (Rual *et al.*, 2005), yeast (Gelperin *et al.*, 2005), and *E. coli* (Arifuzzaman *et al.*, 2006).

 ## 4. Methods

4.1. Protocols for biotinylation of proteins at the C-terminus

4.1.1. *In vitro* protein biotinylation

1. Clone the protein(s) of interest into the vector pTYB1 or pTYB2 following standard molecular biology procedures (Sambrook *et al.*, 1989).
2. Transform the plasmid by heat shock/electroporation into the *E. coli* expression strain ER2566 or any appropriate host that carries a T7 RNA polymerase gene (BL21 (DE3) and its derivatives). Select for positive transformants on LB agar plates containing 100 μg/ml of ampicillin. Incubate plates overnight (10 to 12 h) at 37 °C.
3. Inoculate a single colony in 3 ml of LB-ampicillin (100 μg/ml) media. Grow cultures overnight (12 to 14 h) with constant shaking (200 to 250 rpm) at 37 °C.
4. Dilute the overnight culture 1:100 into 200 ml of fresh LB-ampicillin (100 μg/ml) media. Incubate at 37 °C with constant shaking at 250 rpm till $OD_{600} \approx 0.5$ (approximately 2 to 3 h). Take a sample of 100 μl of culture (uninduced cells), centrifuge at 2500g for 2 min, discard supernatant, and resuspend cells in 50 μl of 1x SDS-PAGE sample dye; vortex and freeze at −20 °C. Add 0.3 to 0.5 mM IPTG to the culture media for induction of protein expression. Incubate cultures with constant shaking for 12 to 16 h at 18 °C.
5. Collect a 100-μl sample of culture (induced cells), spin, discard the supernatant, resuspend in 50 μl of 1x SDS-PAGE sample dye, vortex, and store at −20 °C.
6. Transfer the remaining cultures into Beckman centrifuge tubes. Spin for 10 min (6000g 4 °C) to pellet cells. All subsequent steps should be carried at 4 °C to minimize protein degradation. Alternatively pellets can be stored in −20 °C for up to 1 month but may result in loss or decrease of protein activity.
7. Resuspend cell pellets in 10 ml of ice-cold (4 to 8 °C) lysis buffer (20 mM Tris-HCl, pH 8.0, 500 mM NaCl, 1 mM EDTA and 0.1% Triton X-100. PMSF (20 $\mu$$M$) and 1 m$M$ TCEP may be included).
8. Sonicate cells on ice at 30% amplitude with five 10-s bursts and a 30-s cooling interval between each burst. Monitor release of proteins by Bradford protein assay. If lysis is incomplete, use additional bursts as required.

9. Clarify the lysate by centrifugation at 20,000*g* for 30 min at 4 °C. Transfer the supernatant and save on ice. Add 5 μl of 2x SDS–PAGE sample dye to 5 μl of clarified lysate. Vortex and freeze at −20 °C until further use.

10. (Optional) Resuspend the pellet (from 9) in 5 ml of column buffer to give a suspension of the insoluble matter. Add 5 μl of 2x SDS–PAGE sample dye to 5 μl of the insoluble matter. Vortex and freeze the sample at −20 °C until further use.

11. Pack an Econo-Column chromatography column (1.5 cm diameter × 30 cm length) with 3 ml of chitin beads (15 to 20 ml of beads/1 L of culture). Preequilibrate the column with 30 ml (10 bed volumes) of column buffer (20 m*M* Tris-HCl, 500 m*M* NaCl, 1 m*M* EDTA, pH 8.0) before loading lysate.

12. Slowly load the clarified lysate (flow rate: 0.5 to 1 ml/min) onto the chitin column and allow the column to empty by gravity flow. Collect 10 μl of the flow-through, add 10 μl of 2x SDS–PAGE sample dye, vortex, and freeze at −20 °C.

13. Wash the column with 30 ml (10 bed volumes) of column buffer at a flow rate of 2 ml/min. Remove 10 μl from the wash, add an equal volume of 2x SDS–PAGE sample dye, vortex, and store at −20 °C.

14. Quickly flush the column with 6 ml (2 bed volumes) of column buffer, containing 50 m*M* of MESNA and 1 m*M* of cysteine-biotin. Stop the flow. Take 20 μl of the flow-through from the quick flush, mix with 20 μl of 2x SDS–PAGE sample dye, vortex, and freeze at −20 °C. Incubate the column at 4 °C overnight.

15. Elute the biotinylated protein with 10 ml of column buffer or PBS (8 g/L NaCl, 0.2 g/L KCl, 0.14 g/L Na$_2$HPO$_4$, 0.24 g/L KH$_2$PO$_4$.) or any other desired buffer. Collect 1-ml fractions of the eluate, and measure the protein concentration by Bradford assay and pool fractions that contain high amounts of biotinylated protein (usually the first few fractions).

16. Freeze and lyophilize the pooled fractions and reconstitute in buffer that maintains the activity of the protein. Take 5 μl of the concentrated protein for SDS–PAGE analysis.

17. (Optional) Remove 20 μl of the chitin beads, resuspend in 20 μl of 2x SDS–PAGE sample dye (proteins bound to chitin beads after elution), and vortex.

18. Thaw the samples collected at the different steps (4, 5, 9, 10, 12, 13, 14, 16, and 17), boil for 5 min, and analyze on a 10% SDS–PAGE gel. For each sample, 5 to 10 μl is sufficient to visualize the proteins by Coomassie brilliant blue staining. This allows for monitoring of each step in the biotinylation procedure (Figure 10.5).

19. Confirm the presence of biotinylated proteins by Western blot using HRP-conjugated NeutraVidin (Pierce).

20. (Optional) Incubate the eluted fraction (Step 15) in an appropriate buffer with excess of Streptavidin Magnesphere Paramagnetic particles (SA-

PMPs; Promega) for 1 h at 4 °C with gentle agitation. Using a Magne-Sphere Technology Magnetic Separation stand (Promega), magnetically capture the biotinylated proteins and carefully remove the supernatant. Save supernatant for further analysis. Wash the SA-PMP-biotinylated protein complex with PBS (3 times). Capture the particles magnetically after each wash and remove the wash solution. Resuspend the SA-PMP in 20 μl of PBS. Add 20 μl of 2x SDS-PAGE sample dye. Analyze the different samples (before and after adsorption, supernatant, and washes) by Western blotting using antibody against the target protein/epitope on the target protein. Estimate the extent of biotinylation.

21. (Optional) Regenerate the chitin beads by the following wash steps: wash column with 3 bed volumes of 0.3 M NaOH; allow the resin to soak for 30 min in NaOH; further wash the resin with 7 bed volumes of 0.3 M NaOH before rinsing with 20 bed volumes of water followed by 5 bed volumes of column buffer. Store regenerated resin at 4 °C. For long term storage, add 0.02% sodium azide to the column buffer.

4.1.1.1. *Practical considerations*

A. Different types of vectors are available for expression of intein-fusion proteins (Table 10.2). For *in vitro* biotinylation, the choice of vectors depends on (1) minimal *in vivo* cleavage of the fusion; (2) maximal on-column cleavage of the fusion and high biotinylation efficiencies; (3) whether the intein fusion is at the N- or C-terminus of target protein, as this may affect protein expression and/or solubility; and (4) whether additional vector-derived sequences may affect protein function or activity. Our findings with EGFP suggest that the target protein and/or amino acid at the protein-intein junction significantly affect the extent of *in vivo* cleavage as well as on-column cleavage and biotinylation. C-terminal glycine residue(s) prevents premature *in vivo* cleavage while maximizing on-column biotinylation (Table 10.3 and Fig. 10.6). (Table 10.4).

B. If unfavorable residues are present at the intein splice sites, use pTYB2 (C-terminal) or pTYB12 (N-terminal) to add vector-derived amino acids, and this in turn depends on the choice of restriction sites. Alternatively, introduce the desired amino acids (e.g., Gly) via PCR primers and clone the gene into SapI site of pTYB1/pTYB11, which places the amino acid at the splice junction; however, in both cases, an additional amino acid will be added to the protein sequence either at the C- (pTYB1) or N-terminus (pTYB11).

C. The extent of biotinylation also depends on the temperature and time for which the reaction is allowed to proceed.

Table 10.2 Comparative analysis of the different intein fusion vectors used

Vector	Intein type and length (in amino acids)	Terminus at which intein is fused	Cleavage conditions	Applications
pTYB1/2	*Sce* VMA intein (454)	C-terminus	Thiols	Protein purification; C-terminal thioester for expressed protein ligation
pTYB11/12	Modified *Sce* VMA intein[a] (510)	N-terminal	Thiols	Protein purification; defined N-terminal (e.g. Cys) for pTYB11
pTWIN1	*Ssp* DnaB mini-intein (154) Modified *Mxe* GyrA intein (198)	N-terminal C-terminal	pH & temperature Thiols	Protein purification; cyclization of proteins; generate proteins having defined N-terminal; C-terminal thioester for EPL
pTWIN2	*Ssp* DnaB mini-intein (154) Modified *Mth* RIR1 intein (134)	N-terminal C-terminal	pH & temperature Thiols	Protein purification; cyclization of proteins; generate proteins having defined N-terminal; C-terminal thioester for EPL

[a] *Sce* VMA intein is modified to undergo cleavage at both termini. Splicing and cleavage are blocked *in vivo* but can be induced *in vitro* in presence of thiols (Chong *et al.*, 1998).

Table 10.3 Effect of C-terminal residue on *in vivo* cleavage and subsequent on-column cleavage/biotinylation

C-terminal residue of EGFP	*In vivo* cleavage	On-column cleavage/ biotinylation
Ala	+	+ + +
Arg	+ + +	+ + +
Asn	+	+
Asp	100%	ND
Cys	+	ND
Gln	+	+ + +
Glu	+ + + +	ND
Gly	+	+ + + +
His	+ + +	+ + +
Ile	+	+
Leu	+ +	+ +
Lys	+ +	+ + + +
Met	+ +	+ + +
Phe	+ +	+ + + +
Pro	+	+ +
Ser	+	+ +
Thr	+	+ + +
Trp	+ +	+ + +
Tyr	+ + +	+ + +
Val	+	+

Notes: ND: Not determined; amino acids are indicated by the three letter codes. (+): <25% ;cleavage/ biotinylation; (++): 25 to 50% cleavage/ biotinylation; (+++): 50 to 75% cleavage/biotinylation; (++++): 75 to 100% cleavage/biotinylation.

D. Lysis (Step 7) and column buffer (Step 11) should be devoid on any thiols (DTT, β-mercaptoethanol), as it would result in cleavage of the fusion before affinity purification.

E. Use freshly prepared MESNA (Step 14). MESNA and not DTT should be used to induce on-column cleavage. Even though MESNA is not as efficient as DTT as an on-column cleaving agent, the use of MESNA has been found to favor a higher incorporation of cysteine-biotin. Our data indicates that both MESNA and cysteine-biotin must be present together for biotinylation to occur. It is important that the quick flush does not exceed for more than 30 min.

F. For oxidation-sensitive proteins, include TCEP [tris-(2-carboxyethyl) phosphine] in the lysis buffers, as it specifically reduces disulfide bonds without affecting thiosters.

G. For estimation of extent of biotinylation (Step 20), use antibodies against target protein/epitope on the protein. Use of antibiotin/

Table 10.4 Troubleshooting guidelines for the site-specific protein biotinylation strategies

Problem	Possible reason	Solution
1. In vitro protein biotinylation		
No or low levels of expression	Incorrect reading frame	Check DNA sequence; alter reading frame by site-directed mutagenesis
	Formation of inclusion bodies; protein is toxic to host cells	Try different expression conditions (vary IPTG concentrations and/or lower temperature)
	Incorrect expression host	Use only ER2566 or bacterial expression host having a copy of T7 RNA polymerase (BL21(DE3) and its derivatives)
	Proteolysis of target proteins	Use protease-deficient bacterial strains (BL21 and its derivatives, Turner strains)
	Size of intein fusion protein is too large	Use vectors of the IMPACT–TWIN systems (NEB) which contain mini-intein
Target protein present in flow through	DTT present in SDS–PAGE sample dye	Prepare fresh buffer(s) omitting DTT and other thiol reagents. TCEP (3–5 mM) can be used as reducing agent instead of DTT
	Lysozyme used for cell lysis which digests chitin binding domain	Use sonication or French press for lysis
	Unfavorable residues (Arg, Asp) at the C-terminal of protein resulting in *in vivo* cleavage	Use pTYB2. Introduce extra nucleotides coding for favorable amino acids (like Gly) to the C-terminal of target protein via PCR primers
Intein fusion present in flow through	Insufficient binding of fusion protein to chitin beads	Reduce sample viscosity by adding protease-free DNase (10 μg/ml) plus MgCl$_2$ (5 mM) to lysis buffer; increase the amount of chitin beads

Problem	Possible cause	Solution
Low amount of biotinylated protein in elute fractions	Cleavage of fusion protein before affinity purification	Avoid DTT (or other thiols) in both lysis and column buffer. TCEP/TCCP (0.1-1 mM) may be included in lysis and column buffer for oxidation sensitive proteins
	High *in vivo* cleavage	Use a different intein and/or change amino acid at the protein-intein junction
	Insufficient amount of cysteine-biotin and/or MESNA	Increase the concentration of cysteine-biotin Use both MESNA and cysteine-biotin instead of DTT for on-column cleavage and biotinylation
	Insufficient incubation time	Incubate for longer time (up to 40 hr) at 4 °C
	Target protein becomes insoluble following cleavage	Increase salt concentration (0.5-2 M NaCl) and/or add a non-ionic detergent (0.1-0.5% Triton X-100 or 0.1-0.2% Tween®-20) to the cleavage buffer
Eluted protein is not pure	Washing conditions not optimized	Increase the number of washes and/or use stringent wash conditions. Include high salt (0.5-1 M NaCl) and non-ionic detergents for wash
Fusion protein bound to beads	Insufficient on-column cleavage	Unfavorable residues at the protein-intein junction; introduce additional nucleotides coding for favorable residues via PCR primers or changeamino acids by site-directed mutagenesis
		Increase pH of column buffer (pH 7-9)
		Longer incubation time and higher temperature may be used (16-23 °C for 16-40 hr) provided protein activity is not altered
2. In vivo protein biotinylation		
Insufficient *in vivo* biotinylation	MESNA and/or cysteine-biotin not optimal	Vary MESNA and/or cysteine-biotin concentrations
	Unfavorable amino acid residues at intein-protein junction	Add different amino acids by PCR, alter RE site and/or use different vector

(continued)

Table 10.4 (continued)

Problem	Possible reason	Solution
3. Cell-free protein biotinylation		
No or low level of protein expression	Premature termination of translation	Check DNA sequence; do site-directed mutagenesis to rectify reading frame and/or eliminate secondary structure in mRNA by mutagenesis
	Secondary structure of mRNA inhibits translation initiation; tag has a negative effect on protein folding	Optimize template design; try different N-or C-terminal tags using the RTS E. coli Linear Template Generation Set
	Plasmid not pure	Repeat the plasmid prep. Alternatively, purify DNA by equilibrium centrifugation in CsCl-ethidium bromide gradient. Repeat, if necessary
	Multiple bands in PCR product	Optimize PCR conditions
	Incorrect/not optimal design of DNA template	Optimize sequence using ProteoExpert RTS E. coli HY bioinformatics tool (www.proteoexpert.com)
	Short expression time; not optimal DNA amount	Increase time for expression up to 9 hr; increase DNA amount added to the reaction
Sufficient protein expression but low amount of active protein	Protein aggregates or is insoluble	Add chaperones and/or cofactors
		For proteins with up to 3 disulfide bonds, allow oxidation of the reaction to proceed for 16–24 hr at 2–8 °C
		Alter reaction conditions (lower temperature in the range 4–14 °C)
		Include mild detergents in the reaction mixture

NeutrAvidin results in detection of only biotinylated protein pool and not total proteins will lead to erroneous estimation.

4.1.2. *In vivo* protein biotinylation

1. The plasmids used for *in vitro* biotinylation strategy can also be used for *in vivo* biotinylation. Hence, the initial steps of transformation, bacterial culture, and protein expression with IPTG are essentially the same as described in the *in vitro*–based method (Step 1 to 4).
2. Grow the bacterial culture overnight (12 h) at 23 to 25 °C with constant shaking at 200 rpm.
3. At the end of this period, add 30 mM MESNA and 3 mM cysteine-biotin to the bacterial culture from Step 2. Both MESNA and cysteine-biotin should be added together for optimal *in vivo* cleavage and biotinylation to occur. Incubate the culture for an additional 24 h at 4 °C with constant shaking at 250 rpm.
4. Centrifuge at 6000g for 15 min at 4 °C to obtain cell pellet. Discard supernatant. Wash the cell pellet at least 2 to 3 times with column buffer or PBS to remove excess unreacted cysteine-biotin. After each wash, spin at 6000g for 15 min at 4 °C.
5. Resuspend the cell pellet in 4 ml of ice-cold lysis buffer (4 to 8 °C; 1 ml of lysis buffer/50 ml culture).
6. Sonicate cells on ice at 30% amplitude with five 10-s bursts and a 30-s cooling interval between each burst. Monitor release of proteins by Bradford protein assay. Use additional bursts as required for incomplete lysis.
7. Centrifuge the crude cellular lysate at 20,000g for 30 min at 4 °C. Transfer the clarified cellular lysate to a 15-ml falcon tube. Directly transfer 10 to 20 μl to a 384-well plate for spotting (no additional protein purification steps are required).
8. (Optional) Assess the extent of protein biotinylation using SA-PMPs as described in Step 20 of *in vitro* protein biotinylation procedure.

4.1.3. Cell-free protein expression and biotinylation

1. Transfer 12 μl of *E. coli* lysate, 10 μl of reaction mixture, 12 μl of amino acids, 1 μl of 1 mM methionine, and 5 μl of reconstitution buffer from the RTS 100 *E. coli* kit (Roche Diagnostics) to a PCR tube.
2. Add DNA template (0.1 μg of linear or 0.5 μg of plasmid DNA) to the reaction mixture. In our experiments, pIVEX control vector GFP (Roche Diagnostics) was used as the plasmid DNA template, but other plasmids can also be used. For plasmid DNA templates, OD_{260}/OD_{280} absorbance ratio should be ≥ 1.7, while for PCR templates, remove inhibitory primer-dimers by PCR cleanup kit, and not by agarose gel purification.
3. Add biotin-conjugated puromycin (5′-biotin-dc-Pmn; Dharmacon RNA Technologies) to a final concentration of 25 μM and make up

the reaction volume to 50 μl with RNase-free deionized water. Mix carefully by gently flicking the tubes.

4. Maintain the reaction mixture at 30 °C for 6 h. Use 5 μl to check for biotinylated proteins by Western blotting with HRP-conjugated Neutravidin (Pierce). For proteins in the 20 to 30 kDa range, precipitate the proteins with acetone before adding SDS-sample buffer, as the reaction solution contains polymers that interfere with subsequent SDS-PAGE analysis.

5. Remove any excess unincorporated 5'-biotin-dc-Pmn by using a NAP-5 column. Before sample addition, equilibrate the NAP column with 10 ml of column buffer and allow the column to empty by gravity flow. Add 450 μl of equilibration buffer to the reaction mix from Step 3 and apply it to the preequilibrated column. Allow the entire sample to enter by gravity flow.

6. Elute the biotinylated proteins by adding 1 ml of PBS buffer and estimate the protein concentration by Bradford assay.

7. (Optional) For diluted protein samples, concentrate by freeze-drying. Alternatively, use a Microcon centrifugal filter device to concentrate the protein.

8. Reconstitute the protein in appropriate buffer for protein activity and/or storage and take 10 μl in a 384–well plate for microarray spotting.

9. (Optional) In a microcentrifuge tube, incubate the RTS mixture with excess of Immobilized Neutravidin Biotin Binding Protein for 1 h at 4 °C with gentle agitation. Spin at 2500g for 1 to 2 min and remove supernatant. Save supernatant for further analysis. Wash NeutrAvidin-bound protein complex with 0.5 to 1 ml of PBS, spin at 2500g for 1 to 2 min, and remove supernatant. Repeat wash steps 3 to 4 times and remove the final wash. Boil samples in SDS-PAGE sample dye. Analyze the samples (before and after adsorption and supernatant) by Western blot using antibody against the target protein/epitope on target protein. Estimate the extent of biotinylation by comparing the intensity of the band obtained before and after adsorption using the formula and method indicated in Step 20 of *in vitro* biotinylation.

4.2. Protocol for generation of N-terminal cysteine-containing proteins

The protocol for generation of N-terminal cysteine proteins is essentially the same as that for *in vitro* protein biotinylation, except for some minor modifications, which are listed subsequently.

1. For expression of N-terminal cysteine containing proteins, we made use of the pTWIN vectors (NEB). Because both pTWIN1 and pTWIN2 allow for the fusion of the same *Ssp* DnaB intein, N-terminal to the

target protein, either vector is suitable for this purpose. The target gene was cloned into the SapI and PstI sites. This did not add any vector derived sequences.

2. Following induction with 0.5 mM IPTG at $OD_{600} \approx 0.6$, protein expression was allowed to proceed for overnight at 15 °C. It is important to test out different conditions for induction so as to minimize *in vivo* cleavage.

3. Lysis was done in column buffer (20 mM Tris-HCl, pH 8.5, 500 mM NaCl, 1 mM EDTA). In addition, PMSF (20 μM) and 0.1 to 0.2% Tween 20 can also be added to the buffer.

4. All steps of centrifugation, cellular lysis, preequilibration of chitin column, column loading, and wash must be carried out at 4 °C to minimize cleavage of the intein tag.

5. Cleavage of the intein tag is mediated by a change in pH. For this purpose, equilibrate the column with cleavage buffer (20 mM Tris-HCl, pH 7.0, 500 mM NaCl, 1 mM EDTA) and incubate the column overnight at room temperature with gentle agitation. Monitor each step of the purification by SDS-PAGE analysis. If sufficient cleavage occurs, increase the time of cleavage to 40 h at 4 °C.

4.3. Preparation of slides

1. Make piranha solution (7:3 (v/v) H_2SO_4:H_2O_2) by slowly adding H_2O_2 to H_2SO_4 as the solution becomes extremely hot.

2. Place glass slides (75 mm length × 25 mm width) in a slide-staining rack (695 mm × 86 mm × 21 mm). Transfer them to a slide-staining jar (81 mm × 161 mm × 21 mm) containing piranha solution. Soak for at least 4 h; alternatively, slides can be left in piranha solution until needed. Only resistant plasticware (e.g., polypropylene, polymethylpentene) should be used and, before disposal, piranha solution must be neutralized with base (Na_2CO_3/NaOH).

3. Wash the slides copiously with deionized water, rinse with 95% ethanol, and dry the slides under a stream of pure nitrogen or centrifugation (100g, 1 to 2 min).

4. Soak the freshly cleaned slides in a solution containing 380 ml of 100% ethanol, 8 ml of water, and 12 ml aminopropyltriethoxysilane for 2 h with constant stirring. Wash slides 2 to 3 times with 95% ethanol.

5. Cure slides at 150 °C for at least 2 h or overnight. Rinse slides with ethanol and dry.

6. Place the amine slides (Step 4) in a solution of 180 mM succinic anhydride and 1 M $Na_2B_4O_7$ (pH 9) in DMF. React slides for 15 to 20 min and then soak the slides in boiling water for 2 min. Rinse the slides copiously with 95% ethanol and dry slides under a stream of nitrogen.

7. Activate the carboxylic acid functionalized slides with a solution of TBTU, DIEA, and NHS (1:2:1) in DMF for 3 h. Rinse slides with ethanol and dry.

4.3.1. Thioester slides

A. Place the NHS slides (Step 7) in a solution of 120 mM DIEA and 100 mM benzylmercaptan in DMF and react overnight (Fig. 10.4). Rinse slides with ethanol and dry. The thioester-functionalized slides can now be used for spotting or stored in a dessicator.

4.3.2. Avidin slides

A. To the resulting NHS slides (Step 7), apply 40 to 60 μl of 1 mg/ml avidin in 10 mM NaHCO$_3$ (pH 9). Cover with coverslip (22 mm × 75 mm) for even application and incubate for 30 min at room temperature in a humidified chamber. Care must be taken that no air bubbles are trapped between the coverslip and the slide, as such voids would prevent a homogenous avidin coating.

B. Wash the slides with deionized water in a slide tray. Dry slides by centrifugation (32g, 5 min) or under a stream of pure nitrogen, and air-dry them in a clean environment (e.g., fume hood) until completely dry.

C. Quench the unreacted epoxy groups with 2 mM of aspartic acid solution in 0.5 M of NaHCO$_3$ buffer (pH 9) for 10 min at room temperature and with constant shaking. Use at least 25 ml of solution per 4 slides.

D. Wash the slides with deionized water and dry as described in Step B. Avidin slides can then be stored at 4 °C for extended times, but for best results, always prepare fresh before use. Test functionalization of the slides using a positive control such as FITC-biotin.

4.4. Spotting of slides and detection of proteins

1. Prepare 384-well source plate(s) by dissolving the lyophilized proteins (biotinylated or N-terminal cysteine containing) in PBS or any other appropriate buffer (pH 7.4). Alternatively, use crude cellular lysate containing either the biotinylated proteins or N-terminal cysteine-containing proteins. Spot the proteins onto the slides using a microarray spotter. Incubate for 10 min or longer (if required) in a humidified chamber.

2. Wash the spotted slides once with PBS for 10 min. Rinse well with water. Dry slides by centrifugation (32g, 5 min) or under a stream of pure nitrogen, and air-dry in a clean environment (e.g., fume hood) until completely dry.

3. Block the slide surface with 1% BSA in PBS for at least 1 h. Incubate the slides with fluorescently labeled antibodies against target proteins using a coverslip in a humid chamber for 1 to 2 h, as required.
4. Wash the slides with PBST (PBS containing 0.1% Tween-20); additional wash steps may be necessary to improve signal-to-noise ratio.
5. Rinse with distilled water, dry under a nitrogen stream, and scan using a microarray scanner fitted with the appropriate filters.

5. Concluding Remarks

We hope that the methods presented will enable readers to make informed decisions as to the method that best suits their needs. *In vitro* protein biotinylation, though time-consuming, provides proteins with the highest purity, with biotinylated proteins constituting >95% of the eluted protein pool. Because the C-terminal residue greatly affects the biotinylation efficiencies (and intein cleavage), careful attention must be paid during the design step such that unfavorable residues such as Asp (promotes *in vivo* cleavage), Cys, and Ile (reduces biotinylation efficiencies) can be avoided, while favorable residues like Gly can be incorporated. The guidelines, however, are purely empirical, based on our observations with EGFP, and by no means complete.

In vivo biotinylation is relatively straightforward and, under optimized conditions, as much as 20 to 40% of the proteins can be biotinylated. Crude cellular lysates can be used for spotting without any further intervening protein purification steps. The only other biotinylated protein present in *E. coli* lysate is the biotin carboxyl carrier protein (BCCP) of acetyl-CoA carboxylase (Cronan, 1990). The strategy has also been extended to mammalian cells (Lue *et al.*, 2004; Fig. 10.6) but with limited success. For difficult-to-express proteins, cell-free transcription/translation in conjunction with biotinylated puromycin was devised. Although biotinylated proteins can be obtained within hours, the downside of this approach is the low biotinylation efficiencies (≈15 to 20%). However, the compatibility of this approach with Gateway cloning systems, together with recent developments that indicate increases in the incorporation of puromycin by up to 80% (Agafonov *et al.*, 2006), make this method particularly amenable for high-throughput protein biotinylation in 96-well formats.

Certain limitations exist in the field of protein immobilization using NCL approaches. For example, the ligation step between cysteine and α-thioester is compromised by low-efficiency (Rauh and Waldmann, 2007). To circumvent this, new methodologies have been developed in which amino acids other than cysteine can be used for ligation reactions. One such approach has been the development of cysteine-free NCL via the use of thiol-containing auxiliaries (Bin *et al.*, 2006) and, more recently, through auxiliary-free, TCEP-assisted phenolic ester–derived couplings (Chen *et al.*, 2007). In an alternative method, instead of replacing cysteine, Kent and colleagues have developed efficient kinetically controlled ligation strategies by replacing alkyl thioesters with aryl thioesters (Bang *et al.*, 2006). The difference in reactivity between the two thioesters is attributed to lower pK_a values for the thiol component of aryl thioesters, which makes them better leaving groups (Macmillan, 2006). So far, the variants to NCL have been restricted to the convergent synthesis of glycopeptides and small proteins. It would be exciting to see if these innovative approaches can be extended to the fabrication of protein arrays.

REFERENCES

Agafonov, D. E., Rabe, K. S., Grote, M., Voertler, C. S., and Sprinzl, M. (2006). C-terminal modifications of a protein by UAG-encoded incorporation of puromycin during *in vitro* protein synthesis in the absence of release factor 1. *Chembiochem.* **7**, 330–336.

Arifuzzaman, M., Maeda, M., Itoh, A., Nishikata, K., Takita, C., Saito, R., Ara, T., Nakahigashi, K., Huang, H.-C., Hirai, A., Tsuzuki, K., Nakamura, S., *et al.* (2006). Large scale identification of protein-protein interaction of *Escherichia coli* K-12. *Genome Res.* **16**, 686–691.

Bang, D., and Kent, S. B. H. (2004). A one pot total synthesis of Crambin. *Angew. Chemie. Int. Ed.* **43**, 2534–2538.

Bang, D., Pentelute, B. L., and Kent, S. B. H. (2006). Kinetically controlled ligation for the convergent chemical synthesis of proteins. *Angew. Chemie. Int. Ed.* **45**, 3985–3988.

Bin, W., Chen, J., Warren, J. D., Chen, G., Hua, Z., and Danishefsky, S. J. (2006). Building complex glycopeptides: Development of a cysteine-free native chemical ligation. *Angew. Chemie. Int. Ed.* **45**, 4116–4125.

Camarero, J. A., Kwon, Y., and Coleman, J. A. (2004). Chemoselective attachment of biologically active proteins to surfaces by expressed protein ligation and its application for "protein chip" fabrication. *J. Am. Chem. Soc.* **126**, 14730–14731.

Chattopadhaya, S., Tan, L. P., and Yao, S. Q. (2006). Strategies for site-specific protein biotinylation using *in vitro*, *in vivo* and cell free systems: towards functional protein arrays. *Nat. Protoc.* **1**, 2386–2398.

Chattopadhaya, S., and Yao, S. Q. (2006). *In* "Enzyme Assays: High Throughput Screening, Genetic Selection and Fingerprinting" (Jean-Louis Reymond, ed.), Wiley-VCH, Weinheim.

Chen, G., Wan, Q., Tan, Z., Kan, C., Hua, Z., Ranganathan, K., and Danishefsky, S. J. (2007). Development of efficient methods for accomplishing cysteine-free peptide and glycopeptide coupling. *Angew. Chemie. Int. Ed.* **46**, 7383–7387.

Chong, S., Montello, G. E., Zhang, A., Cantor, E. J., Liao, W., Xu, M.-Q., and Benner, J. (1998). Analysis of yeast protein kinases using protein chips. *Nucleic Acids Res.* **26**, 5109.

Cronan, J. E. (1990). Biotination of proteins *in vivo*. *J. Biol. Chem.* **265**, 10327–10333.

Cronan, J. E., and Reed, K. E. (2000). Biotinylation of proteins *in vivo*: A useful posttranslational modification for protein analysis. *Methods Enzymol.* **326**, 440–458.

de Araújo, A. D., Palomo, J. M., Cramer, J., Köhn, M., Schröder, H., Wacker, R., Niemeyer, C., Alexandrov, K., and Waldmann, H. (2006). Diels-Alder ligation and surface immobilization of proteins. *Angew. Chemie. Int. Ed.* **45**, 296–301.

Dawson, P. E., Muir, T. W., Lewis, I. C., and Kent, S. B. (1994). Synthesis of proteins by native chemical ligation. *Science* **266**, 776–779.

Gelperin, D. M., White, M. A., Wilkinson, M. L., Kon, Y., Kung, L. A., Wise, K. J., Lopez-Hoyo, N., Jiang, L., Piccirillo, S., Yu, H., Gerstein, M., Dumont, M. E., *et al.* (2005). Biochemical and genetic analysis of the yeast proteome with a movable ORF collection. *Genes Dev.* **19**, 2816–2826.

Girish, A., Sun, H., Yeo, D. S. Y., Chen, G. Y. J., Chua, T.-K., and Yao, S. Q. (2005). Site-specific immobilization of proteins in a microarray using intein-mediated protein splicing. *Bioorg. Med. Chem. Lett.* **15**, 2447–2451.

Guschin, D., Yershov, G., Zaslavsky, A., Gemmell, A., Shick, V., Proudnikov, D., Arenkov, P., and Mirzabekov, A. (1997). Manual manufacturing of oligonucleotide, DNA and protein microchips. *Anal. Biochem.* **250**, 203–211.

Hahn, M. E., and Muir, T. W. (2004). Photocontrol of Smad2, a mutiphosphorylated cells signaling protein, through caging of activated phosphoserines. *Angew. Chem. Int. Ed.* **43**, 5800–5803.

Hall, D. A., Ptacek, J., and Synder, M. (2007). Protein Microarray Technology. *Mech. Ageing. Dev.* **128**, 161–167.

Hall, D. A., Zhu, H., Zhu, X., Royce, T., Gerstein, M., and Synder, M. (2004). Regulation of gene expression by a metabolic enzyme. *Science* **306**, 482–484.

Hodneland, C. D., Lee, Y.-S., Min, D.-H., and Mrksich, M. (2002). Selective immobilization of proteins to self-assembled monolayers presenting active site-directed capture ligands. *Proc. Natl. Acad. Sci. USA* **99**, 5048–5052.

Kapust, R. B., and Waugh, D. S. (2000). Controlled intracellular processing of fusion proteins by TEV protease. *Protein. Expr. Purif.* **19**, 312–318.

Kwon, Y., Coleman, M. A., and Camarero, J. A. (2006). Selective Immobilization of protein onto solid supports thorough split-intein-mediated protein trans-splicing. *Angew. Chem. Int. Ed.* **45**, 1726–1729.

LaBear, J., and Ramachandran, N. (2005). Protein microarrays as tools for functional proteomics. *Curr. Opin. Chem. Biol.* **9**, 1–6.

Lesaicherre, M.- L., Lue, R. Y. P., Chen, G. Y. J., Zhu, Q., and Yao, S. Q. (2002). Intein-mediated biotinylation of proteins and its application in a protein microarray. *J. Am. Chem. Soc.* **124**, 8768–8769.

Lin, P.-C., Ueng, S.-H., Tseng, M.-C., Ko, J.-L., Huang, K.-T., Yu, S.-C., Adak, A. K., Chen, Y.-J., and Lin, C.-C. (2006). Site-specific protein modification through Cu^I-catalyzed 1, 2, 3-triazole formation and its implementation in protein microarray fabrication. *Angew. Chemie. Int. Ed.* **45**, 4286–4290.

Lovrinovic, M., Seidel, R., Wacker, R., Schroeder, H., Seitz, O., Engelhard, M., Goody, R. S., and Niemeyer, C. M. (2003). Synthesis of protein-nucleic acid conjugates by expressed protein ligation. *Chem. Commun.* 822–823.

Lue, R. Y. P., Chen, G. Y. J., Hu, Y., Zhu, Q., and Yao, S. Q. (2004). Versatile protein biotinylation strategies for potential high-throughput proteomics. *J. Am. Chem. Soc.* **126**, 1055–1062.

MacBeath, G., and Schreiber, S. L. (2000). Printing proteins as microarrays for high-throughput function determination. *Science* **289**, 1760–1763.

Macmillan, D. (2006). Evolving strategies for protein synthesis converge on native chemical ligation. *Angew. Chemie. Int. Ed.* **45**, 7668–7672.

Miyamoto-Sato, E., Nemoto, N., Kobayashi, K., and Yanagawa, H. (2000). Specific bonding of puromycin to full-length protein at the C-terminus. *Nucleic. Acids. Res.* **28**, 1176–1182.

Muir, T. W. (2003). Semisynthesis of proteins by expressed protein ligation. *Annu. Rev. Biochem.* **72**, 249–289.

Muralidharan, V., Cho, J., Trester-Zedlitz, M., Kowalik, L., Chait, B. T., Raleigh, D. P., and Muir, T. W. (2004). Domain specific incorporation of noninvasive optical probes into recombinant proteins. *J. Am. Chem. Soc.* **126,** 14004–14012.

Muralidharan, V., and Muir, T. W. (2006). Protein ligation: An enabling technology for the biophysical analysis of proteins. *Nat. Methods.* **3,** 429–438.

Paborsky, L. R., Dunn, K. E., Gibbs, C. S., and Dougherty, J. P. (2004). A nickel chelate microtiter plate assay for six histidine-containing proteins. *Anal. Biochem.* **234,** 60–65.

Pellois, J. P., and Muir, T. W. (2005). A ligation and photorelease strategy for the temporal and spatial control of protein function in living cells. *Angew. Chem. Int. Ed.* **44,** 5713–5717.

Ramachandran, N., Hainsworth, E., Bhullar, B., Eisenstein, S., Rosen, B., Lau, A. Y., Walter, J. C., and LaBaer, J. (2004). Self-assembling protein microarrays. *Science* **305,** 86–90.

Rauh, D., and Waldmann, H. (2007). Linking chemistry and biology for the study of protein function. *Angew. Chemie. Int. Ed.* **46,** 826–829.

Sambrook, J., Fritsch, E. F., and Maniatis, T. (1989). *Molecular Cloning: A Laboratory Manual* 2nd edition, Cold Spring Harbor Laboratory Press, New York, USA.

Schwartz, E. C., Saez, L., Young, M. Y., and Muir, T. W. (2007). Post-translational enzyme activation in an animal via optimized conditional protein splicing. *Nat. Chem. Biol.* **3,** 50–54.

Soellner, M. B., Dickson, K. A., Nilsson, B. L., and Raines, R. T. (2003). Site-specific protein immobilization by Staudinger ligation. *J. Am. Chem. Soc.* **125,** 11790–11791.

Sohma, Y., Pentelute, B. L., Whittaker, J., Hua, Q.-X., Whittaker, L. J., Weiss, M. A., and Kent, S. B. H. (2007). Comparative properties of insulin-like growth factor (IGF-1) and [Gly7D-Ala] IGF-1 prepared by total chemical synthesis. *Angew. Chemie. Int. Ed.* DOI.10.1002/anie.200703521.

Tan, L. P., Chen, G. Y. J., and Yao, S. Q. (2004). Expanding the scope of site-specific protein biotinylation strategies using small molecules. *Bioorg. Med. Chem. Lett.* **14,** 5735–5738.

Tan, L. P., Lue, R. Y. P., Chen, G. Y. J., and Yao, S. Q. (2004). Improving the intein-mediated, site-specific protein biotinylation strategies both *in vitro* and *in vivo*. *Bioorg. Med. Chem. Lett.* **14,** 6067–6070.

Tao, S.-C., Chen, C.-S., and Zhu, H. (2007). Applications of Protein Microarray Technology. *Comb. Chem. High. Through. Screen* **10,** 706–718.

Tao, S.-C., and Zhu, H. (2006). Protein chip fabrication by capture of nascent polypeptides. *Nat. Biotechnol.* **24,** 1253–1254.

Tolbert, T. J., and Wong, C.-H. (2000). Intein mediated synthesis of proteins containing carbohydrates and other molecular probes. *J. Am. Chem. Soc.* **122,** 5421–5428.

Tolbert, T. J., and Wong, C.-H. (2002). New methods for proteomic research: preparation of proteins with N-terminal cysteines for labeling and conjugation. *Angew. Chemie. Int. Ed.* **41,** 2171–2174.

Uttamchandani, M., Wang, J., and Yao, S. Q. (2006). Protein and small molecule arrays: Powerful tools for high-throughput proteomics. *Mol. BioSyst.* **2,** 58–68.

Vijayendran, R. A., and Leckband, D. E. (2001). A quantitative assessment of heterogeneity for surface-immobilized proteins. *Anal. Chem.* **73,** 471–480.

Watzke, A., Köhn, M., Rodriguez, M. G., Wacker, R., Schröder, H., Breinbauer, R., Kuhlmann, J., Alexandrov, K., Niemeyer, C. M., Goody, R. S., and Waldmann, H. (2006). Site-selective protein immobilization by Staudinger ligation. *Angew. Chemie. Int. Ed.* **45,** 1408–1412.

Xu, M. -Q., Paulus, H., and Chong, S. (2000). Fusions to self-splicing inteins for protein purification. *Method. Enzymol.* **326,** 376–418.

Yin, J., Liu, F., Li, X., and Walsh, C. T. (2004). Labeling proteins with small molecules by site-specific posttranslational modification. *J. Am. Chem. Soc.* **126,** 7754–7755.

Zhang, K., Diehl, M. R., and Tirrell, D. A. (2005). Artificial polypeptide scaffold for protein immobilization. *J. Am. Chem. Soc.* **127,** 10136–10137.

Zhu, H., Klemic, J. F., Chang, S., Bertone, P., Casamayor, A., Klemic, K. G., Smith, D., Gerstein, M., Reed, M. A., and Synder, M. (2000). Analysis of yeast protein kinases using protein chips. *Nat. Genet.* **26,** 283–289.

Zhu, H., Bilgin, M., Bangham, R., Hall, D. A., Casamayor, A., Bertone, P., Lan, N., Jansen, R., Bidlingmaoer, S., Houfek, T., Mitchell, T., Miller, P., *et al.* (2001). Global analysis of protein activities using proteome chips. *Science* **293,** 2101–2105.

CHAPTER ELEVEN

SEMISYNTHESIS OF UBIQUITYLATED PROTEINS

Robert K. McGinty,* Champak Chatterjee,* *and* Tom W. Muir*

Contents

* Laboratory of Synthetic Protein Chemistry, Rockefeller University, New York, USA

Methods in Enzymology, Volume 462
ISSN 0076-6879, DOI: 10.1016/S0076-6879(09)62011-5

Abstract

Most, if not all, proteins are at one point or another posttranslationally modified so as to regulate their biological function. One of the most common protein modifications is ubiquitylation, in which the small protein ubiquitin is attached to a target protein in a multistep process involving dedicated ubiquitin ligases. Ubiquitylation is best known for its role in protein turnover. In this case, attachment of a polyubiquitin chain to a target protein leads to its eventual destruction by the 26S proteasome. However, attachment of ubiquitin is not always a kiss of death for the recipient protein; it is increasingly clear that the modification plays additional roles, including regulating protein trafficking and protein signaling. Understanding these functions at the molecular level necessitates that we have access to homogenous ubiquitylated proteins, something that has proved very difficult using standard biochemical approaches. In this chapter, we describe the development of synthetic chemistries and protein semisynthesis methods that permit the site-specific ubiquitylation of proteins. The utility of this methodology is illustrated through the synthesis of ubiquitylated histones.

1. INTRODUCTION

The development of recombinant DNA technology and techniques for the heterologous overexpression of polypeptides has allowed for the routine preparation of many proteins. However, these standard techniques are limited to the 20 naturally occurring amino acids. In many cases it is desirable to obtain proteins with posttranslational modifications (PTMs), unnatural amino acids, or biophysical probes—this was the motivation behind the development of the various methods covered by this volume. As illustrated by some of the chapters herein, the study of PTMs has proved a particular hotbed of activity for biochemists armed with these powerful new approaches. Proteins are subject to numerous PTMs that can alter the chemical structure (and hence function) of the molecule in myriad ways; these can range from quite subtle modifications such as the attachment of a methyl group to more dramatic changes such as that resulting from protein splicing (Walsh, 2006). Many of the most common PTMs have been successfully installed into proteins using protein semisynthesis and suppressor mutagenesis (reviewed in Flavell and Muir, 2009; Wang and Schultz, 2005). However, there are a few notable gaps in this coverage. One of these is the attachment of the small protein ubiquitin—a modification know as ubiquitylation (Hochstrasser, 1996).

Posttranslational modification of proteins by side-chain ubiquitylation was first discovered 30 years ago on histone H2A (Ballal *et al.*, 1975) and soon after on histone H2B (West and Bonner, 1980). The 76-amino-acid protein ubiquitin (Ub) is covalently linked to the target protein through the formation of an isopeptide bond between the C-terminal glycine of Ub and an ε-amino group of a lysine side chain on the target protein. Ubiquitin is activated for isopeptide bond formation by a series of enzymes called E1-3 ligases (Hochstrasser, 1996). The E1 ligase first activates Ub through the formation of a ubiquityl-[C(O)S]-E1 thioester in a step that requires ATP and progresses via an Ub-AMP intermediate. Ubiquitin can then be transferred to a family of E2 ligases capable of targeted substrate ubiquitylation. In many cases, a member of the E3 ligase family is required to maximize substrate specificity. In addition to their monoubiquitylation function, E2 and E3 ligases are capable of polyubiquitylating substrates through processive ubiquitylation of ubiquitin lysines. The removal of Ub is performed by dedicated Ub proteases (Ubps). Typically, monoubiquitylation controls substrate function by directing cellular location or by facilitating protein-protein interactions, while polyubiquitylation results in targeted destruction of the substrate by the proteasome (Hicke *et al.*, 2005; Hochstrasser, 1996). In addition to Ub, a series of other ubiquitin-like proteins (Ubls) including the small ubiquitin-like modifier (SUMO), have been discovered. Although the functions of the Ubls are still being characterized, most Ubls perform functions similar to, yet distinct from, monoubiquitylation (reviewed in Seeler and Dejean, 2003).

Our laboratory is interested in the role of histone ubiquitylation in regulating the structure, stability, and function of chromatin (Chatterjee *et al.*, 2007; McGinty *et al.*, 2008). Histone proteins are subject to numerous PTMs, including acetylation, phosphorylation, methylation, ADP-ribosylation, ubiquitylation, and sumoylation (Kouzarides, 2007). Growing evidence for dynamic regulation of these PTMs, position- and modification-specific protein interactions, and biochemical cross-talk between modifications has strengthened the histone code hypothesis, in which histone PTMs are integral to choreographing the expression of the genome (Jenuwein and Allis, 2001). Monoubiquitylation of H2A (uH2A) is associated with transcriptional silencing in higher eukaryotes, as well as in DNA-damage repair (Weake and Workman, 2008). In contrast, monoubiquitylation of H2B (uH2B) is strongly correlated with regions of active transcription, and related to this, with enhanced methylation of lysine residues in histone H3 (Weake and Workman, 2008). Many questions remain unanswered with respect to the precise role of histone ubiquitylation in these processes. One of the major obstacles is the difficulty associated with accessing homogenously ubiquitylated histones for use in biochemical studies—this is a general problem in the ubiquitin field. For instance, *in vitro* reconstitution of H2B with recombinant ubiquitin E1-E3 ligases and associated factors allows for the production of uH2B in limited quantities (Zhu *et al.*, 2005). Because of its natural abundance, uH2B can also be purified from endogenous sources (Davies and Lindsey, 1994; West and Bonner, 1980). However, heterogeneity due to the presence of additional modifications complicates biochemical analyses. To circumvent these problems, we decided to employ expressed protein ligation (EPL) technology to regioselectively ubiquitylate histones, thus bypassing the requirement for the complex cellular ubiquitylation machinery and ensuring chemical homogeneity.

2. Semisynthesis of Ubiquitylated Histone H2B

2.1. Overall synthetic design

Expressed protein ligation allows for the formation of an amide bond between two polypeptides of recombinant or synthetic origins, one containing an α-thioester, and the other an N-terminal cysteine (Muir, 2003). Although EPL has been applied to countless proteins and has allowed for the introduction of a wide range of PTMs into these, the methodology had not been previously applied to the preparation of ubiquitylated proteins. There are a number of challenges here, principally the large size of the PTM (Ub is an ≈8-kDa protein) and the need for a native isopeptide linkage between the C-terminus of Ub and the ε-amino group of the target lysine. In initial studies, we developed a general approach that allows the attachment of Ub to a target lysine within a peptide through a completely native linkage (Chatterjee *et al.*, 2007). Key to our strategy was the use of photocleaveable ligation auxiliary (6), which is coupled to the side chain of a target lysine residue in a synthetic peptide via a glycyl linker. The ligation auxiliary acts as an N-terminal cysteine surrogate to react in an EPL reaction with a recombinant ubiquitin α-thioester generated from a corresponding intein fusion (Fig. 11.1). Auxiliary groups have been used in traceless native chemical ligation (NCL) reactions for a number of years (reviewed in Dawson and Kent, 2000; Muralidharan and Muir, 2006). However, their widespread use has been impeded by the requirement for a sterically undemanding ligation site. Fortunately, such a condition can be met in the case of Ub and Ubls, as the last two C-terminal residues of the protein family are always Gly-Gly, an ideal junction for auxiliary-mediated ligation. Thus, our synthetic scheme calls for ligation of an ubiquitin α-thioester lacking the C-terminal Gly (i.e. residues 1 to 75) and a synthetic peptide containing auxiliary 6 linked to a lysine side chain in the peptide via a glycyl linker (which will ultimately correspond to Gly76 of ubiquitin). Following the ligation, the auxiliary is removed in a traceless manner by photolysis. In a sense, the intein plays the role of the E1 ubiquitin-activating enzyme in our system, while the

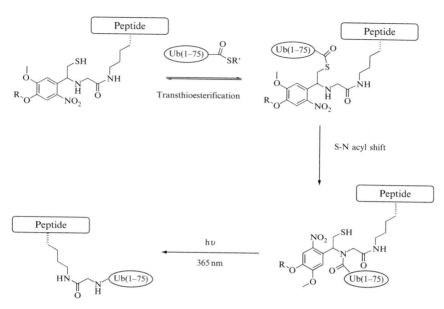

Figure 11.1 Principle of auxiliary-mediated native chemical ligation. Auxiliary 7 is coupled to a lysine side chain in a synthetic peptide via a glycyl linker. This glycine will eventually become Gly76 of ubiquitin. Next, this peptide is ligated to a protein α-thioester, in this case corresponding to the first 75 residues of ubiquitin (Ub). In the last step the auxiliary group is removed by photolysis. Hence the ligation is traceless with respect to the thiol group.

auxiliary is the synthetic equivalent of the E2/E3 ligase enzymes. Using this strategy, we demonstrated that synthetic peptides could not only be ubiquitylated efficiently and specifically but also modified with other Ubls such as SUMO (Chatterjee et al., 2007).

We have recently extended this approach to the preparation of full-length ubiquitylated H2B (McGinty et al., 2008). In this case, the target protein is assembled from three pieces, two recombinant α-thioesters and one synthetic peptide, and employs two traceless orthogonal EPL reactions (Fig. 11.2). The first ligation reaction involves the photolytic ligation auxiliary described previously, whereas the second ligation reaction makes use of a desulphurization reaction that converts the Cys residue at the ligation site back into a native alanine. The net result of all these operations is that the final ubiquitylated H2B product is completely native in structure. Following purification, semisynthetic uH2B could be successfully incorporated into core histone octamers with wild-type recombinant H2A, H3, and H4 that were purified from an optimized E. coli expression system (McGinty et al., 2008). These octamers can subsequently be reconstituted into chemically defined mononucleosomes using 147-bp DNA

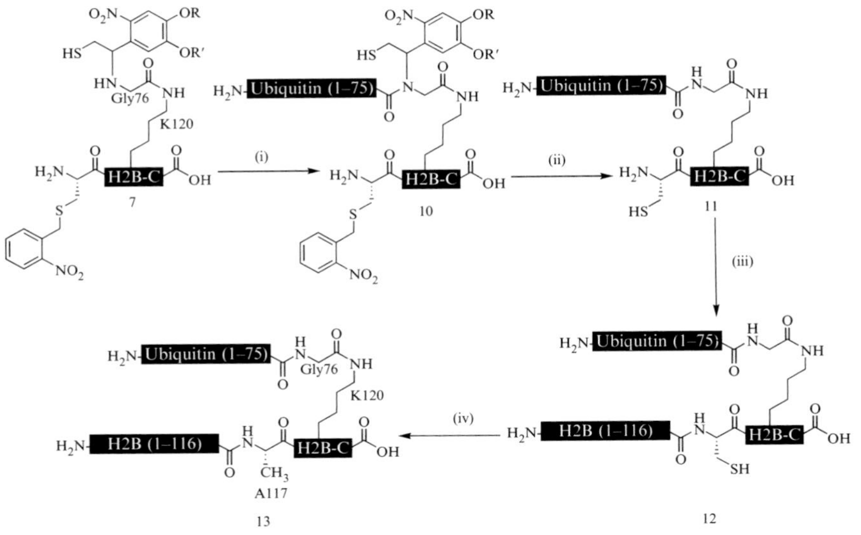

Figure 11.2 Synthesis of ubiquitylated histone 2B via a three-piece sequential EPL strategy. Use of the auxiliary-mediated NCL and a desulfurization step results in the native protein **13**. Step (i); EPL 1 between peptide **7** and Ub(1–75)-MES, **8**. (ii); photolysis at 365 nm. (iii); EPL 2 between branched peptide **11** and H2B(1–116)-MES, **9**. (iv); cysteine desulfurization using Raney nickel.

containing a nucleosomal targeting sequence, so-called 601 DNA (Lowary and Widom, 1998). Experimental details of the auxiliary synthesis and its application to the preparation of uH2B follow.

3. METHODS

3.1. General methods

Amino acid derivatives, preloaded Wang resin, and coupling reagents are purchased from Novabiochem (San Diego, CA). *E. coli* BL21(DE3) and pLysS cells are purchased from Novagen (Madison, WI). Sephacryl S-200 resin is obtained from GE Healthcare (Piscataway, NJ). Restriction enzymes, pTXB1 vector, and chitin resin are obtained from New England Biolabs (Ipswich, MA). Criterion 18% Tris-HCl, and Criterion 5% TBE gels are purchased from Biorad (Hercules, CA). Centricons are purchased from Sartorius (Goettingen, Germany). PCR purification and gel-extraction kits are purchased from Qiagen (Valencia, CA). All other chemical reagents are purchased from Sigma-Aldrich (Milwaukee, WI) or Fisher Scientific (Pittsburgh, PA). Analytical and semipreparative scale reverse-phase HPLC

(RP–HPLC) are performed on a Hewlett-Packard 1100 series instrument using Vydac C18 columns (4 × 150 mm; 10 × 250 mm) at 1 and 4 mL/min, respectively. Unless otherwise noted, all analytical gradients are 0 to 73% B over 30 min (A: 0.1% trifluoroacetic acid (TFA) in water; B: 90% acetronitrile, 0.1% TFA in water). Preparative and process scale RP-HPLC are performed on a Waters DeltaPrep 4000 system connected to a Waters 486 tunable detector using Vydac C18 columns (22 × 250 mm; 50 × 250 mm) at 15 and 30 mL/min, respectively. Size-exclusion chromatography is performed on an AKTA FPLC system from GE Healthcare equipped with a P-920 pump and a UPC-900 monitor. ESI-MS is performed on a SciexAPI-100 single quadrupole mass spectrometer. Primer synthesis and DNA sequencing are performed by Integrated DNA Technologies and Genewiz, respectively.

4. SYNTHESIS OF PHOTOCLEAVABLE LIGATION AUXILIARY

The ligation auxiliary **7** can be prepared in eight steps from vanillin following the basic procedure reported by Pellois and Muir (2005). The synthetic route is shown in Figure 11.3.

4.1. 4-(2-Methoxy-5-nitro-4-vinyl-phenoxy)-butyric acid methyl ester (2)

To a suspension of methyltriphenylphosphonium bromide (6.5 g, 18.2 mmol) in THF (30 mL), a solution of sodium bis(trimethylsilyl) amide (18.2 mL, 1 M in THF, 18.2 mmol) is added dropwise over 30 min at 0 °C and stirred for 1 h. A solution of 4-(4-formyl-2-methoxy-5-nitro-phenoxy)-butyric acid methyl ester (**1**) (available in 2 steps from vanillin (McMinn and Greenberg, 1996)) (4.2 g, 14.0 mmol) in THF (40 mL) is then added dropwise and the mixture stirred at room temperature for an additional 16 h. The solvent is then evaporated and the residue dissolved in chloroform (25 mL). The organic phase is washed with a saturated solution of ammonium chloride (2 × 25 mL), brine (2 × 300mL), dried (Na$_2$SO$_4$) and concentrated *in vacuo*. Flash column chromatography of the residue (silica gel, 30% ethyl acetate in petroleum ether) yields **2** (2.4 g, 8.1 mmol, 58%). Compound **3**: R$_f$ = 0.54 (silica gel, 30% ethyl acetate in petroleum ether); ^{1}H NMR (400 MHz, DMSO-d$_6$): δ 7.56 (s, 1H, H$_{ar}$), 7.19 (s, 1H, H$_{ar}$), 7.09 (dd, J = 17.3, 11 Hz, 1H, H$_2$), 5.88 (d, J = 17.3 Hz, 1H, H$_{1b}$), 5.45 (d, J = 11.1 Hz, 1H, H$_{1a}$), 4.14 (t, J = 6.3 Hz, 2H, OCH$_2$), 3.94 (s, 3H, OCH$_3$), 3.61 (s, 3H, COOCH$_3$), 2.48 (m, 2H, CH$_2$-CO), 2.01 (m, 2H, CH$_2$). ^{13}C NMR (400 MHz, DMSO-d$_6$): δ 173.75, 153.94, 148.17,

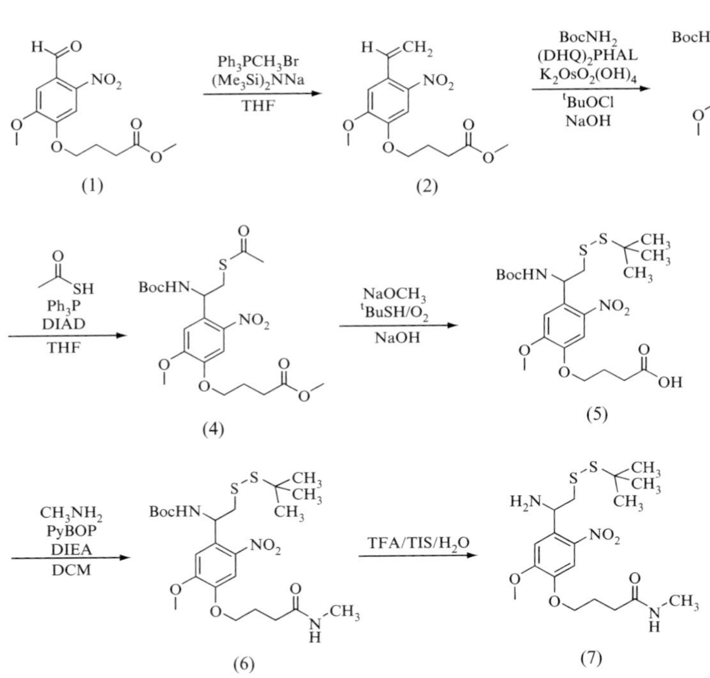

Figure 11.3 Synthetic route to ligation auxiliary 7.

140.78, 133.02, 128.00, 119.15, 110.64, 109.35, 68.74, 57.18, 52.22, 30.69, 24.83. HRMS (FAB); m/z: calculated for $C_{14}H_{18}O_6N$: 296.1134; found: 296.1135 ($[MH^+]$).

4.2. 4-[4-(1-*tert*-Butoxycarbonylamino-2-hydroxy-ethyl)-2-methoxy-5-nitro-phenoxy]-butyric acid methyl ester (3)

To a stirred solution of tert–butylcarbamate (2.0 g, 16.8 mmol) in n-propanol (20mL), sodium hydroxide (33 mL 0.5 M in water, 16.5 mmol) and tert-butyl hypochlorite (1.9 mL) are added and the resulting mixture is stirred for 5 min. The flask is then put in a slurry of ice water and $(DHQ)_2PHAL$ (211 mg, 0.27 mmol, in 20 mL of n- propanol), styrene **2** (1.6 g, 5.4 mmol, in 80 mL of n-propanol) and $K_2OsO_2(OH)_4$ (70 mg, 0.2 mmol) are sequentially added. The resulting mixture is stirred for 16 h at 4 °C. The product is extracted with ethyl acetate (2 × 100 mL), the combined organic extracts washed with brine (2 × 500 mL), dried ($MgSO_4$), and evaporated. Flash column chromatography of the residue (silica gel, 5% methanol in dichloromethane) followed by recrystallization in chloroform/hexanes affords **3** (1.2 g, 2.7 mmol, 50%). Compound **3**: $R_f = 0.38$ (silica gel, 5% methanol in dichloromethane); ^{1}H NMR (400 MHz, CDCl$_3$): δ 7.63 (s, 1H, Har), 7.02 (s, 1H, Har), 5.62 (s, 1H, NH), 5.46 (m, 1H, CHNH), 4.12 (t, J = 6.0 Hz, 2H, OCH$_2$), 3.96 (m, 2H, CH$_2$OH), 3.95 (s, 3H, OCH$_3$), 3.71 (s, 3H, COOCH$_3$), 3.50 (s, 1H, OH), 2.56 (t, J = 7.2 Hz, 2H, CH$_2$-CO), 2.19 (q, J = 6.7 Hz, 2H, CH$_2$), 1.42 (s, 9H, t-Bu). ^{13}C NMR (400 MHz, CDCl$_3$): δ 173.74, 155.81, 154.16, 147.58, 68.66, 65.39, 56.80, 53.68, 52.12, 51.28, 30.76, 28.69, 24.66. HRMS (FAB); m/z: calculated for C$_{19}$H$_{29}$O$_9$N$_2$: 429.1873; found: 429.1872 ([MH^+]).

4.3. 4-[4-(2-Acetylsulfanyl-1-*tert*-butoxycarbonylamino-ethyl)-2-methoxy-5-nitro-phenoxy]-butyric acid methyl ester (4)

Diisopropyl azodicarboxylate (0.92 mL, 4.7 mmol) is added to an ice-cold solution of triphenylphosphine (1.2 g, 4.7 mmol) in THF (30 mL). The mixture is stirred under nitrogen for 30 min, during which time a white precipitate forms. A solution of alcohol 3 (1.0 g, 2.3 mmol) in THF (20 mL) and thiolacetic acid (0.33 mL, 4.7 mmol) is added dropwise and the mixture is stirred for 17 h at room temperature. The solvent is evaporated, the residue dissolved in methanol (200 mL) and left to stand at 0 °C for 16 h. The precipitate is filtered and the solvent evaporated. Flash column chromatography of the residue (silica gel, 40% ethyl acetate in petroleum ether) affords 4 (1.0 g, 2.1 mmol, 91%). Compound 4: R$_f$ = 0.50 (40% ethyl acetate in petroleum ether); ^{1}H NMR (400 MHz, CDCl$_3$): δ 7.63 (s, 1H, Har), 7.01 (s, 1H, Har), 5.69 (s, 1H, NH), 5.00 (m, 1H, CHNH), 4.12 (t, J = 6.0 Hz, 2H, OCH$_2$), 3.98 (s, 3H, OCH$_3$), 3.71 (s, 3H, COOCH$_3$), 3.35 (m, 2H, CH$_2$S), 2.57 (t, J = 7.2 Hz, 2H, CH$_2$-CO), 2.40 (s, 3H, SCOCH$_3$), 2.20 (q, J = 6.7 Hz, 2H, CH$_2$), 1.41 (s, 9H, tBu). ^{13}C NMR (400 MHz, CDCl$_3$): δ 197.78, 173.76, 155.81, 154.34, 147.73, 139.37, 70.53, 68.65, 56.88, 52.13, 33.93, 30.94, 30.76, 28.67, 24.66, 22.34. HRMS (FAB); m/z: calculated for C$_{21}$H$_{31}$O$_9$N$_2$S: 487.1750; found: 487.1749 ([MH^+]).

4.4. 4-[4-(1-*tert*-Butoxycarbonylamino-2-*tert*-butyldisulfanyl-ethyl)-2-methoxy-5-nitro-phenoxy]-butyric acid (5)

To a solution of thioester 4 (500 mg, 1.0 mmol) in methanol (30 mL) is added sodium methoxide (2.0 mL, 0.5 M in methanol, 1.0 mmol). The mixture is stirred under nitrogen for 15 min. 2-methyl-2-propanethiol (3.4 mL, 30 mmol) is added and the mixture stirred under oxygen overnight. A solution of sodium hydroxide (15 mL, 1.0 M in water, 15 mmol) is then added and the mixture stirred for another 3 h. The solution is acidified with hydrochloric acid (20 mL, 1.0 M in water, 20 mmol) and extracted with ethyl acetate (2 × 50 mL). The organic phase is washed with brine

(2 × 25 mL), dried (MgSO$_4$), and concentrated *in vacuo*. Flash column chromatography of the residue (silica gel, 40% ethyl acetate in petroleum ether) affords **5** (467 mg, 0.9 mmol, 90%). Compound **5**: R$_f$ = 0.26 (silica gel, 40% ethyl acetate in petroleum ether); ^{1}H NMR (400 MHz, CDCl$_3$): δ 7.62 (s, 1H, H$_{ar}$), 7.03 (s, 1H, H$_{ar}$), 5.56 (m, 1H, C*H*NH), 4.14 (t, J = 6.0 Hz, 2H, OCH$_2$), 3.95 (s, 3H, OCH$_3$), 3.30 (d, J = 13.0 Hz, 2H, CH$_2$S), 2.60 (t, J = 7.2 Hz, 2H, CH$_2$-CO), 2.20 (m, 2H, CH$_2$), 1.41 (s, 9H, tBu), 1.34 (s, 9H, S-tBu). ^{13}C NMR (400 MHz, CDCl$_3$): δ 177.70, 154.26, 147.56, 140.72, 132.62, 110.48, 100.00, 68.55, 56.84, 51.19, 48.73, 30.60, 30.24, 30.18, 28.66, 24.66. HRMS (FAB); *m/z*: calculated for C$_{22}$H$_{35}$O$_8$N$_2$S$_2$: 519.1835; found: 519.1837 ([*MH*$^+$]).

4.5. 4-[4-(1-Amino-2-tert-butyldisulfanyl-ethyl)-2-methoxy-5-nitro-phenoxy]-N-methyl-butyramide (6)

To a stirred solution of 4-[4-(1-*tert*-Butoxycarbonylamino-2-*tert*-butyldisul-fanyl-ethyl)-2-methoxy-5-nitrophenoxyl-butyric acid (438 mg, 0.844 mmol) in DCM (10 mL) at room temperature is added PyBOP (439 mg, 0.844 mmol) and N,N-diisopropylethylamine (0.43 ml, 2.53 mmol). After an additional 5 min of stirring, methylamine (4.22 mL, 8.44 mmol, 2M in THF) is added to the mixture and the reaction is allowed to proceed for a further 10 h at room temperature. At this stage, the solvent is removed *in vacuo* and the residue dissolved in a mixture of TFA (9.5 mL), triisopro-pylsilane (0.25 mL), and water (0.25 mL). The resulting solution is stirred at room temperature for 1 h to effect complete removal of the Boc–protecting group. The removal of water by lyophilization yields crude amide **6**, which is purified by preparative RP-HPLC (gradient 20 to 45% B, 60 min) to obtain the pure compound in 36% isolated yield over two steps. ^{1}H NMR (400 MHz, CDCl$_3$): δ 7.55 (s, 1H, H$_{ar}$), 7.29 (s, 1H, H$_{ar}$), 6.13 (m, 1H, N*H*CH$_3$), 5.35 (m, 1H, C*H*NH$_2$), 4.12 (t, 2H, OCH$_2$), 4.11 (s, 3H, OCH$_3$), 3.33 (d, 2H, CH$_2$S), 2.78 (d, 3H, C*H*$_3$NH), 2.41 (t, 2H, CH$_2$-CO), 2.15 (m, 2H, CH$_2$), 1.33 (s, 9H, S-tBu). ^{13}C NMR (400 MHz, CDCl$_3$): δ 173.90, 154.51, 148.62, 141.78, 124.74, 111.11, 109.78, 69.06, 57.08, 49.06, 42.46, 32.96, 30.12, 26.72, 25.12, 22.29. HRMS (ESI-FT); m/z calculated for C$_{18}$H$_{29}$O$_5$N$_3$S$_2$ (M+H)$^+$ 432.1622, found 432.1609.

▶ 5. PEPTIDE SYNTHESIS

Ligation auxiliary **6** is compatible with Fmoc-based solid–phase peptide synthesis. It can be attached to a designated lysine chain in a synthetic peptide using an orthogonal protection strategy. Typically, we perform this coupling step on the solid phase after chain assembly and before the cleavage/global deprotection step. As an example, we provide details of the synthesis of synthetic peptide **7**, which is derived from the C-terminus of H2B (residues 117-125). In this case, auxiliary **6** is attached to lysine 120 via a glycyl linker. In addition, the peptide contains an Ala117-to-Cys mutation to facilitate a second regioselective EPL reaction (see Fig. 11.2). Critical to the synthetic design is transient protection of the N-terminal cysteine in **1**, which precludes unwanted double ligation of ubiquitin. The photoremovable S-(o-nitrobenzyl) group is used for this purpose, as it is easily removed by photolysis at the same time as we remove the ligation auxiliary after ligation of Ub.

5.1. Chemical synthesis of peptide 7

The sequence corresponding to residues 117 to 125 of *Xenopus* H2B with an A117C replacement is synthesized on preloaded Wang resin by manual solid-phase peptide synthesis with a 9-fluorenylmethoxycarbonyl (Fmoc) N^α protection strategy and using 2-(1H-benzotriazole-1-yl)-1,1,3,3-tetramethyluronium hexafluorophosphate (HBTU) for amino acid activation. Standard ^{t}butyl side-chain protection is used throughout with the following exceptions; the ε-amino group of K120 is protected with the 4-methyltrityl (Mtt) group, and the thiol group of C117 is protected with an o-nitrobenzyl group. N^α-(^{t}butoxycarbonyl)-S-(o-nitrobenzyl)-L-cysteine (Boc-Cys(ONB)) used in the peptide synthesis is prepared as previously described (McGinty, 2008 #3072). The glycyl linker and ligation auxiliary are installed on the solid phase as follows: (1) The Mtt group on K120 is deprotected by successive incubations of the peptidyl-resin with 1% TFA in DCM containing 1% triisopropylsilane (TIS) for 10-min intervals until no yellow color evolves; (2) bromoacetic acid (222 mg, 1.6 mmol) is coupled to the ε-NH_2 of K120 with DIC (1.6 mmol) for 3 × 1 h; and (3) the ligation auxiliary **6** (32.6 mg, 76 mmol) and DIEA (1.4 mmol) are added to bromoacetylated resin (120 mg, 76 mmol) suspended in 600 mL of DMF. The alkylation is allowed to proceed for 72 h at room temperature, after which, the resin is dried. Following cleavage from the resin with TFA:TIS:H_2O (95:2.5:2.5) for 3 h, peptide **7** is purified by RP-HPLC on a preparative scale using a 28 to 38% B gradient over 45 min, yielding 10.8 mg of peptide. Peptide **7**; ESI-MS $(M+H)^+$ observed: 1606.8 Da; expected: 1607.9 Da.

6. GENERATION OF RECOMBINANT PROTEIN α-THIOESTERS

The sequential EPL strategy for the preparation of uH2B requires the generation of two recombinant protein α-thioester building blocks, Ub(1-75)-α-MES and H2B(1-116)-α-MES. Both proteins can be readily obtained by thiolysis of the corresponding intein fusions using established procedures (Muralidharan and Muir, 2006). Note that, in general, there are several criteria that go into designing an optimal intein fusion construct for use in semisynthesis. A full discussion of these can be found in Muralidharan and Muir (2006), as well as in several of the chapters in this volume, most notably by Szewczuk *et al.*

6.1. Preparation of ubiquitin(1-75)- α-MES (8)

The partial human ubiquitin gene, *ub(1-75)*, can be amplified by PCR using the primers: 5′-GGGAATTCCATATGCAGATCTTCGTGAAGACTC-3′ and 5′-GAATATATGCTCTTCCGCAACCTCTGAGACGGA-3′ with the plasmid pHub76 as the template DNA. The PCR product is purified (QIAquick kit), digested with NdeI and SapI restriction enzymes, and ligated into the identically digested pTXB1 vector (New England BioLabs). The resulting plasmid, pHub(1-75), which encodes Ub(1-75) fused at its C-terminus to the GyrA intein and a chitin-binding domain, is verified by DNA sequencing. *E. coli* BL21(DE3) cells (Novagen, Madison, WI) transformed with the plasmid pHub(1-75) are grown in 6 L of Luria-Bertani medium (100 μg/mL ampicillin) at 37 °C with shaking at 250 rpm until an OD_{600} 0.6 to 0.8. Overexpression of the desired protein is induced by the addition of 0.5 mM IPTG followed by an additional 6 h growth at 25 °C. The cells are then harvested by centrifugation at 10,000g for 30 min and the cell pellet resuspended in column buffer A (50 mM Tris, 200 mM NaCl, 1 mM EDTA, pH 7.2). The cells are lysed by passage through a French press, and the soluble fraction is separated from insoluble cellular debris by centrifugation at 18,500 to 20,000g for 20 min. After filtration through a 0.45-μm filter, the supernatant is bound to a 20-mL chitin column, pre-equilibrated with 10 vol of column buffer A, for 2 h at 4 °C. The resin is washed with 35 column-volumes of buffer A, followed by 3 column-volumes of column buffer B containing 50 mM Tris, 200 mM NaCl, and 1 mM EDTA, pH 7.5. Ub(1-75)-α-MES is cleaved from the intein-CBD fusions by incubation with 1.5 column-volumes of buffer B containing 100 mM of the sodium salt of 2-mercaptoethanesulfonic acid (MESNA) for 87 h. The eluted Ub(1-75)-α-MES thioester is subsequently purified by C18 semi-preparative RP-HPLC (gradient of 30 to 45% B over 45 min (R_t 27.5 min)). This yields 10.2 mg of the Ub(1-75)-α-MES ESI-MS; calculated m/z $(M+H)^+$ 8632 Da, found 8632 ± 2 Da.

6.2. Preparation of H2B(1-116)-α-MES (9)

A truncated *Xenopus* H2B gene, containing residues 1 to 116, is PCR amplified using primers H2B-FP (5′-GGAATTCCATATGCCTGAGC-CAGCCAAGTCCGCTCCAGCCCCG-3′)   and   H2B116-RP   (5′-

GGTGGTT<u>GCTCTTC</u>CGCACTTGGTGCCCTCGGACAC-3′) and a *Xenopus* H2B expression plasmid as a template. Following digestion by NdeI and SapI, the fragment is ligated into a similarly digested pTXB1 vector, and the resulting plasmid, pRMH2B-N, which encodes H2B(1-116) fused at its C-terminus to the GyrA intein and a chitin-binding domain, is verified by DNA sequencing. *E. coli* BL21(DE3) cells transformed with pRMH2B-N are grown in Luria-Bertani medium at 37 °C until mid-log phase, and protein expression is induced by the addition of 0.5 m*M* IPTG and allowed to continue at 25 °C for 16 h. After harvesting the cells by centrifugation at 6,800*g* for 15 min, the cell pellet is resuspended in lysis buffer (50 m*M* Tris, 200 m*M* NaCl, 1 m*M* EDTA, pH 7.5) and frozen at −80 °C. Thawed cells are lysed by passage through a French press, and the insoluble material is removed by centrifugation at 26,000*g* for 30 min. The supernatant is filtered and incubated overnight at 4 °C with chitin resin (35 mL) preequilibrated in lysis buffer. The resin is washed with 200 mL of column buffer A (50 m*M* Tris, 200 m*M* NaCl, 1 m*M* EDTA, pH 7.2) and 700 mL of column buffer C (50 m*M* Tris, 200 m*M* NaCl, 1 m*M* EDTA, pH 7.4). The resin is then incubated with cleavage buffer (50 m*M* Tris, 200 m*M* NaCl, 1 m*M* EDTA, 100 m*M* MESNA, pH 7.4) for 70 h, resulting in thiolysis of the intein fusion, forming H2B(1-116)-α-MES, **9**. The column is eluted and the resin washed with 2 × 25 mL cleavage buffer. The thiolysis reaction is repeated and the combined eluate further purified by preparative RP-HPLC using a 42 to 52% B gradient over 45 min, to yield 4 to 5 mg of lyophilized protein per liter of culture. The identity of the purified protein is verified as **9** by ESI-MS ((M+H)$^+$ observed: 12,991 ± 3 Da, expected: 12,991 Da). Mass spectrometry indicates that the non-native N-terminal methionine used for expression of **9** is processed during recombinant expression, leaving the native N-terminal sequence.

7. Expressed Protein Ligation

The ubiquitylated H2B target is generated using a sequential EPL process (see Fig. 11.2). In the first step, auxiliary-containing peptide **7** is ligated to protein α-thioester **8** to give ubiquitylated peptide **10**. Following photolytic removal of the auxiliary and the N-terminal Cys protecting group, this intermediate, **11**, is ligated to protein α-thioester **9**, to give ubiquitylated H2B mutant, **12**, which harbors an Ala117-to-Cys mutation at the second ligation site. In the final step, Raney Ni mediated desulfurization is performed to convert this Cys back to the native Ala, thereby yielding the final ubiquitylated H2B protein **13**. Several aspects of this process are worth emphasizing. First, the auxiliary-mediated EPL reaction takes several days to reach completion, which is considerably slower than typical NCL or EPL reactions (normally complete within 12-24 hours).

This is likely because the S-to-N acyl shift step in the NCL reaction involves a hindered secondary amine, rather than the normal primary amine of an N-terminal Cys. Second, the use of the S-(o-nitrobenzyl) protecting group on peptide **7** is found to be completely stable during this first, lengthy EPL step, thereby ensuring the regioselectivity of this reaction. In contrast, the thiazolidine-based N-terminal Cys protection strategy, which is commonly used in sequential NCL/EPL schemes (Bang and Kent, 2004), is found to reversible over this period, leading to unwanted double ligation of protein **8** (RKM, CC and TWM, unpublished observations). Last, use of the desulfurization reaction is permissible in this instance, as protein **11** contains only 1 Cys residue, namely at the ligation junction. However, the protein does contain a methionine residue, which also desulfurizes (with loss of CH_3SH), albeit at a much slower rate than Cys. Thus, the reduction reaction must be carefully monitored over time (by LCMS) to obtain optimal cysteine desulfurization over methionine.

7.1. Synthesis of ubiquitylated peptide 10

Purified peptide **7** (1.1 mg, 0.69 μmol) is dissolved in 180 μL of buffer containing 3 M guanidinium chloride, 300 mM sodium phosphate at pH 7.8, and 50 mM tris(2-carboxyethyl)phosphine (TCEP), and is incubated at room temperature for 30 min to remove the S-^{t}butyl protecting group on the ligation auxiliary. The resulting reduced peptide is then added to ubiquitin(1–75)-MES, **8** (17.1 mg, 1.98 μmol) and dissolved in ligation buffer (3 M guanidinium chloride, 300 mM sodium phosphate, 100 mM MES, pH 7.8). The reaction volume is increased to 950 μL with ligation buffer and the pH is adjusted to 7.8 using NaOH. After 120 h at 4 °C, the reaction is quenched with 1 mL of 50% HPLC buffer B containing 100 mM TCEP. Ligation product **10** is purified using preparative HPLC with a 32 to 42% B gradient over 45 min, yielding 4.0 mg of lyophilized protein. ESI-MS is used to verify the identity of the ligation product (($M+H)^+$ observed: 10,008 ± 1 Da, expected: 10,009 Da).

7.2. Photolytic deprotection of 10 to give branched protein 11

In a typical reaction, ligation product **10** (3.5 mg, 0.35 μmol) is dissolved in 1.5 mL of photolysis buffer (25% HPLC buffer B containing 10 mM semicarbazide, 10 mM DTT, 10 mM ascorbic acid, and 2 mM cysteine). The resulting solution is irradiated at 365 nm for 4 h using a collimated light source from Oriel (Stratford, CT) equipped with a 200-W Hg lamp. Irradiance (4 mW/cm^2) is measured with a model 840-c monochromic photometer (Irvine, CA). Selective irradiation at 365 nm is achieved by using an analytical line filter (9.4-nm bandwidth) obtained from Oriel. Irradiation effects simultaneous removal of the ligation auxiliary and the

o-nitrobenzyl protecting group, forming branched protein **11**. Semipreparative RP–HPLC purification is accomplished using a 0 to 73% B gradient over 45 min, yielding 2.5 mg of lyophilized protein **11**; ESI–MS $(M+H)^+$ observed: $9{,}547 \pm 2$ Da, expected: 9,548 Da.

7.3. Synthesis of ubiquitylated H2B mutant 12

In a typical reaction, photolysis product, **11** (2.5 mg, 0.25 μmol), and H2B (1-116)–MES, **9** (5.0 mg, 0.38 μmol), are dissolved in ligation buffer to a final volume of 225 μL. The pH of the resulting solution is increased to 7.8 with NaOH, and the reaction is allowed to proceed for 78 h prior to quenching with 225 μL of 50% HPLC buffer B containing 100 mM TCEP. The ligation product, **12**, is purified using semipreparative RP–HPLC with a 42 to 52% B gradient over 45 min, yielding 3.0 mg of product. The identity of the ligation product is verified by ESI–MS $((M+H)^+$ observed: $22{,}395 \pm 3$ Da, expected: 22,397 Da).

7.4. Desulfurization of 12 to give uH2B 13

Raney nickel reduction is used to convert Cys117 of protein **12** to the native Ala. In a typical reaction, protein **12**, (1.3 mg, 58 nmol) is dissolved in 3 mL of desulfurization buffer (6 M guanidinium chloride, 200 mM sodium phosphate, 35 mM TCEP). Raney nickel is prepared by adding 200 mg of NaBH$_4$ to a stirred solution of 1.2 g of nickel acetate in 6 mL of water. After 5 min, the Raney nickel is filtered, washed with 200 mL water, and added to the solution of **12**. The reaction progress is followed closely by HPLC and ESI–MS. An identical amount of fresh Raney nickel is added after 6 h and the reaction is found to be complete at 8.5 h. Note, much longer reaction times (more than 24 h) led to a second desulfurization reaction on methionine. The Raney nickel is pelleted by centrifugation and washed with 4×0.5 mL desulfurization buffer. The reaction supernatant and washes are combined, added to an equivalent volume of 50% HPLC buffer B, and purified using semipreparative RP–HPLC with a 42 to 52% B gradient over 45 min, yielding 1.1 mg of uH2B, **13**. uH2B is characterized by ESI–MS $((M+H)^+$ observed: $22{,}366 \pm 4$ Da, expected: 22,365 Da).

8. GENERATION OF UBIQUITYLATED MONONUCLEOSOMES

Synthetic access to homogenous uH2B opens the way to a range of biochemical and biophysical investigations (McGinty *et al.*, 2008). For most of these studies, it is necessary to incorporate the ubiquitylated protein into mononucleosomes and even polynucleosome arrays. The reader will recall that the nucleosome is the basic repeating unit of chromatin. The core particle is composed of 1.7 turns of DNA containing 147 bp wrapped around a highly basic octameric protein spool containing two copies of each core histone: H2A, H2B, H3, and H4. The overall molecular weight of this nucleoprotein complex is ≈200 kDa. Fortunately, robust procedures have been developed for the *in vitro* reconstitution of mononucleosomes from purified recombinant histones and double-stranded DNA containing nucleosomal targeting sequences (Luger *et al.*, 1997a; Luger *et al.*, 1997b; Luger *et al.*, 1999). These protocols are readily adapted to the reconstitution of chemically defined ubiquitylated mononucleosomes as detailed subsequently.

8.1. Recombinant histone preparation

Recombinantly expressed *Xenopus* histones H2A, H2B, H3, and H4 are prepared similarly to previously described protocols (Luger *et al.*, 1999). The three N-terminal residues of H2B (PEP) were added to the *Xenopus* H2B expression plasmid. The DNA encoding residues 1 to 125 of H2B is PCR amplified using primers H2B-FP described earlier and H2B-RP (5′-CGGGATCCTTACTTGGCGCTGGTGTACTTG-3′) and using the H2B plasmid described previously as a template. Following digestion with NdeI and BamHI, the H2B gene is ligated into a similarly digested pET vector from the *Xenopus* H2A expression plasmid (Luger *et al.*, 1997b). For protein expression, *E. coli* BL21(DE3)pLysS cells transformed with the appropriate histone expression plasmid are grown in 6 L of 2xTY media at 37 °C until mid-log phase, and protein expression is induced by the addition of 0.5 mM IPTG at 37 °C for 2 to 3 h. Cells are harvested and lysed as described previously for proteins **8** and **9**. The insoluble fractions of the bacterial lysates are washed twice with 20 mL of wash buffer (20 mM Tris, 200 mM NaCl, 1 mM EDTA, 1 mM 2-mercaptoethanol, pH 7.5) and once with 20 mL of triton wash buffer (20 mM Tris, 200 mM NaCl, 1 mM EDTA, 1 mM 2-mercaptoethanol, 1% triton, pH 7.5). DMSO (1 mL) is then added to the pellets and after 15 min, 15 to 50 mL of extraction buffer (7 M guanidinium chloride, 20 mM Tris, 200 mM NaCl, 1 mM EDTA, 1 mM 2-mercaptoethanol, pH 7.5) are added and the suspension stirred for a further 15 to 30 min before clearing by centrifugation at 26,000g for 30 min. The supernatants are purified using a Sephacryl S-200 column (approximately 1 L of bed volume), eluting with extraction buffer. Purified fractions are combined, dialyzed into water containing 2 mM DTT, and lyophilized in aliquots suitable for octamer formation. The identity of all purified histones is verified by ESI-MS.

8.2. Preparation of DNA for nucleosome formation

A plasmid containing 12 copies of 177 bp of the 601 sequence (Dorigo *et al.*, 2003) is purified using a Qiagen Plasmid Giga kit. The 177 bp repeat is digested from the vector using EcoRV sites flanking the segment. The desired segment is selectively precipitated by incrementally increasing the concentration of PEG-6000 from 4 to 8.5%, followed by centrifugation at 26,000*g* to separate precipitated DNA. ScaI digestion and gel purification using a QIAquick Gel Extraction Kit affords a 177 bp fragment of the 601 sequence, 1_177_601. PCR amplification of the central 147 bp region of 1_177_601 is accomplished using primers 147-FP (5′-CTGGA-GAATCCCGGTGCCGAGG-3′) and 147-RP (5′-ACAGGATGTATA-TATCTGACACG-3′) and gel-purified 1_177_601 as a template. Purification of the PCR product using a QIAquick PCR purification kit, yields 1_147_601.

8.3. Ubiquitylated nucleosome formation

Histone octamers are formed as previously described (Luger *et al.*, 1999). Briefly, individual purified histones (H2A, uH2B, H3 and H4) are dissolved in unfolding buffer (7M guanidinium chloride, 20 mM Tris, 10 mM DTT, pH 7.5) at approximately 4 mg/mL. Histones are combined in equimolar amounts (combined protein can range from 0.75 to 12 mg), and the solution is diluted to 1 mg/mL with unfolding buffer. The resulting mixture is dialyzed into refolding buffer (2M NaCl, 10 mM Tris, 1 mM EDTA, 1 mM DTT, pH 7.5) (3 changes of 2 L each). Crude octamer assemblies are then concentrated in Vivaspin 2 and 20 Centricons (3-10 kDa MWCO) and purified using a Superdex 200 10/300 column, eluting with refolding buffer. Octamer quality is verified by 18% SDS-PAGE, followed by Coomassie staining. Octamer samples are stored at −20 °C in 50% glycerol.

Mononucleosomes are formed using a previously described stepwise dilution procedure (Owen-Hughes *et al.*, 1999). Briefly, ubiquitylated octamers and DNA (1_147_601) are combined in 10 μL of high-salt refolding buffer to a final concentration of 3 μM. After incubation at 37 °C for 15 min, 3.3 μL of dilution buffer 1 (10 mM HEPES, 1 mM EDTA, 0.5 mM PMSF, pH 7.9) is added and the temperature dropped to 30 °C. Further dilutions of 6.7, 5, 3.6, 4.7, 6.7, 10, 30, and 20 μL, respectively of buffer 1, are then performed every 15 min. A final dilution with 100 μL of dilution buffer 2 (10 mM Tris, 1 mM EDTA, 0.1% NP-40, 5 mM DTT, 0.5 mM PMSF, 20% glycerol, pH 7.5) is then carried out. After an additional 15 min, the nucleosomes are concentrated using Vivaspin 500 Centricons (3 to 10 kDa MWCO) at 4 °C. Nucleosome formation is verified by separation on a Criterion 5% TBE gel run in 0.5x TBE, followed by ethidium bromide staining.

9. CONCLUSIONS

In this chapter, we have provided detailed experimental procedures for the preparation of monoubiquitylated histone H2B. Access to this modified protein is an important first step toward fully understanding the role of this PTM in regulating the structure and function of chromatin. Along these lines, we have recently shown that H2B ubiquitylation is sufficient to dramatically stimulate intranucleosomal H3K79 methylation by the human methyltransferase hDot1 (McGinty *et al.*, 2008). More generally, the semisynthetic protocols described herein can be readily adapted to the preparation of other ubiquitylated peptides (and following additional ligations, proteins) with a range of applications including antibody development and biochemical pull-down experiments. Moreover, this same technology can be extended to other members of the ubiquitin family, as already demonstrated with sumoylation (Chatterjee *et al.*, 2007).

ACKNOWLEDGMENTS

We would like to thank T.J. Richmond for donating the 12_177_601 plasmid. We would like to thank C.D. Allis for contributing the *Xenopus* histone plasmids for recombinant histone expression. This work was funded by the U.S. National Institutes of Health and the Starr Foundation. R.K.M. was supported by an MSTP grant.

REFERENCES

Ballal, N. R., Kang, Y. J., Olson, M. O. J., and Busch, H. (1975). Changes in Nucleolar Proteins and Their Phosphorylation Patterns during Liver-Regeneration. *J. Biol. Chem.* **250**, 5921–5925.

Bang, D., and Kent, S. B. H. (2004). A one-pot total synthesis of crambin. *Angew. Chemie. Intl. Ed.* **43**, 2534–2538.

Chatterjee, C., McGinty, R. K., and Muir, T. W. (2007). Auxiliary-Mediated Site-Specific Peptide Ubiquitylation. *Angewandte Chemie International Edition* **46**, 2814–2818.

Davies, N., and Lindsey, G. G. (1994). Histone H2b (and H2a) Ubiquitination Allows Normal Histone Octamer and Core Particle Reconstitution. *Biochimica Et Biophysica Acta-Gene Structure and Expression* **1218**, 187–193.

Dawson, P. E., and Kent, S. B. H. (2000). Synthesis of Native Proteins by Chemical Ligation. *Ann. Rev. Biochem.* **69**, 923–960.

Dorigo, B., Schalch, T., Bystricky, K., and Richmond, T. J. (2003). Chromatin fiber folding: Requirement for the histone H4N-terminal tail. *J. Mol. Biol.* **327**, 85–96.

Flavell, R. R., and Muir, T. W. (2009). Expressed protein ligation (EPL) in the study of signal transduction, ion conduction, and chromatin biology. *Accounts Chem. Res.* **42**, 107–116.

Hicke, L., Schubert, H. L., and Hill, C. P. (2005). Ubiquitin-binding domains. *Nature Rev. Mol. Cell Biol.* **6**, 610–621.

Hochstrasser, M. (1996). Ubiquitin-dependent protein degradation. *Ann. Rev. Gen.* **30**, 405–439.

Jenuwein, T., and Allis, C. D. (2001). Translating the histone code. *Science* **293**, 1074–1080.

Kouzarides, T. (2007). Chromatin Modifications and Their Function. *Cell* **128**, 693–705.

Lowary, P. T., and Widom, J. (1998). New DNA sequence rules for high afinity biding to histone octamer and sequence-directed nucleosome positioning. *J. Mol. Biol.* **276**, 19–42.

Luger, K., Mader, A. W., Richmond, R. K., Sargent, D. F., and Richmond, T. J. (1997a). Crystal structure of the nucleosome core particle at 2.8 A resolution. *Nature* **389**, 251–260.

Luger, K., Rechsteiner, T. J., Flaus, A. J., Waye, M. M. Y., and Richmond, T. J. (1997b). Characterization of nucleosome core particles containing histone proteins made in bacteria. *J. Mol. Biol.* **272**, 301–311.

Luger, K., Rechsteiner, T. J., and Richmond, T. J. (1999). Preparation of nucleosome core particle from recombinant histones. *Chromatin.* **304**, 3–19.

McGinty, R. K., Kim, J., Chatterjee, C., Roeder, R. G., and Muir, T. W. (2008). Chemically ubiquitylated histone H2B stimulates hDot1L-mediated intrnucleosomal methylation. *Nature* **453**, 812–816.

McMinn, D. L., and Greenberg, M. M. (1996). *Tetrahedron* **52**, 3827.

Muir, T. W. (2003). Semisynthesis of Proteins by Expressed Protein Ligation. *Ann. Rev. Biochem.* **72**, 249–289.

Muralidharan, V., and Muir, T. W. (2006). Protein Ligation: An enabling technology for the biophysical analysis of proteins. *Nat. Methods* **3**, 429–438.

Owen-Hughes, T., Utley, R. T., Steger, D. J., West, J. M., John, S., Cote, J., Havas, K. M., and Workman, J. L. (1999). Analysis of nucleosome disruption by ATP-driven chromatin remodeling complexes. *Methods Mol. Biol.* **119**, 319–331.

Pellois, J.-P., and Muir, T. W. (2005). A ligate and photorelease strategy for the temporal and spatial control of protein function in living cells. *Angew. Chemie. Int. Ed.* submitted.

Seeler, J. S., and Dejean, A. (2003). Nuclear and unclear functions of SUMO. *Nat. Rev. Mol. Cell Biol.* **4**, 690–699.

Walsh, C. T. (2006). Posttranslation Modification of Proteins: Expanding Nature's Inventory. Roberts and Company Publishers, Greenwood Village.

Wang, L., and Schultz, P. G. (2005). Expanding the genetic code. *Angew. Chemie. Intl. Ed.* **44**, 34–66.

Weake, V. M., and Workman, J. L. (2008). Histone ubiquitination: Triggering gene activity. *Mol. Cell.* **29**, 653–663.

West, M. H. P., and Bonner, W. M. (1980). Histone 2b Can Be Modified by the Attachment of Ubiquitin. *Nucleic Acids Res.* **8**, 4671–4680.

Zhu, B., Zheng, Y., Pham, A. D., Mandal, S. S., Erdjument-Bromage, H., Tempst, P., and Reinberg, D. (2005). Monoubiquitination of human histone H2B: The factors involved and their roles in HOX gene regulation. *Mol. Cell.* **20**, 601–611.

Author Index

Subject Index

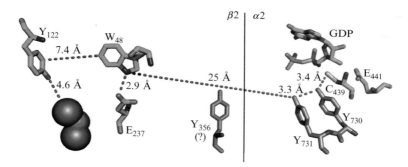

Mohammad R. Seyedsayamdost and JoAnne Stubbe, Figure 3.1 The radical transfer pathway in *E. coli* RNR. Residues Y_{122}, W_{48}, E_{237}, and Y_{356} reside in $\beta2$, while Y_{731}, Y_{730}, C_{439}, and E_{441} reside in $\alpha2$. Residues in gray have been shown to be redox–active using DOPA–$\beta2$ and NH_2Y–$\alpha2$s (see text). Note that the position of Y_{356} is unknown from structural studies.

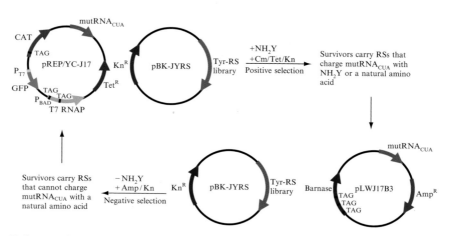

Mohammad R. Seyedsayamdost and JoAnne Stubbe, Figure 3.2 General scheme for selection of NH_2Y-RS. In the positive selection, suppression of a permissive TAG codon in chloramphenicol acetyl transferase (CAT, the Cm resistance gene) allows synthetases that charge $mutRNA_{CUA}$ with any amino acid to survive. The positive selection may also be performed by monitoring green fluorescence stemming from GFPuv, which contains a T7 promoter and is expressed when the TAG codons in the T7 RNA polymerase gene are suppressed. The synthetases are then carried through a negative selection cycle in which those that suppress the TAG codons in the barnase gene in the absence of NH_2Y are eliminated. Therefore, only synthetases that are functional with the host's translation machinery and do not charge $mutRNA_{CUA}$ with a natural amino acid will survive. See text for a description of the features on each plasmid.

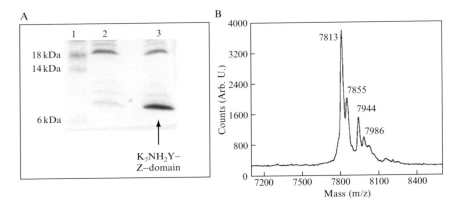

Mohammad R. Seyedsayamdost and JoAnne Stubbe, Figure 3.3 SDS-PAGE and MALDI-TOF MS analysis of K_7NH_2Y-Z-domain. (A) SDS gel of purified Z-domain after expression in the absence (lane 2) or presence (lane 3) of NH_2Y. The arrow designates the band corresponding to K_7NH_2Y-Z-domain. Protein ladder and MWs are shown in lane 1. (B) MALDI-TOF MS of purified K_7NH_2Y-Z-domain obtained under positive ionization mode. For the four main peaks in the spectrum, m/z $[M+H]^+$ are indicated. These correspond to N-terminally cleaved Met form of K_7NH_2Y-Z-domain, $[M+H]^+_{exp} = 7814$; its acetylated form, $[M+H]^+_{exp} = 7856$; full-length K_7NH_2Y-His-Z-domain, $[M+H]^+_{exp} = 7945$; and its acetylated form, $[M+H]^+_{exp} = 7987$.

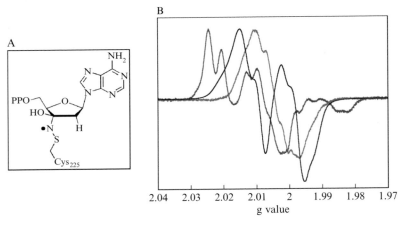

Mohammad R. Seyedsayamdost and JoAnne Stubbe, Figure 3.5 (A) Structure of the active site N-centered radical formed after reaction of RNR with N_3ADP. This radical is formed at the expense of $Y_{122}{}^\bullet$. It is stable on the minute time scale and is abbreviated as $N^\bullet$ in the text. (B) EPR spectra of $N^\bullet$ (green), $Y_{122}{}^\bullet$ (blue), and $NH_2Y_{731}{}^\bullet$ (red). The distinct features in the low-field region between ≈ 2.02 and ≈ 2.03 were used to perform subtractions and quantitations of the three species in N_3ADP assays.

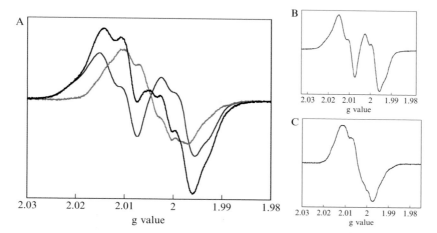

Mohammad R. Seyedsayamdost and JoAnne Stubbe, Figure 3.7 Reaction of
NH_2Y-$\alpha2$/ATP with wild-type $\beta2$/CDP monitored by EPR spectroscopy. (A) The
reaction components were mixed at 25 °C to yield final concentrations of 20 μM
$Y_{731}NH_2Y$-$\alpha2$/$\beta2$ complex, 1 mM CDP, and 3 mM ATP in assay buffer. After 20 s,
the reaction was quenched by hand-freezing in liquid N_2 and the EPR spectrum
recorded at 77 K. Unreacted Y_{122}• (blue trace, 58% of total spin) was subtracted
from the observed spectrum (black trace) to reveal the spectrum of NH_2Y_{731}• (red
trace, 42 % of total spin). (B) Reaction of $Y_{731}NH_2Y$-$\alpha2$/$\beta2$ in the absence of CDP/
ATP. Only the Y_{122}• spectrum is observed in this case. (C) Reaction of $Y_{730}NH_2Y$-
$\alpha2$ with $\beta2$ and CDP/ATP, similar to that described in (A). The subtracted spectrum
of NH_2Y_{730}• is shown.

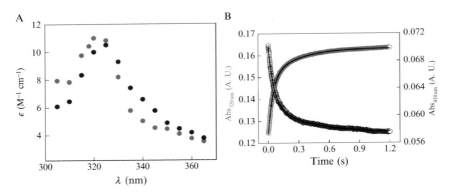

Mohammad R. Seyedsayamdost and JoAnne Stubbe, Figure 3.8 Characterization of NH_2Y-α2s by SF UV-vis spectroscopy. (A) Point-by-point reconstruction of the UV-vis spectrum of $NH_2Y_{730}{}^\bullet$ (blue dots) and $NH_2Y_{731}{}^\bullet$ (red dots). Prereduced $Y_{730}NH_2Y$-α2 and ATP in one syringe were mixed with wild-type β2 and CDP from another syringe, yielding final concentrations of 10 μM, 3 mM, 10 μM, and 1 mM, respectively. With $Y_{731}NH_2Y$-α2, the reaction was carried out at final concentrations of 9 μM $Y_{731}NH_2Y$-α2/β2, 1 mM CDP, and 3 mM ATP. The absorption change was monitored in 5-nm intervals; at each λ, 2 to 4 time courses were averaged and corrected for the absorption of $Y_{122}{}^\bullet$ using previously determined ε in this spectral range. The corrected ΔOD was converted to ε, which was then plotted against λ. (B) Prereduced $Y_{731}NH_2Y$-α2 and CDP in one syringe were mixed in a 1:1 ratio with β2 and ATP from another syringe to yield the same concentrations as in (A). The concentration of $Y_{122}{}^\bullet$ (blue trace) and $NH_2Y_{731}{}^\bullet$ (red trace) were monitored at 410 nm and 320 nm, respectively. A total of 6 traces were averaged at each λ. See text for kinetic parameters.

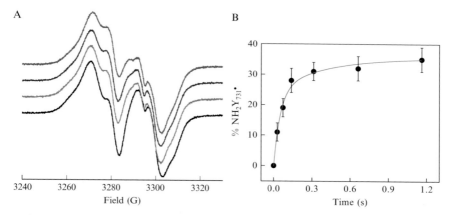

Mohammad R. Seyedsayamdost and JoAnne Stubbe, Figure 3.9 Analysis of $NH_2Y_{731}^{\bullet}$ formation by rapid-freeze quench EPR spectroscopy. (A) $Y_{731}NH_2Y\text{-}\alpha2$ and ATP were rapidly mixed with wild-type $\beta2$ and CDP to yield final concentrations of 19 μM $\beta2/Y_{731}NH_2Y\text{-}\alpha2$, 3 m$M$ ATP and 1 mM CDP. The reaction was quenched at various time points by spraying its contents into a liquid isopentane bath maintained at $\approx -140\ ^\circ C$. The EPR spectra were subsequently recorded at 77 K. Shown are the spectra at 0 ms (black), 72 ms (red), 312 ms (blue), and 1.16 s (green). (B) Kinetic trace obtained after subtracting the $Y_{122}^{\bullet}$ component from each trace and plotting the percentage of $NH_2Y_{731}^{\bullet}$ (relative to total spin) versus time. In this double exponential fit, the rate constant and amplitude for the slow phase have been held constant using the values from the SF UV-vis studies. Therefore, the fit yields the parameters for the fast kinetic phase. See text for kinetic parameters.

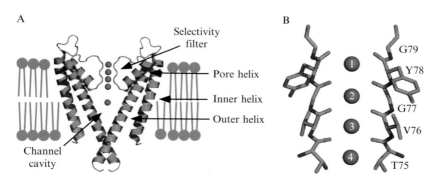

Alexander G. Komarov et al., Figure 7.1 The K^+ channel pore. (A) The KcsA channel is shown (pdb:1k4c). Only two opposite subunits of the tetrameric protein are shown. The selectivity filter, residues 75 to 79, is colored red. K^+ ions in the channel are depicted as magenta spheres. (B) Close-up view of the selectivity filter in stick representation with the front and the back subunits removed. The four K^+-binding sites in the selectivity filter are shown and the ion-binding sites are numbered 1 through 4.

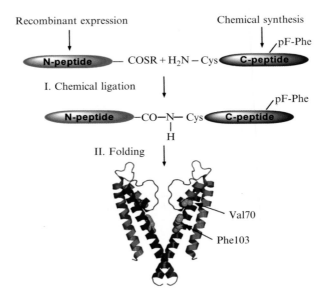

Alexander G. Komarov *et al.*, Figure 7.3   Semisynthesis of KcsA. The KcsA poly-peptide residues (1 to 123) is obtained by the expressed protein ligation of a recombi-nantly expressed N-peptide thioester (1 to 69) and a chemically synthesized peptide (70 to 123) with an N-terminal cysteine. The KcsA polypeptide obtained by the ligation reaction is folded to the native tetrameric state. Two subunits of the KcsA tetramer are shown. Val70, the ligation site, and Phe103, which is substituted with pF-Phe is depicted in space fill.

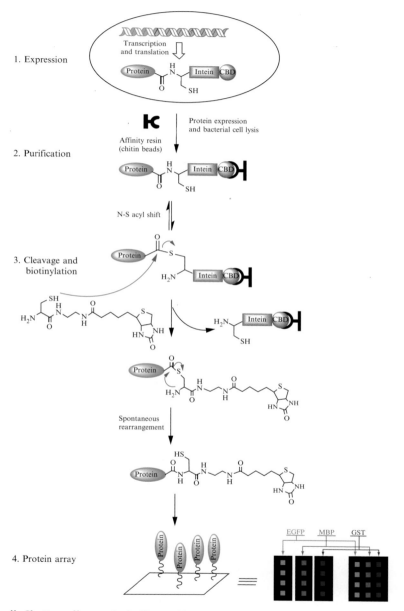

1. Expression

2. Purification

3. Cleavage and biotinylation

4. Protein array

Souvik Chattopadhaya *et al.*, Figure 10.1 Schematic representation of the different steps of *in vitro* purification, on–column biotinylation, and immobilization on avidin slides to generate functional protein arrays.

A

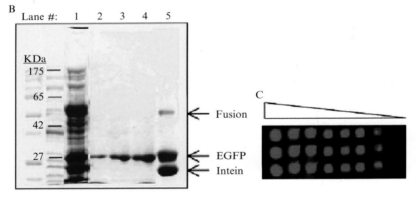

B

Lane #: 1   2   3   4   5

KDa
175 —
65 —
42 —
27 —

Fusion
EGFP
Intein

C

Souvik Chattopadhaya *et al.*, Figure 10.4 (A) Scheme showing the preparation of thioester slides. (B) Expression and *in vitro* purification of N-terminal cysteine-containing EGFP by intein-mediated cleavage on a chitin column. Lane 1: Total cell lysate containing expressed CBD-intein-EGFP fusion. Lane 2 to 4: First three elute fractions. Lane 5: Proteins retained on the column after elution. (C) Detection of N-terminal cysteine containing EGFP on thioester slides. (L to R) 1, 0.5, 0.2, 0.1, 0.05, 0.02, 0.01 mg/ml of EGFP was spotted (in triplicates) and detection was done using Cy5-labeled anti-EGFP antibody.